Silvia Bertoluzza • Ricardo H. Nochetto
Alfio Quarteroni • Kunibert G. Siebert
Andreas Veeser

Multiscale and Adaptivity: Modeling, Numerics and Applications

C.I.M.E. Summer School,
Cetraro, Italy 2009

Editors:
Giovanni Naldi
Giovanni Russo

Springer

FONDAZIONE CIME ROBERTO CONTI

Silvia Bertoluzza
CNR
Istituto di Matematica Applicata
e Tecnologie Informatiche
Pavia
Italy

Ricardo H. Nochetto
University of Maryland
Department of Mathematics
College Park, MD
USA

Alfio Quarteroni
École Polytechnique Fédérale
de Lausanne
Chaire de Modelisation
et Calcul Scientifique (CMCS)
Lausanne
Switzerland

Kunibert G. Siebert
Universität Stuttgart
Fakultät für Mathematik und Physik
Stuttgart
Germany

Andreas Veeser
Università degli Studi di Milano
Dipartimento di Matematica
Milano
Italy

ISBN 978-3-642-24078-2 e-ISBN 978-3-642-24079-9
DOI 10.1007/978-3-642-24079-9
Springer Heidelberg Dordrecht London New York

Lecture Notes in Mathematics ISSN print edition: 0075-8434
ISSN electronic edition: 1617-9692

Library of Congress Control Number: 2011943495

Mathematics Subject Classification (2010): 65M50, 65N50, 65M55, 65T60, 65N30, 65M60, 76MXX

Printed on acid-free paper

Springer is part of Springer Science+Business Media (www.springer.com)

Preface

The CIME-EMS Summer School in applied mathematics on "Multiscale and Adaptivity: Modeling, Numerics and Applications" was held in Cetraro (Italy) from July 6 to 11, 2009. This course has focused on mathematical methods for systems that involve multiple length/time scales and multiple physics. The complexity of the structure of these systems requires suitable mathematical and computational tools. In addition, mathematics provides an effective approach toward devising computational strategies for handling multiple scales and multiple physics. This course brought together researchers and students from different areas such as partial differential equations (PDEs), analysis, mathematical physics, numerical analysis, and scientific computing to address the challenges present in these issues. Physical, chemical, and biological processes for many problems in computational physics, biology, and material science span length and time scales of many orders of magnitude. Traditionally, scientists and research groups have focused on methods that are particularly applicable in only one regime, and knowledge of the system at one scale has been transferred to another scale only indirectly. Microscopic models, for example, have been often used to find the effective parameters of macroscopic models, but for obvious computational reasons, microscopic and macroscopic scales have been treated separately.

The enormous increase in computational power available (due to the improvement both in computer speed and in efficiency of the numerical methods) allows in some cases the treatment of systems involving scales of different orders of magnitude, arising, for example, when effective parameters in a macroscopic model depend on a microscopic model, or when the presence of a singularity in the solution produces a continuum of length scales. However, the numerical solution of such problems by classical methods often leads to an inefficient use of the computational resources, even up to the point that the problem cannot be solved by direct numerical simulation. The main reasons for this are that the necessary resolution of a fine scale entails an over-resolution of coarser scales, the position of the singularity is not known beforehand, the gap between the scales is too big for a treatment in the same framework. In other cases, the structure of the mathematical models that treat the system at the different scales varies a lot, and therefore new mathematical techniques are required to treat systems described by different mathematical models. Finally, in many cases one is interested in the accurate treatment of a small portion of a large system, and it is too expensive to treat the whole system at the required accuracy. In such cases, the region of interest is modeled and discretized with great accuracy, while the remaining parts of the system are described by some reduced model, which enormously simplifies the calculation, still providing reasonable boundary conditions for the region of interest, allowing the required level of detail in such region.

The outstanding and internationally renowned lecturers have themselves contributed in an essential way to the development of the theory and techniques that constituted the subjects of the courses. The selection of the five topics of the CIME-EMS Course was not an easy task because of the wide spectrum of recent developments in multiscale methods and models. The six world leading experts illustrated several aspects of the multiscale approach.

Silvia Bertoluzza, from IMATI-CNR Pavia, described the concept of nonlinear sparse wavelet approximation of a given (known) function. Next she showed how the tools just introduced can be applied in order to write down efficient adaptive schemes for the solution of PDEs.

Bjorn Engquist, from ICES University of Texas at Austin, gradually guided the audience toward the realm of "Multiscale Modeling," by providing mathematical ground for state-of-the-art analytical and numerical multiscale problems.

Alfio Quarteroni, from EPFL, Lausanne, and Politecnico di Milano, considered adaptivity in mathematical modeling for the description and simulation of complex physical phenomena. He showed that the combination of hierarchical mathematical models can be set up with the aim of reducing the computational complexity in the real life problems.

Ricardo H. Nochetto, from University of Maryland, and Andreas Veeser, from Università di Milano, in their joint course started with an overview of the a posteriori error estimation for finite element methods, and then they exposed recent results about the convergence and complexity of adaptive finite element methods.

Kunibert G. Siebert, from Universität Duisburg-Essen, described the implementation of adaptive finite element methods using toolbox ALBERTA (created by Alfred Schmidt and Kunibert G. Siebert, which is freely available).

The main "senior" lecturers were complemented by four young speakers, who gave account of detailed examples or applications during an afternoon session dedicated to them. Matteo Semplice, Università dell'Insubria, has spoken about "Numerical entropy production and adaptive schemes for conservation laws," Tiziano Passerini, from Emory University, about "A 3D/1D geometrical multiscale model of cerebral vasculature," Loredana Gaudio, MOX Politecnico di Milano, about "Spectral element discretization of optimal control problems," and Carina Geldhauser, Universität Tuebingen, described "A discrete-in-space scheme converging to an unperturbed Cahn–Hilliard equation." Both the lectures and the active interactions with and within the audience contributed to the scientific success of the course, which was attended by about 60 people of various nationality (14 different countries), ranging from first year PhD students to full professors. The present volume collects the expanded version of the lecture notes by Silvia Bertoluzza, Alfio Quarteroni (with Marco Discacciati and Paola Gervasio as coauthors), Ricardo H. Nochetto, Andreas Veeser, and Kunibert G. Siebert. We are grateful to them for such high quality scientific material.

As editors of these Lecture Notes and as scientific directors of the course, we would like to thank the many persons and Institutions that contributed to the success of the school. It is our pleasure to thank the members of the Scientific Committee of CIME for their invitation to organize the School; the Director, Prof. Pietro Zecca, and the Secretary, Prof. Elvira Mascolo, for their efficient support during the organization and their generous help during the school. We were particularly pleased by the fact that the European Mathematical Society (EMS) chose to cosponsor this CIME course as one of its Summer School in applied mathematics for 2009. Our special thanks go to the lecturers for their early preparation of the material to be distributed to the participants, for their excellent performance in teaching the courses and their stimulating scientific contributions. All the participants contributed to the creation of an exceptionally friendly atmosphere in the beautiful environment around the School. We also wish to thank Dipartimento di Matematica of the Università degli Studi di Milano, and Dipartimento di Matematica ed Informatica of the Università degli Studi di Catania for their financial support.

Catania *Giovanni Naldi*
Milano *Giovanni Russo*

Contents

Adaptive Wavelet Methods 1
Silvia Bertoluzza
1 Introduction 1
2 Multiresolution Approximation and Wavelets 2
2.1 Riesz Bases 2
2.2 Multiresolution Analysis 3
2.3 Examples 9
2.4 Beyond $L^2(\mathbb{R})$ 17
3 The Fundamental Property of Wavelets 21
3.1 The Case $\Omega = \mathbb{R}$: The Frequency Domain Point of View vs. the Space Domain Point of View 22
4 Adaptive Wavelet Methods for PDE's: The First Generation 34
4.1 The Adaptive Wavelet Collocation Method 37
5 The New Generation of Adaptive Wavelet Methods 40
5.1 A Posteriori Error Estimates 41
5.2 Nonlinear Wavelet Methods for the Solution of PDE's 46
5.3 The CDD2 Algorithm 48
5.4 Operations on Infinite Matrices and Vectors 51
References 54

Heterogeneous Mathematical Models in Fluid Dynamics and Associated Solution Algorithms 57
Marco Discacciati, Paola Gervasio, and Alfio Quarteroni
1 Introduction and Motivation 57
2 Variational Formulation Approach 67
2.1 The Advection–Diffusion Problem 67
2.2 Variational Analysis for the Advection–Diffusion Equation 68
2.3 Domain Decomposition Algorithms for the Solution of the Reduced Advection–Diffusion Problem 72
2.4 Numerical Results for the Advection–Diffusion Problem 77
2.5 Navier–Stokes/Potential Coupled Problem 80

2.6 Asymptotic Analysis of the Coupled Navier–Stokes/ Darcy Problem ... 82
2.7 Solution Techniques for the Navier–Stokes/Darcy Coupling ... 85
2.8 Numerical Results for the Navier–Stokes/Darcy Problem ... 90
3 Virtual Control Approach ... 94
3.1 Virtual Control Approach Without Overlap for AD Problems ... 95
3.2 Domain Decomposition with Overlap ... 105
3.3 Virtual Control Approach with Overlap for the Advection–Diffusion Equation ... 108
3.4 Virtual Control with Overlap for the Stokes–Darcy Coupling ... 114
3.5 Coupling for Incompressible Flows ... 119
References ... 120

Primer of Adaptive Finite Element Methods ... 125
Ricardo H. Nochetto and Andreas Veeser
1 Piecewise Polynomial Approximation ... 125
1.1 Classical vs Adaptive Pointwise Approximation ... 126
1.2 The Sobolev Number: Scaling and Embedding ... 127
1.3 Conforming Meshes: The Bisection Method ... 129
1.4 Finite Element Spaces ... 133
1.5 Polynomial Interpolation in Sobolev Spaces ... 134
1.6 Adaptive Approximation ... 139
1.7 Nonconforming Meshes ... 143
1.8 Notes ... 145
1.9 Problems ... 146
2 Error Bounds for Finite Element Solutions ... 148
2.1 Model Boundary Value Problem ... 148
2.2 Galerkin Solutions ... 149
2.3 Finite Element Solutions and A Priori Bound ... 150
2.4 A Posteriori Upper Bound ... 151
2.5 Notes ... 157
2.6 Problems ... 158
3 Lower A Posteriori Bounds ... 159
3.1 Local Lower Bounds ... 160
3.2 Global Lower Bound ... 166
3.3 Notes ... 167
3.4 Problems ... 168
4 Convergence of AFEM ... 170
4.1 A Model Adaptive Algorithm ... 171
4.2 Convergence ... 172
4.3 Notes ... 178
4.4 Problems ... 179
5 Contraction Property of AFEM ... 180
5.1 Modules of AFEM for the Model Problem ... 180
5.2 Basic Properties of AFEM ... 182

5.3 Contraction Property of AFEM 185
5.4 Example: Discontinuous Coefficients 189
5.5 Extensions and Restrictions 191
5.6 Notes 193
5.7 Problems 193
6 Complexity of Refinement 194
6.1 Chains and Labeling for $d = 2$ 195
6.2 Recursive Bisection 197
6.3 Conforming Meshes: Proof of Theorem 1 199
6.4 Nonconforming Meshes: Proof of Lemma 3 204
6.5 Notes 205
6.6 Problems 206
7 Convergence Rates 206
7.1 The Total Error 207
7.2 Approximation Classes 208
7.3 Quasi-Optimal Cardinality: Vanishing Oscillation 212
7.4 Quasi-Optimal Cardinality: General Data 215
7.5 Extensions and Restrictions 218
7.6 Notes 221
7.7 Problems 221
References 223

Mathematically Founded Design of Adaptive Finite Element Software 227
Kunibert G. Siebert
1 Introduction 227
1.1 The Variational Problem 229
1.2 The Basic Adaptive Algorithm 230
2 Triangulations and Finite Element Spaces 232
2.1 Triangulations 232
2.2 Finite Element Spaces 234
2.3 Basis Functions and Evaluation of Finite Element Functions 240
2.4 ALBERTA Realization of Finite Element Spaces 244
3 Refinement By Bisection 246
3.1 Basic Thoughts About Local Refinement 246
3.2 Bisection Rule: Bisection of One Single Simplex 248
3.3 Triangulations and Refinements 252
3.4 Refinement Algorithms 255
3.5 Complexity of Refinement By Bisection 260
3.6 ALBERTA Refinement 262
3.7 Mesh Traversal Routines 263
4 Assemblage of the Linear System 268
4.1 The Variational Problem and the Linear System 269
4.2 Assemblage: The Outer Loop 272

4.3 Assemblage: Element Integrals 276
4.4 Remarks on Iterative Solvers 283
5 The Adaptive Algorithm and Concluding Remarks 285
5.1 The Adaptive Algorithm 286
5.2 Concluding Remarks 294
6 Supplement: A Nonlinear and a Saddlepoint Problem 297
6.1 The Prescribed Mean Curvature Problem in Graph Formulation 297
6.2 The Generalized Stokes Problem 301
References 308

List of Participants 311

Adaptive Wavelet Methods

Silvia Bertoluzza

Abstract Wavelet bases, initially introduced as a tool for signal and image processing, have rapidly obtained recognition in many different application fields. In this lecture notes we will describe some of the interesting properties that such functions display and we will illustrate how such properties (and in particular the simultaneous good localization of the basis functions in both space and frequency) allow to devise several adaptive solution strategies for partial differential equations. While some of such strategies are based mostly on heuristic arguments, for some other a complete rigorous justification and analysis of convergence and computational complexity is available.

1 Introduction

Wavelet bases were introduced in the late 1980s as a tool for signal and image processing. Among the applications considered at the beginning we recall applications in the analysis of seismic signals, the numerous applications in image processing – image compression, edge-detection, denoising, applications in statistics, as well as in physics. Their effectiveness in many of the mentioned fields is nowadays well established: as an example, wavelets are actually used by the US *Federal Bureau of Investigation* (or FBI) in their fingerprint database, and they are one of the ingredient of the new MPEG media compression standard. Quite soon it became clear that such bases allowed to represent objects (signals, images, turbulent fields) with singularities of complex structure with a low number of degrees of freedom, a property that is particularly promising when thinking of an application to the numerical solution of partial differential equations: many PDEs have in fact

S. Bertoluzza (✉)
Istituto di Matematica Applicata e Tecnologie Informatiche del CNR, v. Ferrata 1, Pavia, Italy
e-mail: silvia.bertoluzza@imati.cnr.it

S. Bertoluzza et al., *Multiscale and Adaptivity: Modeling, Numerics and Applications*, Lecture Notes in Mathematics 2040, DOI 10.1007/978-3-642-24079-9_1,

solutions which present singularities, and the ability to represent such solution with as little as possible degrees of freedom is essential in order to be able to implement effective solvers for such problems. The first attempts to use such bases in this framework go back to the late 1980s and early 1990s, when the first simple adaptive wavelet methods [38] appeared. In those years the problems to be faced were basic ones. The computation of integrals of products of derivative of wavelets – object which are naturally encountered in the variational approach to the numerical solution of PDEs – was an open problem (solved later by Dahmen and Micchelli in [24]). Moreover, wavelets were defined on $\mathbb{R}$ and on $\mathbb{R}^n$. Already solving a simple boundary value problem on $(0, 1)$ (the first construction of wavelets on the interval [19] was published in 1993) posed a challenge.

Many steps forward have been made since those pioneering works. In particular *thinking in terms of wavelets* gave birth to some new approaches in the numerical solution of PDEs. The aim of this course is to show some of these new ideas. In particular we want to show how one key property of wavelets (the possibility of writing equivalent norms for the scale of Besov spaces) allows to write down some new adaptive methods for solving PDE's.

2 Multiresolution Approximation and Wavelets

2.1 *Riesz Bases*

Before starting with defining wavelets, let us recall the definition and some properties of Riesz bases [14], which will play a relevant role in the following. Let H denote an Hilbert space and let $V \subseteq H$ denote a subspace. A basis $\mathscr{B} = \{\varphi_k, k \in I\}$ ($I \subseteq \mathbb{N}$ index set) of V is a Riesz basis if and only if the following norm equivalence holds:

$$\| \sum_k c_k e_k \|_H^2 \simeq \sum_k |c_k|^2.$$

Here and in the following we use the notation $A \simeq B$ to signify that there exist positive constants c and C, independent of any relevant parameter, such that $cB \leq A \leq CB$. Analogously we will use the notation $A \lesssim B$ (resp. $A \gtrsim B$), meaning that $A \leq CB$ (resp. $A \geq cB$).

Letting $P : H \rightarrow V$ be any projection operator ($P^2 = P$), it is not difficult to realize that there exist a sequence $\mathscr{G} = \{g_k, k \in I\}$ such that for all $f \in H$ we have the identity

$$Pf = \sum_{k \in I} \langle f, g_k \rangle \varphi_k.$$

The sequence g_k is *biorthogonal* to the basis $\mathscr{B}$, that is we have that

$$\langle g_k, \varphi_i \rangle = \delta_{i,k}.$$

Moreover the sequence $\mathscr{G}$ is a Riesz basis for the subspace P^*H (P^* denoting the adjoint operator to P), and P^* takes the form

$$P^* f = \sum_{k \in I} \langle f, \varphi_k \rangle g_k.$$

2.2 Multiresolution Analysis

We start by introducing the general concept of multiresolution analysis in the univariate case [39].

Definition 1. A *Multiresolution Analysis* (MRA) of $L^2(\mathbb{R})$ is a sequence $\{V_j\}_{j \in \mathbb{Z}}$ of closed subspaces of $L^2(\mathbb{R})$ verifying:

(i) The subspaces are nested: $V_j \subset V_{j+1}$ for all $j \in \mathbb{Z}$.
(ii) The union of the spaces is dense in $L^2(\mathbb{R})$ and the intersection is null:

$$\overline{\cup_{j \in \mathbb{Z}} V_j} = L^2(\mathbb{R}), \quad \cap_{j \in \mathbb{Z}} V_j = \{0\}. \tag{1}$$

(iii) There exists a *scaling function* $\varphi \in V_0$ such that $\{\varphi(\cdot - k), k \in \mathbb{Z}\}$ is a Riesz basis for V_0.
(iv) $f \in V_0$ implies $f(2^j \cdot) \in V_j$.

Several properties descend directly from this definition. First of all it is not difficult to check that the above properties imply that for all j the set $\{\varphi_{j,k}\ k \in \mathbb{Z}\}$, with

$$\varphi_{j,k} = 2^{j/2}\varphi(2^j \cdot - k) \tag{2}$$

is Riesz basis for V_j, yielding, uniformly in j, a norm equivalence between the L^2 norm of a function in V_j and the ℓ^2 norm of the sequence of its coefficients.

Moreover, the inclusion $V_0 \subset V_1$ implies that the scaling function φ can be expanded in terms of the basis of V_1 through the following *refinement equation*:

$$\varphi(x) = \sum_{k \in \mathbb{Z}} h_k \varphi(2x - k), \tag{3}$$

with $(h_k)_k \in \ell^2(\mathbb{Z})$. The function φ is then said to be a *refinable function* and the coefficients h_k are called *refinement coefficients*.

Let now $f \in L^2(\mathbb{R})$. We can consider approximations $f_j \in V_j$ to f at different levels j. Since $V_j \subset V_{j+1}$ it is not difficult to realize that the approximation f_{j+1} of a given function f at level $j+1$ must "contain" more information on f than f_j. The idea underlying the construction of wavelets is the one of somehow encoding the "loss of information" that we have when we go from f_{j+1} to f_j. Let us for instance consider $f_j = P_j f$, where $P_j : L^2(\mathbb{R}) \to V_j$ denotes the $L^2(\mathbb{R})$-orthogonal projection onto V_j. Remark that $P_{j+1}P_j = P_j$ (a direct consequence of the nestedness of the spaces V_j). Moreover, we have that $P_j P_{j+1} = P_j$: f_{j+1} contains in this case all information needed to retrieve f_j. We can in this case introduce the orthogonal complement $W_j \subset V_{j+1}$ ($W_j \perp V_j$ and $V_{j+1} = V_j \oplus W_j$).

A similar construction can be actually carried out in a more general framework, in which P_j is not necessarily the orthogonal projection. To be more general, let us start by choosing a sequence of uniformly bounded (not necessarily orthogonal) projectors $P_j : L^2(\mathbb{R}) \to V_j$ verifying the following properties:

$$P_j P_{j+1} = P_j, \tag{4}$$

$$P_j(f(\cdot - k2^{-j}))(x) = P_j f(x - k2^{-j}), \tag{5}$$

$$P_{j+1} f((2\cdot))(x) = P_j f(2x). \tag{6}$$

Remark again that the inclusion $V_j \subset V_{j+1}$ guarantees that $P_{j+1} P_j = P_j$. On the contrary, property (4) is not verified by general non-orthogonal projectors and expresses the fact that the approximation $P_j f$ can be derived from $P_{j+1} f$ without any further information on f. Equations (5) and (6) require that the projector P_j respects the translation and dilation invariance properties (iii) and (iv) of the MRA.

Since $\{\varphi_{0,k}\}$ is a Riesz basis for V_0 there exists a biorthogonal sequence $\{\tilde{\varphi}_{0,k}\}$ of $L^2(\mathbb{R})$ functions such that

$$P_0 f = \sum_k \langle f, \tilde{\varphi}_{0,k} \rangle \varphi_{0,k}.$$

Property (5) implies that the biorthogonal basis has itself a translation invariant structure, as stated by the following proposition.

Proposition 1. *Letting $\tilde{\varphi} = \tilde{\varphi}_{0,0}$ we have that*

$$\tilde{\varphi}_{0,k}(x) = \tilde{\varphi}(x - k). \tag{7}$$

Proof. We observe that

$$P_0(f(\cdot + n))(x) = \sum_k \langle f(\cdot + n), \tilde{\varphi}_{0,k} \rangle \varphi(x - k) = \sum_k \langle f(\cdot), \tilde{\varphi}_{0,k}(\cdot - n) \rangle \varphi(x - k),$$

$$P_0 f(x + n) = \sum_k \langle f(\cdot), \tilde{\varphi}_{0,k} \rangle \varphi(x - k + n).$$

Thanks to (5) we have that

$$\sum_k \langle f(\cdot), \tilde{\varphi}_{0,k}(\cdot - n) \rangle \varphi(x - k) = \sum_k \langle f(\cdot), \tilde{\varphi}_{0,k} \rangle \varphi(x - k + n),$$

and, since $\{\varphi_{0,k}\}$ is a Riesz basis for V_0, implying that the coefficients (and in particular the coefficient of $\varphi_{0,0} = \varphi(x)$) are uniquely determined, this implies, for all $f \in L^2(\mathbb{R})$,

$$\langle f, \tilde{\varphi}_{0,0}(\cdot - n)\rangle = \langle f, \tilde{\varphi}_{0,n}\rangle,$$

that is, by the arbitrariness of f, $\tilde{\varphi}_{0,n} = \tilde{\varphi}_{0,0}(\cdot - n)$. □

In an analogous way, thanks to property (6) it is not difficult to prove the following Proposition.

Proposition 2. *We have*

$$P_j f = \sum_k \langle \tilde{\varphi}_{j,k}, f\rangle \varphi_{j,k} \quad \textit{with} \quad \tilde{\varphi}_{j,k}(x) = 2^{j/2}\tilde{\varphi}(2^j x - k). \tag{8}$$

Moreover the set $\{\tilde{\varphi}_{j,k},\ k \in \mathbb{Z}\}$ *forms a Riesz basis for the subspace* $\tilde{V}_j = P_j^*(L^2(\mathbb{R}))$ *(where* P_j^* *denotes the adjoint of* P_j*).*

Finally, property (4) implies that the sequence $\tilde{V}_j$ is nested.

Proposition 3. *The sequence* $\{\tilde{V}_j\}$ *satisfies* $\tilde{V}_j \subset \tilde{V}_{j+1}$.

Proof. Property (4) implies that $P_{j+1}^* P_j^* f = P_j^* f$. Now we have $f \in \tilde{V}_j$ implies $f = P_j^* f = P_{j+1}^* P_j^* f \in \tilde{V}_{j+1}$. □

Corollary 1. *The function* $\tilde{\varphi} = \tilde{\varphi}_{0,0}$ *is refinable.*

The above construction derives from the a priori choice of a sequence P_j, $j \in \mathbb{Z}$, of (oblique) projectors onto the subspaces V_j. A trivial choice is to define P_j as the $L^2(\mathbb{R})$ orthogonal projector. It is easy to see that all the required properties are satisfied by such a choice. In this case, since the $L^2(\mathbb{R})$ orthogonal projector is self adjoint, we have $\tilde{V}_j = V_j$, and the biorthogonal function $\tilde{\varphi}$ belongs itself to V_0. Clearly, in the case that $\{\varphi_{0,k}, k \in \mathbb{Z}\}$ is an orthonormal basis for V_0 (as in the Haar basis case of the forthcoming Example I, or as for Daubechies MRA's) we have that $\tilde{\varphi} = \varphi$. Another possibility would be to choose P_j to be the Lagrangian interpolation operator. This choice, which we will describe later on, does however fall outside of the framework considered here, since interpolation is not an $L^2(\mathbb{R})$ bounded operator.

Infinitely many other choices are possible in theory but quite difficult to construct in practice. The solution is then to go the other way round, constructing the function $\tilde{\varphi}$ directly and defining the projectors P_j by (8) [18]. We then introduce the following definition:

Definition 2. A refinable function

$$\tilde{\varphi} = \sum_k \tilde{h}_k \tilde{\varphi}(2\cdot - k) \in L^2(\mathbb{R}) \tag{9}$$

is dual to φ if

$$\langle \varphi(\cdot - k), \tilde{\varphi}(\cdot - l)\rangle = \delta_{k,l} \quad k, l \in \mathbb{Z}.$$

It is possible to prove that the translates of the dual refinable function are a Riesz basis for the subspace that they span.

Assuming then that we have a refinable function $\tilde{\varphi}$ dual to φ, we can define the projector P_j using (8).

$$P_j f = \sum_{k \in \mathbb{Z}} \langle f, \tilde{\varphi}_{j,k} \rangle \varphi_{j,k}.$$

The operator P_j is bounded and it is indeed a projector: it is not difficult to check that $f \in V_j \Rightarrow P_j f = f$.

Remark 1. As it happened for the projector P_j, the dual refinable function $\tilde{\varphi}$ is not uniquely determined, once φ is given. Different projectors correspond to different dual functions. It is worth noting that P.G. Lemarié [37] proved that if φ is compactly supported then there exists a dual function $\tilde{\varphi} \in L^2(\mathbb{R})$ which is itself compactly supported.

The dual of P_j

$$P_j^* f = \sum_{k \in \mathbb{Z}} \langle f, \varphi_{j,k} \rangle \tilde{\varphi}_{j,k}$$

is also an oblique projector onto the space $Im(P_j^*) = \tilde{V}_j$, where

$$\tilde{V}_j = span < \tilde{\varphi}_{j,k}, \ k \in \mathbb{Z} > .$$

It is not difficult to see that since $\tilde{\varphi}$ is refinable then the $\tilde{V}_j$'s are nested.

Remark 2. The two different ways of defining the dual MRA are equivalent. A third approach yields also an equivalent structure. In fact, assume that we have a sequence $\tilde{V}_j$ of spaces such that the following inf-sup conditions hold uniformly in j:

$$\inf_{v_j \in V_j} \sup_{w_j \in \tilde{V}_j} \frac{\langle v_j, w_j \rangle}{\|v_j\|_{L^2(\mathbb{R})} \|w_j\|_{L^2(\mathbb{R})}} \gtrsim 1, \qquad \inf_{w_j \in \tilde{V}_j} \sup_{v_j \in V_j} \frac{\langle v_j, w_j \rangle}{\|v_j\|_{L^2(\mathbb{R})} \|w_j\|_{L^2(\mathbb{R})}} \gtrsim 1. \tag{10}$$

Then it is possible to define a bounded projector $P_j : L^2(\mathbb{R}) \to V_j$ as $P_j v = v_j$, v_j being the unique element of V_j such that

$$\langle v_j, w_j \rangle = \langle v, w_j \rangle \quad \forall w_j \in \tilde{V}_j.$$

It is not difficult to see that if the sequence $\tilde{V}_j$ is a multiresolution analysis (that is, if it satisfies the requirements of Definition 1) then the projector P_j satisfies properties (4)–(6). Conversely the uniform boundedness of the projector P_j and of its adjoint $\tilde{P}_j$ actually implies the validity of the two inf-sup conditions (10).

2.2.1 Wavelets

Whatever the way chosen to introduce the dual multiresolution analysis, we have now a natural way to define a space W_j which complements V_j in V_{j+1}. More precisely we set

$$V_{j+1} = V_j \oplus W_j, \qquad W_j = Q_j(V_{j+1}), \qquad Q_j = P_{j+1} - P_j. \tag{11}$$

Remark that $Q_j^2 = Q_j$, that is Q_j is indeed a projector on W_j. W_j can also be defined as the kernel of P_j in V_{j+1}. Iterating for j decreasing the splitting (11) we obtain a *multiscale decomposition* of V_{j+1} as

$$V_{j+1} = V_0 \oplus W_0 \oplus \cdots \oplus W_j.$$

By construction we also have, for all $f \in L^2(\mathbb{R})$, the decomposition

$$P_{j+1} f = P_0 f + \sum_{m=0}^{j} Q_m f.$$

In other words the approximation $P_{j+1} f$ is decomposed as a coarse approximation at scale 0 plus a sequence of fluctuations at intermediate scales 2^{-m}, $m = 0, \ldots, j$.

If we are to express the above identity in terms of a Fourier expansion, we need bases for the spaces W_j. Depending on the nature of the spaces considered such bases might be readily available (see for instance the construction of interpolating wavelets). However this is not, in general, the case. A general procedure to construct a suitable basis for W_j is the following [30]: define two sets of coefficients:

$$g_k = (-1)^k \tilde{h}_{1-k}, \qquad \tilde{g}_k = (-1)^k h_{1-k}, \qquad k \in \mathbb{Z}$$

and introduce a pair of *dual wavelets* $\psi \in V_1$ and $\tilde{\psi} \in \tilde{V}_1$

$$\psi(x) = \sum_k g_k \varphi(2x - k) \qquad \tilde{\psi}(x) = \sum_k \tilde{g}_k \tilde{\varphi}(2x - k). \tag{12}$$

The following theorem holds [18]:

Theorem 1. *The integer translates of the wavelet functions ψ and $\tilde{\psi}$ are orthogonal to $\tilde{\varphi}$ and φ, respectively, and they form a couple of biorthogonal sequences. More precisely, they satisfy*

$$\langle \psi, \tilde{\psi}(\cdot - k) \rangle = \delta_{0,k} \qquad \langle \psi(\cdot - k), \tilde{\varphi} \rangle = \langle \tilde{\psi}(\cdot - k), \varphi \rangle = 0. \tag{13}$$

The projection operator Q_j can be expanded as

$$Q_j f = \sum_k \langle f, \tilde{\psi}_{j,k} \rangle \psi_{j,k}$$

and the functions $\psi_{j,k}$ constitute a Riesz basis of W_j.

For any function $f \in L^2(\mathbb{R})$, $P_j f$ in V_j can be expressed as

$$P_j f = \sum_k c_{j,k}\varphi_{j,k} = \sum_k c_{0,k}\varphi_{0,k} + \sum_{m=0}^{j-1}\sum_k d_{m,k}\psi_{m,k}, \tag{14}$$

with $d_{m,k} = \langle f, \tilde{\psi}_{m,k}\rangle$ and $c_{m,k} = \langle f, \tilde{\varphi}_{m,k}\rangle$.
Both $\{\varphi_{j,k}, k \in \mathbb{Z}\}$ and $\{\varphi_{0,k}, k \in \mathbb{Z}\} \cup_{0\le m<j-1} \{\psi_{m,k}, k \in \mathbb{Z}\}$ are bases for V_j and (14) expresses a change of basis. Thanks to the density property (1) for $j \to +\infty$ $P_j f$ converges to f in $L^2(\mathbb{R})$. Then, taking the limit for $j \to +\infty$ in (14), we obtain

$$f = \sum_k c_{0,k}\varphi_{0,k} + \sum_{m=0}^{+\infty}\sum_k d_{m,k}\psi_{m,k}.$$

We will see in the following that, under quite mild assumptions, the convergence is unconditional.

2.2.2 The Fast Wavelet Transform

The idea is now to design an algorithm allowing to compute efficiently the coefficients $c_{j-1,k}(f)$ and $d_{j-1,k}(f)$ directly from the coefficients $c_{j,k}(f)$, which uniquely identify $P_j f$. The key is the refinement equation (9), which gives us a "fine to coarse" discrete projection algorithm:

$$\begin{aligned} c_{j-1,k} &= \langle f, \tilde{\varphi}_{j-1,k}\rangle = 2^{(j-1)/2}\langle f, \tilde{\varphi}(2^{j-1}\cdot - k)\rangle \\ &= 2^{(j-1)/2}\langle f, \sum_n \tilde{h}_n \tilde{\varphi}(2^j \cdot -2k-n)\rangle = \frac{1}{\sqrt{2}}\sum_n \tilde{h}_n c_{j,2k+n}. \end{aligned}$$

An analogous relation holds for $d_{j-1,k}$. On the other hand, thanks to (3), given the projection $P_{j-1}f = \sum_k c_{j-1,k}\varphi_{j-1,k}$ we are able to express it in terms of basis functions at the finer scale

$$\begin{aligned} P_{j-1}f &= 2^{(j-1)/2}\sum_k c_{j-1,k}\varphi(2^{j-1}\cdot - k) \\ &= 2^{(j-1)/2}\sum_k c_{j-1,k}\sum_n h_n \varphi(2^j \cdot -2k-n) \\ &= \frac{1}{\sqrt{2}}\sum_k \Big[\sum_n h_{k-2n} c_{j-1,n}\Big]\varphi_{j,k}. \end{aligned}$$

Analogously we have

$$Q_{j-1}f = \frac{1}{\sqrt{2}} \sum_k \Big[\sum_n g_{k-2n} d_{j-1,n} \Big] \varphi_{j,k}.$$

Since $P_j f = P_{j-1} f + Q_{j-1} f$ we immediately get

$$P_j f = \sum_k \frac{1}{\sqrt{2}} \left[\sum_n h_{k-2n} c_{j-1,n} + \sum_n g_{k-2n} d_{j-1,n} \right] \varphi_{j,k}.$$

In summary, the one level decomposition algorithm reads

$$c_{j,n} = \frac{1}{\sqrt{2}} \sum_k \tilde{h}_{k-2n} c_{j+1,k} \qquad d_{j,n} = \frac{1}{\sqrt{2}} \sum_k \tilde{g}_{k-2n} c_{j+1,k}$$

while its inverse, the one level reconstruction algorithm can be written as

$$c_{j+1,k} = \frac{1}{\sqrt{2}} \Big[\sum_n h_{k-2n} c_{j,n} + \sum_n g_{k-2n} d_{j,n} \Big].$$

Once the one level decomposition algorithm is given, giving the coefficient vectors $(c_{j,k})_k$ and $(d_{j,k})_k$ in terms of the coefficient vector $(c_{j+1,k})_k$, we can iterate it to obtain $(c_{j-1,k})_k$ and $(d_{j-1,k})_k$ and so on until we get all the coefficients for the decomposition (14).

2.3 Examples

2.3.1 Example I: Daubechies Wavelets

The Haar basis: The first, simplest, example of a wavelet basis is the Haar basis, which was introduced in 1909 by Alfred Haar as an example of a countable orthonormal system for $L^2(\mathbb{R})$. In the Haar wavelet case V_j is defined to be the space of piecewise constant functions with uniform mesh size $h = 2^{-j}$:

$$V_j = \{w \in L^2(\mathbb{R}) \text{ such that } w|_{I_{j,k}} \text{ is constant}\},$$

where we denote by $I_{j,k}$ the dyadic interval $I_{j,k} := (k2^{-j}, (k+1)2^{-j})$. It is not difficult to see that the sequence $\{V_j,\ j \in \mathbb{Z}\}$ is indeed a multiresolution analysis. In particular an orthonormal basis for V_j is given by the family

$$\varphi_{j,k} := 2^{j/2} \varphi(2^j \cdot -k) \quad \text{with} \quad \varphi = \chi|_{(0,1)}. \tag{15}$$

Letting $P_j : L^2(\mathbb{R}) \to V_j$ be the $L^2(\mathbb{R})$-orthogonal projection onto V_j, clearly we have

$$P_j f = \sum_k c_{j,k}(f)\varphi_{j,k}, \qquad c_{j,k}(f) = \langle f, \varphi_{j,k}\rangle,$$

and the dual multiresolution analysis $\{\tilde{V}_j\}$ coincides with $\{V_j\}$. The space W_j is then the orthogonal complement of V_j in V_{j+1}:

$$V_{j+1} = W_j \oplus V_j, \quad W_j \bot V_j,$$

and the $L^2(\mathbb{R})$-orthogonal projection $Q_j := P_{j+1} - P_j$ onto W_j verifies

$$Q_j f|_{I_{j+1,2k}} = P_{j+1} f|_{I_{j+1,2k}} - (P_{j+1} f|_{I_{j+1,2k}} + P_{j+1} f|_{I_{j+1,2k+1}})/2$$

$$= P_{j+1} f|_{I_{j+1,2k}}/2 - P_{j+1} f|_{I_{j+1,2k+1}}/2, \tag{16}$$

$$Q_j f|_{I_{j+1,2k+1}} = P_{j+1} f|_{I_{j+1,2k+1}} - (P_{j+1} f|_{I_{j+1,2k}} + P_{j+1} f|_{I_{j+1,2k+1}})/2$$

$$= -P_{j+1} f|_{I_{j+1,2k}}/2 + P_{j+1} f|_{I_{j+1,2k+1}}/2. \tag{17}$$

It is then not difficult to realize that we can expand $Q_j f$ as

$$Q_j f = \sum_k d_{j,k}(f)\psi_{j,k} \quad \text{with} \quad \psi_{j,k} = 2^{j/2}\psi(2^j \cdot -k),$$

where

$$\psi := \chi_{(0,1/2)} - \chi_{(1/2,1)}.$$

Since the functions $\psi_{j,k}$ at fixed j are an orthonormal system, they do constitute an orthonormal basis for W_j. We have then

$$d_{j,k}(f) = \langle f, \psi_{j,k}\rangle.$$

Daubechies' compactly supported orthonormal wavelets: In her 1988 groundbreaking paper [29] Ingrid Daubechies managed to generalize the Haar basis and construct a class of MRA's such that both φ and ψ have arbitrarily high regularity R, are supported in $(0, L)$ and they generate by translations and dilations orthonormal bases for the spaces V_j and W_j. The projectors P_j are, as in the Haar case, L^2 orthogonal projectors. Also in this case the scaling function and the dual function coincide and we have $\tilde{V}_j = V_j$, $\tilde{W}_j = W_j$, $\tilde{\varphi} = \varphi$ and $\tilde{\psi} = \psi$ (see Fig. 1 for an example of scaling and wavelet functions in this framework).

A characteristic of Daubechies' wavelets is that, unlike the Haar basis, the spaces V_j and the function φ are not given directly. By giving an algorithm to construct them, Daubechies characterizes all the sequences $h = (h_k)_k$ for which a unique solution φ to the refinement equation (3) exists and is orthogonal to its integer translates and smooth. Once $(h_k)_k$ is built, the spaces V_j are then defined as the span of $\{\varphi_{j,k}, k \in \mathbb{Z}\}$ with $\varphi_{j,k}$ defined by (2). The function φ is, by construction, refinable, and the sequence $\{V_j,\ j \in \mathbb{Z}\}$ is a multiresolution analysis. The algorithm to construct the refinement coefficients and the proof that for a given sequence $(h_k)_k$ satisfying suitable conditions (3) has indeed a solution with the required properties is quite technical and it is beyond our scope here to give more details about such a construction. We refer the interested reader to [30]. The coefficient sequences themselves are available, already computed, in table form at different resource sites over the web (see www.wavelet.org).

It is worth noting that the refinement equation is quite powerful and that it is possible to derive from it a lot of information on the function φ. For instance, if we need to plot the function φ we will need access to point values of such a function. Since the φ is supported in $(0, L)$ we have that $\varphi(n) = 0$ for all $n \in \mathbb{Z}, n \notin (0, L)$. For the remaining integers we can write

$$\varphi(n) = \sum_k h_k \varphi(2n - k) = \sum_\ell h_{2n-\ell}\varphi(\ell).$$

If we consider the $L - 1 \times L - 1$ matrix $H = (h_{n,\ell})$ with $h_{n,\ell} = h_{2n-\ell}$, clearly the vector $(\varphi(1), \dots, \varphi(L-1))$ is an eigenvector of H for the eigenvalue 1, which turns out to be unique. Once the values in the integers are computed by any eigenvector computation algorithm, the values in dyadic points are computed recursively thanks again to the refinement equation, which gives us

$$\varphi\left(\frac{n}{2^j}\right) = \sum_k h_k \varphi\left(\frac{n - 2^{j-1}k}{2^{j-1}}\right).$$

Analogous algorithms are available for computing many quantities which are needed for the application in the numerical solution of PDEs, like for instance point values of derivatives, integrals and integrals of product of derivatives.

2.3.2 Example II: B-Splines

Many applications of wavelets to PDEs are based on the multiresolution analysis generated by the spaces V_j:

$$V_j = \{f \in L^2 \cap C^{N-1} \ :\ f|_{I_{j,k}} \in \mathbb{P}^N\}.$$

A basis for V_j whose elements are compactly supported can be constructed by defining the B-spline B_N of degree N recursively by

$$B_0 := \chi_{[0,1]}, \quad B_N := B_0 * B_{N-1} = (*)^{N+1}\chi_{[0,1]},$$

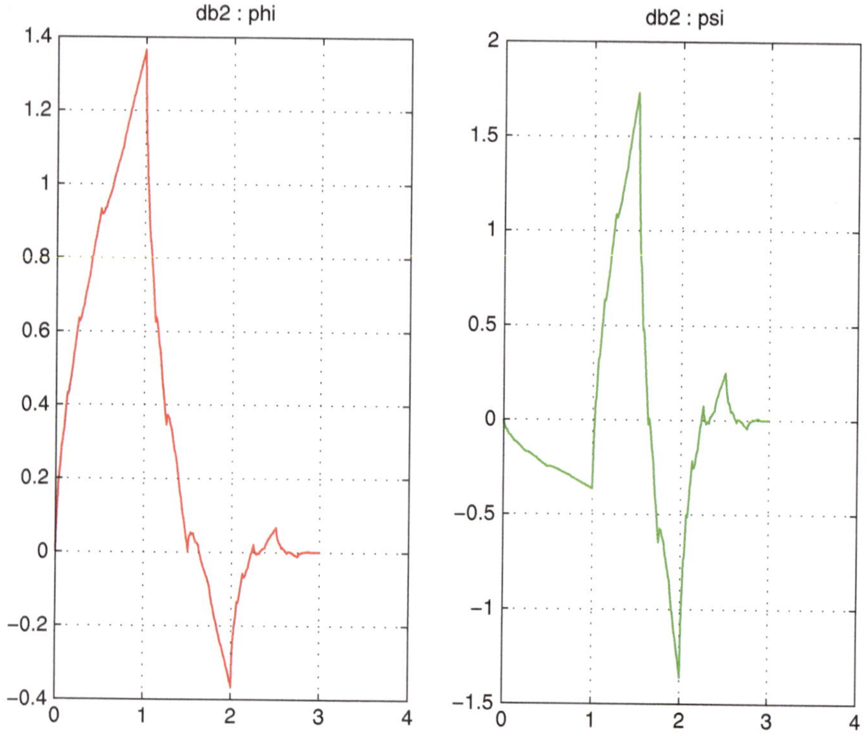

Fig. 1 The scaling and wavelet functions φ and ψ generating a Daubechies' orthonormal wavelet basis

where $*$ denotes the convolution product. The function B_N is supported in $[0, N+1]$, it is refinable and the corresponding scaling coefficients are defined by

$$h_k = \begin{cases} 2^{-N} \binom{N+1}{k}, & 0 \leq k \leq N+1, \\ 0 & \text{otherwise.} \end{cases}$$

The integer translates of the function $\varphi = B_N$ form a Riesz basis for V_j. For N given it is possible to construct an infinite class of compactly supported refinable functions dual to B_N. More precisely, analogously to what is done for Daubechies' wavelets, it is possible [18] to characterize and construct a family of sequences $(\tilde{h}_k)_k$ for which the solution to the refinement equation (9) exists, is dual to the B-spline B_N, has compact support and arbitrarily high smoothness $\tilde{R}$. Figures 2 and 3 show the functions φ, $\tilde{\varphi}$, ψ and $\tilde{\psi}$ for $N = 1$, $\tilde{R} = 0$ and $N = 1$, $\tilde{R} = 1$, respectively.

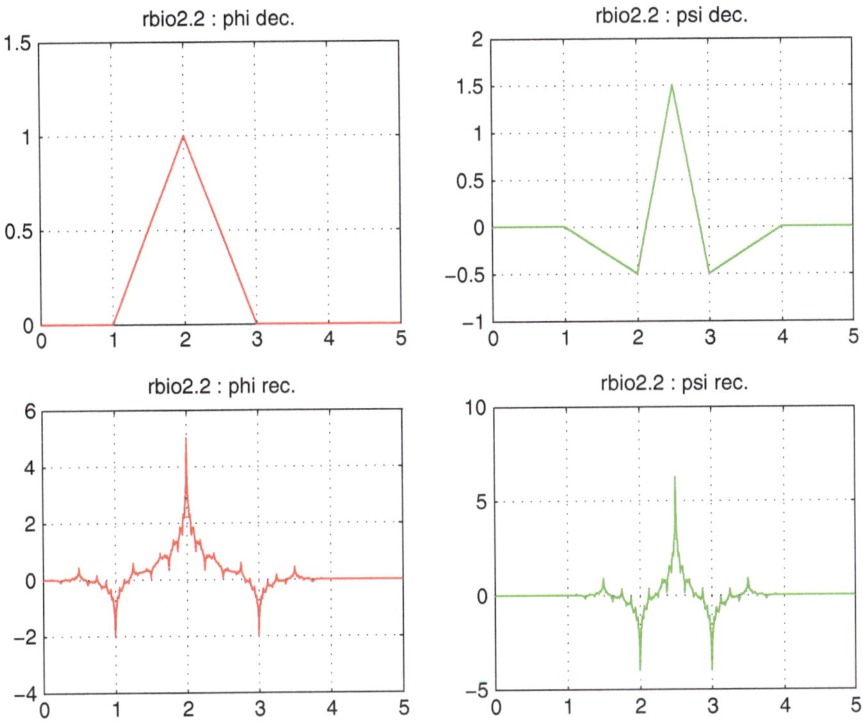

Fig. 2 Scaling and wavelet functions φ and ψ for decomposition (*top*) and the duals $\tilde{\varphi}$ and $\tilde{\psi}$ for reconstruction (*bottom*) corresponding to the biorthogonal basis B2.2

2.3.3 Interpolating Wavelets

It is also interesting to consider an example where L_j is a Lagrangian interpolation operator (see [5, 33]). Clearly, Lagrangian interpolation is not an L^2 bounded operator, consequently interpolating wavelets do not entirely fall in the framework described up to here. However they have some quite useful characteristics that make them particularly well suited for an application to the numerical solution of PDE's.

The Schauder piecewise linear basis: As a first example let us consider the multiresolution analysis generated by the spaces V_j of continuous piecewise linear functions on a uniform mesh with meshsize 2^{-j}

$$V_j = \{w \in C^0(\mathbb{R}) : w \text{ is linear on } I_{j,k},\ k \in \mathbb{Z}\}.$$

We can easily construct a basis for V_j out of the dilated and translated of the "hat function":

$$V_j = span\{\vartheta_{j,k},\ k \in \mathbb{Z}\} \quad \text{with} \quad \vartheta_{j,k} := 2^{j/2}\vartheta(2^j \cdot - k),$$

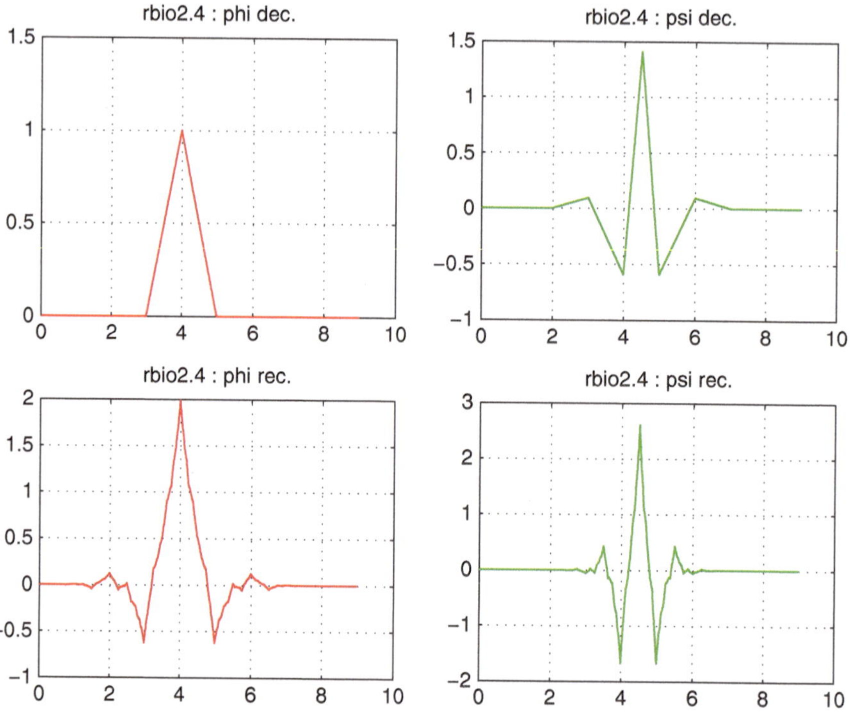

Fig. 3 Example of biorthogonal wavelet basis. Scaling and wavelet functions φ and ψ for decomposition (*top*) and the duals $\tilde{\varphi}$ and $\tilde{\psi}$ for reconstruction (*bottom*) corresponding to the basis B2.4. Remark that the scaling function for decomposition is the same as for the basis B2.2. In both cases V_j is the space of piecewise linears

$$\vartheta(x) = \max\{0, 1 - |x|\}. \tag{18}$$

This basis is a Riesz basis. Remark that the hat function ϑ is the B-spline of order one. The multiresolution analysis V_j itself falls then in the framework described in Sect. 2.3.2 and there exist a whole family of dual multiresolution analyses and of associated wavelets (in Fig. 2 we see one of the possible dual functions). We consider here instead a more straightforward approach. We observe that $f_j \in V_j$ is uniquely determined by its point values at the mesh points $k2^{-j}$. Assuming that f is sufficiently regular we can consider the interpolant $f_j = L_j f$, with L_j denoting the Lagrange interpolation operator: $L_j : C^0(\mathbb{R}) \to V_j$ is defined by

$$L_j f(k2^{-j}) = f(k2^{-j}).$$

It is not difficult to realize that

$$L_j f = \sum_k c_{j,k}(f)\varphi_{j,k}, \quad c_{j,k}(f) = 2^{-j/2} f(2^{-j}k).$$

Remark that L_j is a "projector" ($f \in V_j$ implies $L_j f = f$) but not an $L^2(\mathbb{R})$ bounded projector (it is not even well defined in L^2). Clearly we cannot find an L^2 function $\tilde{\varphi}$ allowing to write L_j in the form (8). However, if we allow ourselves to take $\tilde{\varphi}$ to be the Dirac's delta in the origin ($\tilde{\varphi} = \delta_{x=0}$, see [5]) we see that the basic structure of the whole construction is preserved. Once again $V_j \subset V_{j+1}$ and $L_j f$ can be derived from $L_{j+1} f$ by interpolation

$$2^{j/2} c_{j,k}(f) = f(k2^{-j}) = f(2k2^{-(j+1)}) = 2^{(j+1)/2} c_{j+1,2k}(f).$$

Also in this case we can compute the details that we loose in projecting $L_{j+1} f$ onto V_j by introducing the difference operator $Q_j = L_{j+1} - L_j$.

We observe that the details $Q_j f$ at level j vanish at the mesh points at level j. In fact $L_j f(k2^{-j}) = f(k2^{-j}) = L_{j+1} f(k2^{-j})$ implies

$$Q_j f(k2^{-j}) = 0.$$

We can then expand $Q_j f$ as

$$Q_j f = \sum d_{j,k}(f) \psi_{j,k} \quad \text{with} \quad \psi_{j,k} = 2^{j/2} \psi(2^j \cdot -k),$$

where

$$\psi(x) = \vartheta(2x - 1).$$

This time, the "wavelets" $\psi_{j,k}$ are then simply those nodal functions at level $j+1$ associated to nodes that belong to the fine but not to the coarse grid.

Remark that for j going to infinity, $L_j f$ converges uniformly to f provided f is uniformly continuous. Then, if f uniformly continuous and compactly supported, it can be expressed as the uniform (but not unconditional) limit

$$f = \sum_k c_{0,k}(f) \varphi_{0,k} + \sum_{0 \le j} \sum_k d_{j,k}(f) \psi_{j,k}.$$

Donoho's interpolating wavelets: The Donoho's Interpolating wavelets generalize the example of the Schauder piecewise linear basis. The observation that the hat function (18) can be obtained as autocorrelation of the box function $\varphi = \chi_{(0,1)}$ ($\vartheta = \varphi(\cdot) * \varphi(-\cdot)$) and that the interpolation property $\vartheta(n) = \delta_{0,n}$ is a consequence of the orthogonality of the box functions to its integer translates,

$$\vartheta(n) = \int_{\mathbb{R}} \varphi(x) \varphi(n - x) = \delta_{0,n},$$

suggests to define, in general, an interpolating scaling function ϑ as the autocorrelation of a Daubechies' function φ [5]. Let in fact φ be a C^R normalized compactly supported refinable function, orthogonal to its integer translates. Its autocorrelation

$$\vartheta = \varphi(\cdot) * \varphi(-\cdot)$$

satisfies (as it is easily verified) the following properties:

(a) ϑ is compactly supported (if supp $\varphi \subseteq [0, N]$ then supp $\vartheta \subseteq [-N, N]$)
(b) $\vartheta \in C^{2R}$
(c) ϑ is refinable: this is quite easily seen by observing that the refinement equation (3) rewrites, in the Fourier domain, as $\hat{\varphi}(\xi) = m_0(\xi/2)\hat{\varphi}(\xi/2)$ with $m_0 = \sum_k h_k e^{-ik\xi}$ being a trigonometric polynomial (since for the Daubechies function the refinement coefficients sequence h has finite length); observing that $\widehat{\vartheta} = |\widehat{\varphi}|^2$, we immediately see that

$$\widehat{\vartheta}(\xi) = |m_0(\xi/2)|^2 \widehat{\vartheta}(\xi/2)$$

(d) ϑ satisfies the following interpolation property:

$$\vartheta(n) = \int_{\mathbb{R}} \varphi(y)\varphi(y-n)\,dy = \delta_{n,0}$$

Remark 3. If we take φ to be one of the so called minimal phase Daubechies scaling function, the corresponding function ϑ turns out to be, as pointed out by Beylkin and Saito [40], a *Deslaurier–Dubuc* interpolating function, which was originally introduced by Deslaurier and Dubuc [31] as the limit function of an interpolatory subdivision scheme.

Let us now introduce the functions $\vartheta_{j,k}(x) = 2^{j/2}\vartheta(2^j x - k)$ and the spaces

$$V_j = span\{\vartheta_{j,k},\ k \in \mathbb{Z}\}.$$

The refinability of ϑ implies that the sequence $\{V_j\}$ is nested. Moreover it is possible to prove that for all j the functions $\vartheta_{j,k}$ constitute a Riesz basis for V_j [5], and that the union of the V_j's is dense in L^2. The sequence V_j is then a multiresolution analysis. In order to form complement spaces we follow the same approach as for the Schauder basis, that is we define L_j to be the interpolation operator, that, thanks to the interpolation property of ϑ takes the form

$$L_j f = \sum_k 2^{-j/2} f(k/2^{-j})\vartheta_{j,k}.$$

We can then define Q_j as $Q_j = L_{j+1} - L_j$ and W_j as $W_j = Q_j V_{j+1}$. As in the piecewise linear Schauder basis case, it is not difficult to see that, setting

$$\psi(x) = \vartheta(2x-1), \quad \psi_{j,k} = 2^{j/2}\psi(2^j x - k) = 2^{j/2}\vartheta(2^{j+1}x - (2k+1)), \quad (19)$$

the set $\{\psi_{j,k},\ k \in \mathbb{Z}\}$ is a Riesz basis for W_j and that uniformly continuous and compactly supported functions f can be expressed as the uniform (but not unconditional) limit

$$f = \sum_k c_{0,k}(f)\varphi_{0,k} + \sum_{0 \leq j} \sum_k d_{j,k}(f)\psi_{j,k}.$$

2.4 Beyond $L^2(\mathbb{R})$

What we built in Sect. 2.2 for the space $L^2(\mathbb{R})$ is a complex structure consisting in:

(a) Two coupled multiresolution analyses V_j and $\tilde{V}_j$ (possibly coinciding, in the orthogonal case)
(b) Two sequences of adjoint projectors $P_j : L^2(\mathbb{R}) \to V_j$ and $\tilde{P}_j = P_j^* : L^2(\mathbb{R}) \to \tilde{V}_j$, both verifying a commutativity property of the form (4)
(c) Two dual refinable functions φ and $\tilde{\varphi}$ (the scaling functions) which, by translation and dilation generate biorthogonal bases for the V_j's and the $\tilde{V}_j$'s, respectively, and that allow to write the two projectors P_j and $\tilde{P}_j$ in the form (8)
(d) A sequence of complement spaces W_j (and it is easy to build a second sequence $\tilde{W}_j$ of spaces complementing $\tilde{V}_j$ in $\tilde{V}_{j+1}$)
(e) Two functions ψ and $\tilde{\psi}$ which, by contraction and dilation generate biorthogonal bases for the W_j's and the $\tilde{W}_j$'s
(f) A fast change of basis algorithm, allowing to go back and forth from the coefficients of a given function in V_j with respect to the nodal basis $\{\varphi_{j,k},\ k \in \mathbb{Z}\}$ to the coefficients of the same function with respect to the hierarchical wavelet basis $\{\varphi_{0,k}, k \in \mathbb{Z}\} \cup_{m=0}^{j-1} \{\psi_{m,k}, k \in \mathbb{Z}\}$

In view of the use of wavelets for the solution of PDE's, we would like to have a similar structure for more general domains, also in dimension greater than one. Actually, wavelets for $L^2(\mathbb{R}^d)$ are quite easily built by tensor product and we have basically the same structure as in dimension one (see e.g. [15]). If, on the other hand, we want to build wavelets defined on general, possibly bounded, domains, it is clear that we have to somehow loosen the definition of what a wavelet is. In particular it is clear that for bounded domains we cannot ask for the translation and dilation invariance properties of the spaces V_j and the bases cannot possibly be constructed by contracting and translating a single function φ.

Let us then see which elements and properties of the above structure it is possible to maintain when replacing the domain $\mathbb{R}$ with a general domain $\Omega \subseteq \mathbb{R}^d$. As we did for $\mathbb{R}$, we will start with a nested sequence $\{V_j\}_{j\geq 0}$, $V_j \subset V_{j+1}$, of closed subspaces of $L^2(\Omega)$, corresponding to discretizations with mesh-size 2^{-j}. We will still assume that the union of the V_j's is dense in $L^2(\Omega)$:

$$L^2(\Omega) = \overline{\cup_j V_j}. \tag{20}$$

We will also assume that we have a Riesz basis for V_j of the form $\{\varphi_\mu,\ \mu \in K_j\}$ such that

$$V_j = span < \varphi_\mu,\ \mu \in K_j >,$$

where $K_j \subseteq \{(j,k),\ k \in \mathbb{Z}^d\}$ will denote a suitable set of multi-indexes (for $\Omega = \mathbb{R}$ $K_j = \{(j,k),\ k \in \mathbb{Z}\}$). Clearly, as already observed, it will not be possible to assume the existence of a single function φ such that all the basis function φ_μ are obtained by dilating and translating φ. However remark that a great number of MRA's in bounded domains are built starting from an MRA for $L^2(\mathbb{R}^d)$ with scaling function φ compactly supported [19]. In such case, all the basis functions of the original MRA for $L^2(\mathbb{R}^d)$ whose support is strictly embedded in Ω are retained as basis functions for the V_j on Ω.

We now want to build a wavelet basis. To this aim we will need to introduce either a sequence of bounded projectors $P_j : L^2(\Omega) \to V_j$ satisfying $P_j P_{j+1} = P_j$ (note that $V_j = P_j(L^2(\Omega))$ and that $V_j \subset V_{j+1}$ implies $P_{j+1} P_j = P_j$) or, equivalently, a nested sequence of dual spaces $\tilde{V}_j$ satisfying the two inf-sup conditions of the form (10). Remark that, as it happens in the $L^2(\mathbb{R})$ case, choosing P_j is equivalent to choosing $\tilde{V}_j$. The existence of a biorthogonal Riesz basis $\{\tilde{\varphi}_\mu,\ \mu \in K_j\}$ such that

$$\tilde{V}_j = P_j^*(L^2(\Omega)) = span < \tilde{\varphi}_\mu,\ \mu \in K_j >,$$

and such that

$$P_j f = \sum_{\mu \in K_j} \langle f, \tilde{\varphi}_\mu \rangle \varphi_\mu, \qquad P_j^* f = \sum_{\mu \in K_j} \langle f, \varphi_\mu \rangle \tilde{\varphi}_\mu$$

is easily deduced as in the $L^2(\mathbb{R})$ case. Again, it will not generally be possible to obtain the basis functions $\tilde{\varphi}_\mu$ by dilation and translation of a single function $\tilde{\varphi}$.

As we did for $\mathbb{R}$ we can then introduce the difference spaces

$$W_j = Q_j(L^2(\Omega)), \quad Q_j = P_{j+1} - P_j.$$

We next need to construct a basis for W_j. This is in general a quite technical task, heavily depending on the particular characteristics of the spaces V_j and $\tilde{V}_j$. It's worth mentioning that, once again, if the MRA for Ω is built starting from an MRA for $L^2(\mathbb{R}^d)$ with compactly supported scaling function φ and if the wavelets themselves are compactly supported, then the basis for W_j will include all those wavelet functions on $\mathbb{R}^d$ whose support, as well as the support of the corresponding dual, are included in Ω. It is well beyond the scope of this paper to go into the details of one or another construction of the basis for W_j. In any case, independently of the particular approach used, we will end up with a Riesz basis for W_j of the form $\{\psi_\lambda,\ \lambda \in \nabla_j\}$:

$$W_j = span < \psi_\lambda,\ \lambda \in \nabla_j >,$$

where ∇_j is again a suitable multi-index set with, in the case of bounded domains,

$$\#(\nabla_j) + \#(K_j) = \#(K_{j+1}).$$

At the same time we will end up with a Riesz basis for the dual space $\tilde{W}_j = (P_{j+1}^* - P_j^*)(L^2(\Omega))$:

$$\tilde{W}_j = span\{\tilde{\psi}_\lambda,\ \lambda \in \nabla_j\}.$$

The two bases can be chosen in such a way that they satisfy a biorthogonality relation

$$\langle \psi_\mu, \tilde{\psi}_{\mu'} \rangle = \delta_{\mu,\mu'},\ \mu, \mu' \in \nabla_j,$$

so that the projection operator Q_j can be expanded as

$$Q_j f = \sum_{\lambda \in \nabla_j} \langle f, \tilde{\psi}_\lambda \rangle \psi_\lambda.$$

Moreover it is not difficult to check that we have an orthogonality relation across scales:

$$\lambda \in \nabla_j,\ \lambda' \in \nabla_{j'},\ j \neq j' \Rightarrow \langle \psi_\lambda, \tilde{\psi}_{\lambda'} \rangle = 0, \qquad \mu \in K_{j'},\ j' \leq j \Rightarrow \langle \psi_\lambda, \varphi_\mu \rangle = 0.$$

In summary we have a multiscale decomposition of V_j as

$$V_j = V_0 \oplus W_0 \oplus \cdots \oplus W_{j-1},$$

and for any $f \in L^2(\Omega)$ we can then write

$$P_j f = \sum_{\mu \in K_j} c_\mu \varphi_\mu = \sum_{\mu \in K_0} c_\mu \varphi_\mu + \sum_{m=0}^{j-1} \sum_{\lambda \in \nabla_m} d_\lambda \psi_\lambda$$

with $d_\lambda = \langle f, \tilde{\psi}_\lambda \rangle$ and $c_\mu = \langle f, \tilde{\varphi}_\mu \rangle$. Since the density property (20) implies that

$$\lim_{j \to +\infty} \| f - P_j f \|_{L^2(\Omega)} = 0,$$

taking the limit as j goes to $+\infty$ allows us to write

$$f = \sum_{\mu \in K_0} c_\mu \varphi_\mu + \sum_{j \geq 0} \sum_{\lambda \in \nabla_j} \langle f, \tilde{\psi}_\lambda \rangle \psi_\lambda. \tag{21}$$

Remark 4. A general strategy to build bases with the required characteristics for $]0,1[^d$ out of the bases for $\mathbb{R}^d$ has been proposed in several papers [19, 35]. To actually build wavelet bases for general bounded domains, several strategies have been considered. Following the same strategy as for the construction of wavelet bases for cubes, *wavelet frames* [14] for $L^2(\Omega)$ (Ω Lipschitz domain) can be constructed according to [20]. The most popular approach nowadays is domain decomposition: the domain Ω is split as the disjoint union of tensorial subdomains Ω_ℓ and a wavelet basis for Ω is constructed by suitably assembling wavelet bases for the Ω_ℓ's [12,16,26]. The construction is quite technical, since it is not trivial to retain in the assembling procedure all the relevant properties of the wavelets. Alternatively we can think of building wavelets for general domains directly, without starting from a construction on $\mathbb{R}$. This is for instance the case of finite element wavelets (see e.g. [27]).

2.4.1 Interpolating Wavelets on Cubes

In view of an application in the framework of an adaptive collocation method let us consider in some more detail the construction of interpolating wavelets on the unit square. We start by constructing an interpolating MRA on the unit interval. Let a Deslaurier–Dubuc interpolating scaling function ϑ be given (see Remark 3). Following Donoho [33], we introduce the Lagrange interpolation polynomials

$$l_k^\flat = \prod_{i=0, i\neq k}^{N} \frac{x - k2^{-j}}{2^{-j}(i-k)}, \quad l_k^\sharp = \prod_{i=2^j-N, i\neq k}^{2^j} \frac{x - k2^{-j}}{2^{-j}(i-k)},$$

and for $k = 0, \dots, 2^j$ we define

$$\vartheta_{j,k}^{\Box} = \vartheta_{j,k} + \sum_{\ell=-N}^{-1} l_k^\flat(2^{-j}\ell)\vartheta_{j,\ell}, \quad k = 0, \dots, N,$$

$$\vartheta_{j,k}^{\Box} = \vartheta_{j,k}, \quad k = N+1, \dots, 2^j - (N+1),$$

$$\vartheta_{j,k}^{\Box} = \vartheta_{j,k} + \sum_{\ell=2^j+1}^{2^j+N} l_k^\sharp(2^{-j}\ell)\vartheta_{j,\ell}, \quad k = 2^j - N, \dots, 2^j.$$

We set $V_j^{\Box} = span < \vartheta_{j,k}^{\Box}, k = 0, \dots, 2^j >$. The sequence $V_j^{\Box}$ forms indeed a MRA for $(0,1)$ and the basis functions are, by construction, interpolatory.

By tensor product we then easily define a multiresolution on the square: introducing the two dimensional scaling functions $\Theta_{j,\mathbf{k}}$, $\mathbf{k} = (k_1, k_2) \in \{0, \dots, 2^j\}^2$ defined as

$$\Theta_{j,\mathbf{k}} = \vartheta_{j,k_1}^{\Box} \otimes \vartheta_{j,k_2}^{\Box},$$

we set

$$V_j = span < \Theta_\mu, \ \mu \in K_j >$$

with $K_j = \{(j,\mathbf{k}), \ \mathbf{k} \in \{0, \dots, 2^j\}^2\}$. It is immediate to define an interpolation operator $L_j : C^0([0,1]^2) \to V_j$

$$L_j f = \sum_{\mu=(j,\mathbf{k})\in K_j} 2^{-j} f(\zeta_\mu)\Theta_\mu, \quad \zeta_{j,\mathbf{k}} = \mathbf{k}/2^j.$$

The wavelet basis functions for the complement space

$$W_j = (L_{j+1} - L_j)V_{j+1}$$

are the functions

$$\Psi_{j,\mathbf{k}}^{(1,0)} = \vartheta^{\Box}_{j+1,2k_1-1} \otimes \vartheta^{\Box}_{j,k_2}, \tag{22}$$

$$\Psi_{j,\mathbf{k}}^{(0,1)} = \vartheta^{\Box}_{j,k_1} \otimes \vartheta^{\Box}_{j+1,2k_2-1}, \tag{23}$$

$$\Psi_{j,\mathbf{k}}^{(1,1)} = \vartheta^{\Box}_{j+1,k_1-1} \otimes \vartheta^{\Box}_{j+1,2k_2-1}. \tag{24}$$

In the following we will need to handle the grid points corresponding to the basis functions. To this aim the following notation will be handy: the grid-points corresponding to the wavelets will be indicated as

$$\xi_{j,\mathbf{k}}^{(1,0)} = ((2k_1 - 1)2^{-(j+1)}, k_2 2^{-j}),$$

$$\xi_{j,\mathbf{k}}^{(0,1)} = (k_1 2^{-j}, (2k_2 - 1)2^{-(j+1)}),$$

$$\xi_{j,\mathbf{k}}^{(1,1)} = ((2k_1 - 1)2^{-(j+1)}, (2k_2 - 1)2^{-(j+1)}).$$

Letting

$$\nabla_j = \{(\eta, j, \mathbf{k}),\ \eta \in \{0,1\}^2 \backslash \{0,0\}, \mathbf{k} \in \mathbb{Z}^2 \text{ such that } \xi^{\eta}_{j,\mathbf{k}} \in [0,1]^2\}$$

and $\Lambda = \cup_{j \geq j_0} \nabla_j$, the same arguments as for the interpolating wavelets on $\mathbb{R}$ will allow us to expand any function $f \in C^0([0,1]^2)$ as

$$f = \sum_{\mu \in K_{j_0}} \beta_\mu \Theta_\mu + \sum_{\lambda \in \Lambda} \alpha_\lambda \Psi_\lambda.$$

3 The Fundamental Property of Wavelets

In the previous section we saw in some detail what a couple of biorthogonal multiresolution analyses is, and how this structure allows to build a wavelet basis. However we have yet to introduce the one property that makes of wavelets the powerful tool that they are and that is probably their fundamental characteristics: the *simultaneous good localization in both space and frequency*.

We put ourselves in the framework described in Sect. 2.4. Let us start by writing the wavelet expansion in an even more compact form, by introducing the notation

$$\nabla_{-1} = K_0, \quad \text{and for} \quad \lambda \in \nabla_{-1}, \quad \psi_\lambda := \varphi_\lambda, \quad \tilde{\psi}_\lambda := \tilde{\varphi}_\lambda,$$

which allows us to rewrite the expansion (21) as

$$f = \sum_{j=-1}^{+\infty} \sum_{\lambda \in \nabla_j} \langle f, \tilde{\psi}_\lambda \rangle \psi_\lambda.$$

We will see in the next section that, under quite mild assumptions on φ, $\tilde{\varphi}$, ψ and $\tilde{\psi}$, the convergence in the expansion (21) turns out to be unconditional. This will allow us to introduce a global index set $\Lambda = \cup_{j=-1}^{+\infty} \nabla_j$ and to write

$$f = \sum_{\lambda \in \Lambda} \langle f, \tilde{\psi}_\lambda \rangle \psi_\lambda = \sum_{\lambda \in \Lambda} \langle f, \psi_\lambda \rangle \tilde{\psi}_\lambda. \tag{25}$$

Such formalism will also be valid for the case $\Omega = \mathbb{R}$, where, for $j \geq 0$, $\nabla_j = K_j = \{(j,k),\ k \in \mathbb{Z}\}$. In this case for $\lambda = (j,k)$ we will have

$$\varphi_\lambda = \varphi_{j,k} = 2^{j/2}\varphi(2^j x - k), \qquad \tilde{\varphi}_\lambda = \tilde{\varphi}_{j,k} = 2^{j/2}\tilde{\varphi}(2^j x - k),$$

$$\psi_\lambda = \psi_{j,k} = 2^{j/2}\psi(2^j x - k), \qquad \tilde{\psi}_\lambda = \tilde{\psi}_{j,k} = 2^{j/2}\tilde{\psi}(2^j x - k).$$

3.1 The Case $\Omega = \mathbb{R}$: The Frequency Domain Point of View vs. the Space Domain Point of View

As we saw in the previous section, in the classical construction of wavelet bases for $L^2(\mathbb{R})$ [39], all basis functions φ_λ, $\lambda \in K_j$ and ψ_λ, $\lambda \in \nabla_j$ with $j \geq 0$, as well as their duals $\tilde{\varphi}_\lambda$ and $\tilde{\psi}_\lambda$, are constructed by translation and dilation of a single *scaling function* φ and a single *mother wavelet* ψ (resp. $\tilde{\varphi}$ and $\tilde{\psi}$). Clearly, the properties of the function ψ will transfer to the functions ψ_λ and will imply properties of the corresponding wavelet basis.

To start with, we will then restrict our framework by making some additional assumptions on φ and ψ as well as on their duals $\tilde{\varphi}$ and $\tilde{\psi}$. The first assumption deals with *space localization*. In view of an application to the numerical solution of PDE's we make such an assumption in quite a strong form: we ask that there exists an $L > 0$ and an $\tilde{L} > 0$ such that (with $\lambda = (j,k)$)

$$\text{supp}\varphi \subseteq [-L, L] \quad \Longrightarrow \quad \text{supp}\varphi_\lambda \subseteq [(k-L)/2^j, (k+L)/2^j], \tag{26}$$

$$\text{supp}\tilde{\varphi} \subseteq [-\tilde{L}, \tilde{L}] \quad \Longrightarrow \quad \text{supp}\tilde{\varphi}_\lambda \subseteq [(k-\tilde{L})/2^j, (k+\tilde{L})/2^j], \tag{27}$$

$$\text{supp}\psi \subseteq [-L, L] \quad \Longrightarrow \quad \text{supp}\psi_\lambda \subseteq [(k-L)/2^j, (k+L)/2^j], \tag{28}$$

$$\text{supp}\tilde{\psi} \subseteq [-\tilde{L}, \tilde{L}] \quad \Longrightarrow \quad \text{supp}\tilde{\psi}_\lambda \subseteq [(k-\tilde{L})/2^j, (k+\tilde{L})/2^j], \tag{29}$$

that is, both the wavelet ψ_λ ($\lambda = (j,k)$) and its dual $\tilde{\psi}_\lambda$ will be supported around the point $x_\lambda = k/2^j$, and the size of their support will be of the order of 2^{-j}.

Now let us consider the Fourier transform $\hat{\psi}(\xi)$ of $\psi(x)$. Since ψ is compactly supported, by the Heisenberg indetermination principle, $\hat{\psi}$ cannot be itself compactly supported. However we assume that it is localized in some weaker sense around the frequency 1. More precisely we assume that there exist an integer $M > 0$ and an integer $R > 0$, with $M > R$, such that for $n = 0, \ldots, M$ and for s such that $0 \leq s \leq R$ one has

$$\text{(a)} \quad \frac{d^n\hat{\psi}}{d\xi^n}(0) = 0 \qquad \text{and} \qquad \text{(b)} \quad \int_{\mathbb{R}} (1+|\xi|^2)^s |\hat{\psi}(\xi)|^2 \, d\xi \lesssim 1. \tag{30}$$

Analogously, for $\tilde{\psi}$ we assume that there exist integers $\tilde{M} > \tilde{R} > 0$ such that for $n = 0, \ldots, \tilde{M}$ and for s such that $0 \leq s \leq \tilde{R}$ one has

$$\text{(a)} \quad \frac{d^n\hat{\tilde{\psi}}}{d\xi^n}(0) = 0 \qquad \text{and} \qquad \text{(b)} \quad \int_{\mathbb{R}} (1+|\xi|^2)^s |\hat{\tilde{\psi}}(\xi)|^2 \, d\xi \lesssim 1. \tag{31}$$

The frequency localisation property (30) can be rephrased directly in terms of the function ψ, rather than in terms of its Fourier transform: in fact (30) is equivalent to

$$\int_{\mathbb{R}} x^n \psi(x) \, dx = 0, n = 0, \ldots, M \quad \text{and} \quad \|\psi\|_{H^s(\mathbb{R})} \lesssim 1, 0 \leq s \leq R, \tag{32}$$

which, by a simple scaling argument implies for $\lambda \in \nabla_j$

$$\int_{\mathbb{R}} x^n \psi_\lambda(x) \, dx = 0, n = 0, \ldots, M \quad \text{and} \quad \|\psi_\lambda\|_{H^s(\mathbb{R})} \lesssim 2^{js}, 0 \leq s \leq R. \tag{33}$$

Analogously, we can write for $\tilde{\psi}_\lambda$

$$\int_{\mathbb{R}} x^n \tilde{\psi}_\lambda(x) \, dx = 0, n = 0, \ldots, \tilde{M} \quad \text{and} \quad \|\tilde{\psi}_\lambda\|_{H^s(\mathbb{R})} \lesssim 2^{js}, 0 \leq s \leq \tilde{R}. \tag{34}$$

In the following we will require that also the functions φ and $\tilde{\varphi}$ have some frequency localization property or, equivalently, some smoothness. More precisely we will ask that for all s and $\tilde{s}$ such that, respectively, $0 \leq s \leq R$ and $0 \leq \tilde{s} \leq \tilde{R}$ we have that

$$\text{(a)} \int_{\mathbb{R}} (1+|\xi|^2)^s |\hat{\varphi}(\xi)|^2 \, d\xi \lesssim 1 \text{ and (b)} \int_{\mathbb{R}} (1+|\xi|^2)^{\tilde{s}} |\hat{\tilde{\varphi}}(\xi)|^2 \, d\xi \lesssim 1, \tag{35}$$

or, equivalently, that

$$\text{(a)} \quad \varphi \in H^R(\mathbb{R}) \qquad \text{and} \qquad \text{(b)} \quad \tilde{\varphi} \in H^{\tilde{R}}(\mathbb{R}). \tag{36}$$

Remark 5. Heisenberg's indetermination principle states that a function cannot be arbitrarily well localized both in space and frequency. More precisely, introducing the *position uncertainty* Δx_λ and the *momentum uncertainty* $\Delta \xi_\lambda$ defined by

$$\Delta x_\lambda := \left(\int (x - x_\lambda)^2 |\psi_\lambda(x)|^2 \, dx \right)^{1/2},$$

$$\Delta \xi_\lambda := \left(\int (\xi - \xi_\lambda)^2 |\hat{\psi}_\lambda(\xi)|^2 \, d\xi \right)^{1/2},$$

with $x_\lambda = x_{j,k} = k/2^j$ and $\xi_\lambda = \xi_{j,k} \sim 2^j$ defined by $\xi_\lambda = \int_{\mathbb{R}} \xi |\hat{\psi}_\lambda(\xi)|^2 \, d\xi$, one necessarily has $\Delta x_\lambda \cdot \Delta \xi_\lambda \geq 1$. In our case $\Delta x_\lambda \cdot \Delta \xi_\lambda \lesssim 1$, that is wavelets are simultaneously localized in space and frequency nearly as well as possible.

The frequency localization property of wavelets (30) and (32) can be rephrased in yet a third way as a *local polynomial reproduction* property [15].

Lemma 1. *Let (26)–(29) hold. Then (31) holds if and only if for all polynomial p of degree $d \leq \tilde{M}$ we have*

$$p = \sum_k \langle p, \tilde{\varphi}_{j,k} \rangle \varphi_{j,k}. \tag{37}$$

Analogously (30) holds if and only if for all polynomials p of degree $d \leq M$ we have

$$p = \sum_k \langle p, \varphi_{j,k} \rangle \tilde{\varphi}_{j,k}. \tag{38}$$

Remark that the expressions on the right hand side of both (37) and (38) are well defined pointwise thanks the support compactness property of φ and $\tilde{\varphi}$.

Before going on in seeing what the space-frequency localisation properties of the basis function ψ (and consequently of the wavelets ψ_λ's) imply, let us consider functions with a stronger frequency localisation. Let us then for a moment drop the assumption that ψ and φ are compactly supported and assume instead that their Fourier transform verify

$$\text{supp}(\hat{\psi}) \subseteq [-2, -1] \cup [1, 2], \quad \text{supp}(\hat{f}) \subseteq [-1, 1] \quad \forall f \in V_0.$$

Since for $\lambda \in \nabla_j$ one can then easily check that $\text{supp}(\hat{\psi}_\lambda) \subset [-2^{j+1}, -2^j] \cup [2^j, 2^{j+1}]$, one immediately obtains the following equivalence: letting $f = \sum_\lambda f_\lambda \psi_\lambda$

$$\|f\|^2_{H^s(\mathbb{R})} = \int_{\mathbb{R}} (1 + |\xi|^2)^s |\hat{f}(\xi)|^2 \, d\xi \simeq \sum_j 2^{2js} \| \sum_{\lambda \in \nabla_j} f_\lambda \hat{\psi}_\lambda \|^2_{L^2(\mathbb{R})}. \tag{39}$$

By taking the inverse Fourier transform on the right hand side we immediately see that

$$\|f\|^2_{H^s(\mathbb{R})} \simeq \sum_j 2^{2js} \| \sum_{\lambda\in\nabla_j} f_\lambda \psi_\lambda \|^2_{L^2(\mathbb{R})}.$$

If $\{\psi_\lambda,\ \lambda \in \nabla_j\}$ is a Riesz basis for W_j, (39) implies then

$$\|f\|^2_{H^s(\mathbb{R})} = \|\sum_{\lambda\in\Lambda} f_\lambda\psi_\lambda\|^2_{H^s(\mathbb{R})} \simeq \sum_{j\geq -1} 2^{2js} \sum_{\lambda\in\nabla_j} |f_\lambda|^2. \tag{40}$$

If we only consider partial sums, we easily derive direct and inverse inequalities, namely:

$$\|\sum_{j=-1}^{J}\sum_{\lambda\in\nabla_j} f_\lambda\psi_\lambda\|_{H^s(\mathbb{R})} \lesssim 2^{Js} \|\sum_{j=-1}^{J}\sum_{\lambda\in\nabla_j} f_\lambda\psi_\lambda\|_{L^2(\mathbb{R})} \tag{41}$$

and

$$\|\sum_{j=J+1}^{\infty}\sum_{\lambda\in\nabla_j} f_\lambda\psi_\lambda\|_{L^2(\mathbb{R})} \lesssim 2^{-Js} \|\sum_{j=J+1}^{\infty}\sum_{\lambda\in\nabla_j} f_\lambda\psi_\lambda\|_{H^s(\mathbb{R})}. \tag{42}$$

Properties (41) and (42), which, as we saw, are easily proven if $\hat{\psi}$ is compactly supported, go on holding, though their proof is less evident, in the case of ψ compactly supported, provided (30) and (32) hold. The same is true for property (40). More precisely, by exploiting the polynomial reproduction properties (37) and (38) it is possible to prove the following inequalities [15].

Theorem 2. *For s with $0 \leq s \leq \tilde{M} + 1$, $f \in H^s(\mathbb{R})$ implies*

$$\|f - P_j f\|_{L^2(\mathbb{R})} \lesssim 2^{-sj} |f|_{H^s(\mathbb{R})}. \tag{43}$$

Analogously, for $0 \leq s \leq M + 1$, $f \in H^s(\mathbb{R})$ implies

$$\|f - \tilde{P}_j f\|_{L^2(\mathbb{R})} \lesssim 2^{-sj} |f|_{H^s(\mathbb{R})}. \tag{44}$$

Applying the above theorem to $g = f - P_j f$ and observing that $g - P_j g = g$ we immediately obtain the bound

$$\|(I - P_j)f\|_{L^2(\mathbb{R})} \lesssim 2^{-js} \|(I - P_j)f\|_{H^s(\mathbb{R})}. \tag{45}$$

The following theorem also holds under the assumptions made at the beginning of this section.

Theorem 3 (Inverse inequality). *For all $f \in V_j$ and for all r with $0 \leq r \leq R$ it holds that*

$$\|f\|_{H^r(\mathbb{R})} \lesssim 2^{jr} \|f\|_{L^2(\mathbb{R})}. \tag{46}$$

Analogously for all $f \in \tilde{V}_j$ and for all r with $0 \leq r \leq \tilde{R}$ we have

$$\|f\|_{H^r(\mathbb{R})} \lesssim 2^{jr} \|f\|_{L^2(\mathbb{R})}. \tag{47}$$

Remark that all functions in W_j and $\tilde{W}_j$ verify both direct and inverse inequality

$$f \in W_j \quad \Rightarrow \quad \|f\|_{H^r(\mathbb{R})} \simeq 2^{jr} \|f\|_{L^2(\mathbb{R})}, \quad r \in [0, R],$$

$$f \in \tilde{W}_j \quad \Rightarrow \quad \|f\|_{H^r(\mathbb{R})} \simeq 2^{jr} \|f\|_{L^2(\mathbb{R})}, \quad r \in [0, \tilde{R}].$$

By a duality argument it is not difficult to prove that, for $f \in W_j$ and $f \in \tilde{W}_j$ similar inequalities hold for negative values of s. More precisely, for $f \in W_j$ and $s \in [0, \tilde{R}]$ we have, using the identity $\tilde{Q}_j = \tilde{P}_{j+1}(I - \tilde{P}_j)$ and the direct inequality (44)

$$\|f\|_{H^{-s}(\mathbb{R})} = \sup_{g \in H^s(\mathbb{R})} \frac{\langle f, g\rangle}{\|g\|_{H^s(\mathbb{R})}} = \sup_{g \in H^s(\mathbb{R})} \frac{\langle f, \tilde{Q}_j g\rangle}{\|g\|_{H^s(\mathbb{R})}}$$

$$\lesssim \sup_{g \in H^s(\mathbb{R})} \frac{\|f\|_{L^2(\mathbb{R})} \|(I - \tilde{P}_j) g\|_{L^2(\mathbb{R})}}{\|g\|_{H^s(\mathbb{R})}} \lesssim 2^{-js} \|f\|_{L^2(\mathbb{R})}.$$

Conversely we can write

$$\|f\|_{L^2(\mathbb{R})} = \sup_{g \in L^2(\mathbb{R})} \frac{\langle f, \tilde{Q}_j g\rangle}{\|g\|_{L^2(\mathbb{R})}} \lesssim \frac{\|f\|_{H^{-s}(\mathbb{R})} \|\tilde{Q}_j g\|_{H^s(\mathbb{R})}}{\|g\|_{L^2(\mathbb{R})}} \lesssim 2^{js} \|f\|_{H^{-s}(\mathbb{R})}.$$

In summary we have

Corollary 2.

$$f \in W_j \Rightarrow \|f\|_{H^s(\mathbb{R})} \simeq 2^{js} \|f\|_{L^2(\mathbb{R})}, \quad s \in [-\tilde{R}, R], \tag{48}$$

$$f \in \tilde{W}_j \Rightarrow \|f\|_{H^s(\mathbb{R})} \simeq 2^{js} \|f\|_{L^2(\mathbb{R})}, \quad s \in [-R, \tilde{R}]. \tag{49}$$

Remark 6. Note that an inequality of the form (46) is satisfied by all functions whose Fourier transform is supported in the interval $[-2^J, 2^J]$, while an inequality of the form (45) is verified by all functions whose Fourier transform is supported in $(-\infty, -2^J] \cup [2^J, \infty)$. Such inequalities are inherently bound to the frequency localisation of the functions considered, or, to put it in a different way, to their more or less oscillatory behavior. Saying that a function is "low frequency" means that such function does not oscillate too much. This translates in an inverse type inequality. On the other hand, saying that a function is "high frequency" means that it is purely oscillating, that is it is locally orthogonal to polynomials (where the meaning of "locally" is related to the frequency); this translates in a direct inequality. In many applications the two relations (45) and (46) can actually replace the information on the localisation of the Fourier transform. In particular this will be the case when we deal with functions defined on a bounded set Ω, for which the concept of Fourier transform does not make sense. Many of the things that can be proven for the case $\Omega = \mathbb{R}$ by using Fourier transform techniques, can be proven in an analogous way for bounded Ω by suitably using inequalities of the form (45) and (46).

The most important consequence of the validity of properties (48) and (49) is the possibility of characterizing the regularity of a function through its wavelet coefficients. Since all the functions $\tilde{\psi}_\lambda$ have a certain regularity, namely $\tilde{\psi}_\lambda \in H^{\tilde{R}}(\mathbb{R})$, the Fourier development (25) makes sense (at least formally), provided f has enough regularity for $\langle f, \tilde{\psi}_\lambda \rangle$ to make sense, at least as a duality product, that is provided $f \in H^{-\tilde{R}}(\mathbb{R})$. Moreover it is not difficult to prove that Q_j and $\tilde{Q}_j$ can be extended to operators acting on $H^{-\tilde{R}}$ and H^{-R}, respectively.

The properties of wavelets imply that, given any function $f \in H^{-\tilde{R}}(\mathbb{R})$, by looking at behavior of the $L^2(\mathbb{R})$ norm of $Q_j f$ as j goes to infinity and, more in detail, by looking at the absolute values of the wavelet coefficients $\langle f, \tilde{\psi}_\lambda \rangle$, it is possible to establish whether or not a function belongs to certain function spaces, and it is possible to write an equivalent norm for such function spaces in terms of the wavelet coefficients. More precisely we have the following theorem (see [23, 39]).

Theorem 4. *Let assumptions (26)– (31), and (35) hold. Let $f \in H^{-\tilde{R}}(\mathbb{R})$ and let $s \in]-\tilde{R}, R[$. Then $f \in H^s(\mathbb{R})$ if and only if*

$$\|f\|_s^2 = \sum_{\mu \in K_0} |\langle f, \tilde{\varphi}_\mu \rangle|^2 + \sum_j \sum_{\lambda \in \nabla_j} 2^{2js} |\langle f, \tilde{\psi}_\lambda \rangle|^2 < +\infty. \tag{50}$$

Moreover $\|\cdot\|_s$ is an equivalent norm for $H^s(\mathbb{R})$.

Proof. Thanks to the fact that the functions φ_μ, $\mu \in K_0$ and ψ_λ, $\lambda \in \nabla_j$ constitute Riesz bases for V_0 and W_j, respectively, (50) is equivalent to

$$\|P_0 f\|^2_{L^2(\mathbb{R})} + \sum_{j \geq 0} 2^{2js} \|Q_j f\|^2_{L^2(\mathbb{R})} < +\infty. \tag{51}$$

We will at first show that if (51) holds then $\sum_j Q_j f \in H^s(\mathbb{R})$. We start by observing that the bilinear form $\langle (1-\Delta)^{s/2} \cdot, (1-\Delta)^{s/2} \cdot \rangle = \langle (1-\Delta)^{s/2+\varepsilon} \cdot, (1-\Delta)^{s/2-\varepsilon} \cdot \rangle$ is a scalar product for $H^s(\mathbb{R})$ (this is for instance easily seen in the Fourier domain). We can write

$$\|\sum_j Q_j f\|^2_{H^s(\mathbb{R})} \leq 2 \sum_j \sum_{k>j} \|Q_j f\|_{H^{s+2\varepsilon}(\mathbb{R})} \|Q_k f\|_{H^{s-2\varepsilon}(\mathbb{R})} + \sum_j \|Q_j f\|^2_{H^s(\mathbb{R})}.$$

Thanks to the inverse inequalities we can then bound

$$\|\sum_j Q_j f\|^2_{H^s(\mathbb{R})} \leq 2 \sum_j \sum_{k>j} 2^{js} \|Q_j f\|_{L^2(\mathbb{R})} 2^{ks} \|Q_k f\|_{L^2(\mathbb{R})} 2^{-2\varepsilon|j-k|} + \sum_j 2^{2js} \|Q_j f\|^2_{L^2(\mathbb{R})}.$$

The second sum is finite by assumption and the first sum can be bound by recalling that the convolution product is a bounded operator from $\ell^1 \times \ell^2$ to ℓ^2.

Let now $f \in H^s(\mathbb{R})$. We have, for N arbitrary

$$\Big[\|P_0 f\|^2_{L^2(\mathbb{R})} + \sum_{j=1}^{N} 2^{2js}\|Q_j f\|^2_{L^2(\mathbb{R})}\Big]^2 = \langle f, \tilde{P}_0 P_0 f + \sum_{j=1}^{N} 2^{2sj}\tilde{Q}_j Q_j f\rangle^2$$
$$\lesssim \|f\|^2_{H^s(\mathbb{R})}\|\tilde{P}_0 P_0 f + \sum_{j=1}^{N} 2^{2sj}\tilde{Q}_j Q_j f\|^2_{H^{-s}(\mathbb{R})}. \tag{52}$$

Using the first part of the theorem we get

$$\|\tilde{P}_0 P_0 f + \sum_{j=1}^{N} 2^{2sj}\tilde{Q}_j Q_j f\|^2_{H^{-s}(\mathbb{R})} \lesssim \|\tilde{P}_0 P_0 f\|^2_{L^2(\mathbb{R})} + \sum_{j=1}^{N} 2^{-2sj}2^{4sj}\|\tilde{Q}_j Q_j f\|^2_{L^2(\mathbb{R})}$$
$$\lesssim \|P_0 f\|^2_{L^2(\mathbb{R})} + \sum_{j=1}^{N} 2^{2sj}\|Q_j f\|^2_{L^2(\mathbb{R})}.$$

Dividing both sides of (52) by $\|P_0 f\|^2_{L^2(\mathbb{R})} + \sum_{j=1}^{N} 2^{2sj}\|Q_j f\|^2_{L^2(\mathbb{R})}$ we obtain

$$\|P_0 f\|^2_{L^2(\mathbb{R})} + \sum_{j=1}^{N} 2^{2sj}\|Q_j f\|^2_{L^2(\mathbb{R})} \lesssim \|f\|^2_{H^s(\mathbb{R})}.$$

The arbitrariness of N yields the thesis. □

For $s = 0$ we immediately obtain the following Corollary.

Corollary 3. *If the assumptions of Theorem 4 hold then $\{\psi_\lambda,\ \lambda \in \Lambda\}$ is a Riesz basis for $L^2(\mathbb{R})$.*

A more general result actually holds. In fact, letting $B^{s,p}_q(\mathbb{R}) := B^s_q(L^p(\mathbb{R}))$ denote the Besov space of smoothness order s with summability in L^p, q being a fine tuning index (see e.g. [41]), we have the following theorem [23, 39].

Theorem 5. *Let (26)–(31) and (35) hold. Let $f \in H^{-\tilde{R}}$ and let $s \in]-\tilde{R}, R[$, $0 < p$, $q < +\infty$. Then, setting*

$$\|f\|^q_{s,p,q} = (\sum_{\mu \in K_0} |\langle f, \tilde{\varphi}_\mu\rangle|^p)^{\frac{q}{p}} + \sum_j (\sum_{\lambda \in \nabla_j} 2^{pjs}2^{p(\frac{1}{2}-\frac{1}{p})j}|\langle f, \tilde{\psi}_\lambda\rangle|^p)^{\frac{q}{p}}, \tag{53}$$

$f \in B^{s,p}_q(\mathbb{R})$ if and only if $\|f\|_{s,p,q} < +\infty$. Moreover $\|\cdot\|_{s,p,q}$ is an equivalent norm for $B^{s,p}_q(\mathbb{R})$. An analogous result, in which the ℓ^p (resp. ℓ^q) norms are replaced by the ℓ^∞ norm, holds for either $p = +\infty$ or $q = +\infty$ or both.

3.1.1 The General Case: Ω Domain of $\mathbb{R}^d$

Let us now consider the general case of Ω being a (possibly bounded) Lipschitz domain of $\mathbb{R}^d$. The property of space localization is almost immediately stated also for wavelet bases on Ω.
Localisation in space: For each $\lambda \in \nabla_j$ we have that

$$diam(\operatorname{supp}\varphi_\lambda) \lesssim 2^{-j} \quad \text{and} \quad diam(\operatorname{supp}\tilde{\varphi}_\lambda) \lesssim 2^{-j}, \tag{54}$$

$$diam(\operatorname{supp}\psi_\lambda) \lesssim 2^{-j} \quad \text{and} \quad diam(\operatorname{supp}\tilde{\psi}_\lambda) \lesssim 2^{-j}, \tag{55}$$

and for all $\mathbf{k} = (k_1, k_2, \ldots, k_d) \in \mathbb{Z}^d$ there are at most K (resp. $\tilde{K}$) values of $\lambda \in \nabla_j$ such that

$$\operatorname{supp}\psi_\lambda \cap \square_{j,\mathbf{k}} \neq \emptyset \qquad (\text{resp. } \operatorname{supp}\tilde{\psi}_\lambda \cap \square_{j,\mathbf{k}} \neq \emptyset\ ), \tag{56}$$

where $\square_{j,\mathbf{k}}$ denotes the cube of center $\mathbf{k}/2^j$ and side 2^{-j}.

Remark 7. The last requirement is equivalent to asking that the basis functions at j fixed are uniformly distributed over the domain of definition. It avoids, for instance, that they accumulate somewhere.

Clearly, when working on bounded domains we do not have at our disposal tools like the Fourier transform. Still, we can ask that the basis functions for $L^2(\Omega)$ have the same property as the ones for $L^2(\mathbb{R})$ in term of oscillations. We will then assume that they satisfy an analogous relation to (33). More precisely, using for $x \in \Omega$ and $\alpha \in \mathbb{N}^d$ the notation $x^\alpha = (x_1, \cdots, x_d)^{(\alpha_1, \cdots, \alpha_d)} = x_1^{\alpha_1} x_2^{\alpha_2} \cdots x_d^{\alpha_d}$, we assume that the basis functions ψ_λ verify for all s, $0 \leq s \leq R$ and all $\alpha \in \mathbb{N}^d$ with $\sum_i \alpha_i \leq M$

$$\|\psi_\lambda\|_{H^s(\Omega)} \lesssim 2^{js} \quad \text{and} \quad \int_\Omega x^\alpha \psi_\lambda(x)\, dx = 0. \tag{57}$$

A similar relation is assumed to hold for the dual basis: for all s, $0 \leq s \leq \tilde{R}$ and all $\alpha \in \mathbb{N}$ with $\sum_i \alpha_i \leq \tilde{M}$

$$\|\tilde{\psi}_\lambda\|_{H^s(\Omega)} \lesssim 2^{js} \quad \text{and} \quad \int_\Omega x^\alpha \tilde{\psi}_\lambda(x)\, dx = 0. \tag{58}$$

Also in this case it is possible to prove that (57) and (58) together with the property of space localization imply that for all polynomials p of degree less or equal than $\tilde{M}$ (resp. less or equal that M) we have

$$p = \sum_{\mu \in K_j} \langle p, \tilde{\varphi}_\mu \rangle \varphi_\mu \qquad (\text{resp. } p = \sum_{\mu \in K_j} \langle p, \varphi_\mu \rangle \tilde{\varphi}_\mu). \tag{59}$$

Exactly as in the case $\Omega = \mathbb{R}$ this property allows us to prove a direct type inequality.

Theorem 6 (Direct inequality). *Assume that (54)–(58) hold. Then for all* s, $0 < s \leq \tilde{M} + 1$, $f \in H^s(\Omega)$ *implies*

$$\|f - P_j f\|_{L^2(\Omega)} \lesssim 2^{-js} \|f\|_{H^s(\Omega)}. \tag{60}$$

Analogously for all s, $0 < s \leq M + 1$, $f \in H^s(\Omega)$ *implies*

$$\|f - \tilde{P}_j f\|_{L^2(\Omega)} \lesssim 2^{-js} \|f\|_{H^s(\Omega)}. \tag{61}$$

With a proof which is not much different from the one used for proving the analogous result on $\mathbb{R}$ it is also not difficult to prove that an inverse inequality holds. More precisely we have the following theorem.

Theorem 7 (Inverse inequality). *Assume that (54)–(58) hold. Then for* r *with* $0 < r \leq R$ *it holds that for all* $f \in V_j$

$$\|f\|_{H^r(\Omega)} \lesssim 2^{jr} \|f\|_{L^2(\Omega)}. \tag{62}$$

Analogously, for r *with* $0 < r \leq \tilde{R}$ *it holds that for all* $f \in \tilde{V}_j$

$$\|f\|_{H^r(\Omega)} \lesssim 2^{jr} \|f\|_{L^2(\Omega)}. \tag{63}$$

Analogously to what happens for $\mathbb{R}$ Theorems 6 and 7 allow us to prove a norm equivalence for $H^s(\Omega)$ in term of a suitable weighted ℓ^2 norm of the sequence of wavelet coefficients. More precisely the following theorem, which is proven similarly to Theorem 5, holds [23].

Theorem 8. *Assume that (54)–(58), hold. Let* $f \in (H^{\tilde{R}}(\Omega))'$ *and let* $-\tilde{R} < s < R$. *Let*

$$\|f\|_s^2 = \sum_{\mu \in K_0} |\langle f, \tilde{\varphi}_\mu \rangle|^2 + \sum_j \sum_{\lambda \in \nabla_j} 2^{2js} |\langle f, \tilde{\psi}_\lambda \rangle|^2. \tag{64}$$

Then, for $s \geq 0$, $\|f\|_s$ *is an equivalent norm for the space* $H^s(\Omega)$ *and* $f \in H^s(\Omega)$ *if and only if* $\|f\|_s$ *is finite; for negative* s, $\|f\|_s$ *is an equivalent norm for the space* $(H^{-s}(\Omega))'$ *and* $f \in (H^{-s}(\Omega))'$ *if and only if* $\|f\|_s$ *is finite.*

An analogous result holds for the dual multiresolution analysis, which allows to characterize $H^s(\Omega)$, $s \geq 0$ and $(H^{-s}(\Omega))'$, $s < 0$, for $-R < s < \tilde{R}$. Also for Ω a characterization result for Besov spaces holds, as stated by the following theorem (see once again [23]).

Theorem 9. *Let all the assumption of Theorem 8 hold. Let* $f \in (H^{\tilde{R}}(\Omega))'$ *and let* $s \in]-\tilde{R}, R[$, $0 < p, q < +\infty$. *Then, setting*

$$\|f\|_{s,p,q}^q = (\sum_{\mu \in K_0} |\langle f, \tilde{\varphi}_\mu \rangle|^p)^{\frac{q}{p}} + \sum_j (\sum_{\lambda \in \nabla_j} 2^{pjs} 2^{p(\frac{d}{2} - \frac{d}{p})j} |\langle f, \tilde{\psi}_\lambda \rangle|^p)^{\frac{q}{p}}, \tag{65}$$

$f \in B^{s,p}_q(\Omega)$ if and only if $\|f\|_{s,p,q} < +\infty$. Moreover $\|\cdot\|_{s,p,q}$ is an equivalent norm for $B^{s,p}_q(\Omega)$. An analogous result, in which the ℓ^p (resp. ℓ^q) norms are replaced by the ℓ^∞ norm, holds for either $p = +\infty$ or $q = +\infty$ or both.

Thanks to these norm equivalences we can then evaluate the Sobolev norms for spaces with negative and/or fractionary indexes by using simple operations, namely the evaluation of $L^2(\Omega)$ scalar products and the evaluation of an (infinite) sum.

Remark 8. All the wavelets mentioned until now (with the exception of interpolating wavelet bases) satisfy the assumptions of Theorem 8 for suitable values of M, $\tilde{M}$, R, $\tilde{R}$. A result similar to (65) holds, however, also for interpolating wavelet, which allow to characterize those Sobolev and Besov spaces that are embedded in $C^0(\bar{\Omega})$.

3.1.2 Nonlinear vs. Linear Wavelet Approximation

Let us now consider the problem of approximating a given function $f \in L^2(\Omega)$, $\Omega \subseteq \mathbb{R}^d$ domain in $\mathbb{R}^d$, with N degrees of freedom (that is with a function which we can identify with N scalar coefficients). We distinguish between two approaches:

The first approach is the usual *linear approximation*: a space V_h of dimension N is fixed a priori and the approximation f_h is the $L^2(\Omega)$ projection of f on V_h (example: finite elements on an uniform grid with mesh size h related to N, in which case in dimension d the mesh size and the number of degrees of freedom verify $h \sim N^{-1/d}$).

It is well known that the behavior of the linear approximation error is generally linked to the Sobolev regularity of f. In particular we cannot hope for a high rate of convergence if f has poor smoothness. Several remedies are available in this last case, like for instance adaptive approximation by performing a mesh refinement around the singularities of f. We have then the *non linear approximation* approach: a class of spaces X is chosen a priori. We then choose a space $V_N(f)$ of dimension N in X well suited to f. The approximation f_h is finally computed as an L^2 projection of f onto $V_N(f)$ (example: finite elements with N free nodes). In other words we look for an approximation to f in the *non linear* space

$$\Sigma_N = \cup_{V_N \in X} V_N.$$

Three questions are of interest [32]:

- Which is the relation between the performance of non linear approximation and some kind of smoothness of the function to be approximated.
- How do we compute the non linear approximation of a given function f.
- How do we compute the non linear approximation of an unknown function u (solution of a PDE).

In the following we will give some idea on how these questions are answered in the wavelet framework.

3.1.3 Non Linear Wavelet Approximation

We want to approximate f with an element of the non linear space

$$\Sigma_N = \{u = \sum_\lambda u_\lambda \psi_\lambda : \ \#\{\lambda : u_\lambda \neq 0\} \leq N\}.$$

In order to construct an approximation to f in Σ_N define a *non linear projection* [32].

Definition 3. Let $f = \sum_\lambda f_\lambda \psi_\lambda$. The non linear projector $P_N : L^2 \to \Sigma_N$ is defined as

$$P_N f := \sum_{n=1}^{N} f_{\lambda_n} \psi_{\lambda_n},$$

where

$$|f_{\lambda_1}| \geq |f_{\lambda_2}| \geq |f_{\lambda_3}| \geq \dots |f_{\lambda_n}| \geq |f_{\lambda_{n+1}}| \geq \dots$$

is a decreasing reordering of the wavelet coefficients of f.

Remark 9. If the wavelet basis is orthonormal $\|f\|_{L^2(\Omega)} = \|(f_\lambda)_\lambda\|_{\ell^2}$ and then $P_N f$ is the best N-term approximation (the norm of the error is the ℓ^2 norm of the sequence of discarded coefficients, which is minimized by the projection P_N). In any case, since $\|f\|_{L^2(\Omega)} \simeq \|(f_\lambda)_\lambda\|_{\ell^2}$, we have that $P_N f$ is the best approximation in an L^2 equivalent norm.

We have the following theorem [32], see also [15].

Theorem 10. *$f \in B^{s,q}_q(\Omega)$ with $0 < s < R$ and $q : d/q = d/2 + s$ implies*

$$\|f - P_N f\|_{L^2(\Omega)} \lesssim \|f\|_{B^{s,q}_q(\Omega)} N^{-s/d}.$$

Proof. With the choice q such that $d/q = d/2 + s$, the characterization of the Besov norm in terms of the wavelet coefficients yields

$$\|\sum_\lambda f_\lambda \psi_\lambda\|_{B^{s,q}_q(\Omega)} \simeq \|(f_\lambda)_\lambda\|_{\ell^q}.$$

Let us consider the decreasing reordering of the coefficients:

$$|f_{\lambda_1}| \geq |f_{\lambda_2}| \geq |f_{\lambda_3}| \geq \dots |f_{\lambda_n}| \geq |f_{\lambda_{n+1}}| \geq \dots.$$

We can easily see that

$$n|f_{\lambda_n}|^q \leq \sum_{k \leq n} |f_{\lambda_k}|^q \leq \sum_\lambda |f_\lambda|^q \leq \|f\|^q_{B^{s,q}_q(\Omega)},$$

that is

$$|f_{\lambda_n}| \leq n^{-1/q} \| f \|_{B_q^{s,q}(\Omega)}.$$

Now we can write

$$\| f - P_N f \|_{L^2(\Omega)} = \| \sum_{n>N} f_{\lambda_n} \psi_{\lambda_n} \|_{L^2(\Omega)} \simeq \left(\sum_{n>N} |f_{\lambda_n}|^2 \right)^{1/2}$$

$$\lesssim \| f \|_{B_q^{s,q}(\Omega)} \left(\sum_{n>N} n^{-2/q} \right)^{1/2} \lesssim \| f \|_{B_q^{s,q}(\Omega)} N^{-1/q+1/2}.$$

□

Remark 10. Remark that, for $q < 2$ the space $B_q^{s,q}(\Omega) \supset H^s(\Omega)$. In particular there exist a wide class of functions which are not in $H^s(\Omega)$ but that belong to $B_q^{s,q}(\Omega)$. For such functions, the non linear approximation will be of higher order than linear approximation.

Remark 11. Theorem 10 still holds under the weaker assumption that $\mathbf{f} = (f_\lambda)_\lambda$ verifies $\|\mathbf{f}\|_{\ell_w^q} = \sup_n n |f_{\lambda_n}|^q < +\infty$.

3.1.4 Nonlinear Approximation in H^s

If we want to approximate f in $H^s(\Omega)$, rather than in $L^2(\Omega)$, we simply rescale the basis functions by setting, for $\lambda \in \nabla_j$, $\check{\psi}_\lambda = 2^{-js} \psi_\lambda$, so that $f = \sum \check{f}_\lambda \check{\psi}_\lambda$ with $\| f \|_{H^s(\Omega)} = \|(\check{f}_\lambda)_\lambda\|_{\ell^2}$. In this case for $q : d/q = d/2 + r$ we have that

$$\| f \|_{B_q^{s+r,q}(\Omega)} \simeq \|(\check{f}_\lambda)_\lambda\|_{\ell^q}.$$

We then apply the same procedure to the sequence $(\check{f}_\lambda)_\lambda$. In particular we define this time the non linear projector as

Definition 4. Let $f = \sum_\lambda \check{f}_\lambda \check{\psi}_\lambda$. The non linear projector $P_N : H^s(\Omega) \to \Sigma_N$ is defined as

$$P_N f := \sum_{n=1}^{N} \check{f}_{\lambda_n} \check{\psi}_{\lambda_n}, \tag{66}$$

where

$$|\check{f}_{\lambda_1}| \geq |\check{f}_{\lambda_2}| \geq |\check{f}_{\lambda_3}| \geq \ldots |\check{f}_{\lambda_n}| \geq |\check{f}_{\lambda_{n+1}}| \geq \ldots$$

is a decreasing reordering of the rescaled wavelet coefficients.

We have the following theorem, whose proof is identical to the proof of Theorem 10.

Theorem 11. *$f \in B_q^{s+r,q}(\Omega)$ with $0 < r < R-s$ and $q : d/q = d/2+r$, implies*

$$\|f - P_N f\|_{H^s(\Omega)} \lesssim \|f\|_{B_q^{s+r,q}(\Omega)} N^{-r/d}.$$

Remark 12. By abuse of notation we will also indicate by $P_N : \ell^2 \to \ell^2$ the operator mapping the coefficient vector $\mathbf{f} = (f_\lambda)$ of the function f to the coefficient vector of its nonlinear projection $P_N f$.

3.1.5 The Issue of Boundary Conditions

When aiming at using wavelet bases for the numerical solution of PDE's, one has, of course, to take into account the issue of boundary conditions. If, for instance, in the equation considered, essential boundary conditions (for example $u = 0$) need to be imposed on a portion Γ_e of the boundary, we will want the basis functions ψ_λ, $\lambda \in \Lambda$, to satisfy themselves the corresponding homogeneous boundary conditions on Γ_e. Depending on the projectors P_j, the dual wavelets $\tilde{\psi}_\lambda$ will not however need to satisfy themselves the same homogeneous boundary conditions (see [25]), though this might be the case (if for instance the projector P_j is chosen to be the $L^2(\Omega)$ orthogonal projector). Depending on whether the ψ_λ and the $\tilde{\psi}_\lambda$ satisfy or not some homogeneous boundary conditions, the same boundary conditions will be incorporated in the spaces that we will be able to characterize through such functions. It is not the aim of this paper to go into details. Let us just give an idea on the kind of results that hold. To fix the ideas let us consider for example the case of homogeneous Dirichlet boundary condition $u = 0$ on $\partial\Omega$ and let us concentrate on the characterization of Sobolev spaces. If, for all $\lambda \in \Lambda$, $\psi_\lambda = 0$ on Γ, then (64) will hold provided f belongs to the $H^s(\Omega)$ closure of $H^s(\Omega) \cap H_0^1(\Omega)$, that we will denote $\mathscr{H}_0^s(\Omega)$. If in turn the $\tilde{\psi}_\lambda$'s satisfy $\tilde{\psi}_\lambda = 0$ on Γ, we clearly will not be able to characterize (through scalar products with such functions) the space $(H^s(\Omega))'$, but only the space $(\mathscr{H}_0^s(\Omega))'$. Moreover it is clear that, if for all $\lambda \in \Lambda$ the ψ_λ's satisfy an homogeneous boundary condition, we can expect a direct inequality of the form (60) to hold only if we assume that the function f to approximate satisfy itself the same homogeneous boundary conditions. For more details see [15].

4 Adaptive Wavelet Methods for PDE's: The First Generation

As we saw in the previous section the wavelet decomposition of a function provides a quite straightforward *smoothness analysis* tool, that has been successfully applied, for example in edge detection algorithms. It is then quite natural to think of using such a tool also in the framework of adaptive wavelet methods for PDE's. The first heuristic idea was to design an adaptive method where, in order to drive the refining/coarsening strategy, wavelet coefficients are used instead of error indicators. The heuristic underlying such a strategy is that big wavelet coefficients are indicators of a potential lack of smoothness, which requires refinement, while small wavelet coefficients indicate that an unnecessary refinement has been performed and the corresponding basis function are superfluous. This idea led Maday et al. [38], to propose a simple adaptive wavelet algorithm for the solution of Burger's equation in dimension one:

$$u_t - \nu u_{xx} + u u_x = f, \quad u(x,0) = u_0(x).$$

The algorithm proposed, based on an implicit/explicit Euler time discretization scheme (adapting the idea to more effective time discretization methods is straightforward), has the following form:

- `initialize: expand` u_0 `as` $u_0 = \sum_{\lambda \in \Lambda} u_\lambda^0 \psi_\lambda$
- $u^n \sim u(n\Delta t) \to u^{n+1} \sim u((n+1)\Delta t)$:
  - ⋆ $\Lambda^{n+1} = \cup_{\lambda : |u_\lambda^n| > \varepsilon} \mathcal{N}(\lambda)$, `where for` $\lambda = (j,k)$

$$\mathcal{N}(\lambda) = \{(j,k+1),(j,k),(j,k-1),(j+1,2k-1),(j+1,2k)\}$$

  - ⋆ $V_h^{n+1} = span < \psi_\lambda, \ \lambda \in \Lambda^{n+1} >$
  - ⋆ `solve, by a Galerkin projection on` V_h^{n+1}

$$u^{n+1} - \nu \Delta t u_{xx}^{n+1} = u^n + \Delta t f((n+1)\Delta t) - \Delta t u^n u_x^n.$$

At each iteration the set Λ^{n+1} contains all the immediate neighbors (at the same level and at the next finer level) of those indexes λ corresponding to coefficients u_λ sufficiently big. The strength of this algorithm lies in its extreme simplicity. The decision on whether to refine/derefine or not is taken by simply looking at the (already available) wavelet coefficients of the solution at the previous time step, so that no auxiliary computation of an error indicator is needed. Moreover the refining and the derefining procedures are themselves much simpler than in the finite element case: adding or removing one basis function results (at the linear system level) in adding or removing the corresponding line and column from the mass and stiffness matrices.

The algorithm has however some drawbacks and some serious limitations. First of all, it is not difficult to realize that the proposed refining strategy, which for each relevant wavelet coefficient only adds two neighbors at only two subsequent levels, implicitly assumes that the time step is small enough. And even with such an assumption there is little rigorous theoretical analysis guaranteeing convergence of such a method (see [9]), as it happens for more refined wavelet methods, which we will describe and partially analyze in Sect. 5. Moreover, while implementing both refining and coarsening operation is extremely simple, the implementation of the wavelet Galerkin method is far from being straightforward. Just to give an example of the kind of difficulties that have to be faced, let us for instance consider the task of computing the entries of a stiffness matrix $R = (r_{\lambda,\mu})$

$$r_{\lambda,\mu} = \int \psi'_\lambda \psi'_\mu.$$

Unfortunately, we generally do not have a close expression for ψ: we only know that it is defined in terms of the scaling function φ by a relation of the form (12). We can then reduce the computation of $r_{\lambda,\mu}$ to

$$r_{\lambda,\mu} = 4 \sum_{\ell,\ell'} g_\ell g_{\ell'} \int 2^{3(j+m)/2} \varphi'(2^{j+1}x - (2k+\ell))\varphi'(2^{m+1}x - (2n+\ell')). \quad (67)$$

If $j = m$, a simple change of variable allows us to write

$$\int \varphi'(2^{j+1}x - a)\varphi'(2^{j+1}x - b) = 2^{-(j+1)} \int \varphi'(x)\varphi'(x - (b-a)).$$

For $a - b$ integer, the integral on the right hand side can be computed by observing that, thanks to the refinement equation (3), for all integers n we have that

$$\int \varphi'(x)\varphi'(x-n) = 4 \sum_{\ell,\ell'} h_\ell h_{\ell'} \int \varphi'(2x-\ell)\varphi'(2x-(2n+\ell'))$$

$$= 2 \sum_{\ell,\ell'} h_\ell h_{\ell'} \int \varphi'(x)\varphi'(x-(2n+\ell'-\ell)).$$

Analogously to what we did for the point values of Daubechies' scaling functions, the values of the integrals of the form $\int \varphi'(x)\varphi'(x-n)$ can then be computed, up to a multiplicative constant (whose value can also be computed (see [24])), by solving, once and for all, an eigenvalue/eigenvector problem. If $m \neq j$ the only way of computing the integrals on the righthand side of (67) is to reduce them to a linear combination of integrals of products of functions at the same level, by taking advantage of the injection $V_j \subset V_m$ (if $m > j$) or $V_m \subset V_j$ (if ($j > m$). Clearly in this case the computation of the integral can turn out to be quite expensive and/or quite cumbersome (see [10] for the efficient computation of stiffness matrix entries in the adaptive wavelet framework). Clearly the implementation of the wavelet Galerkin method is even more difficult if variable coefficients or nonlinear expressions are considered. In particular as far as nonlinear problems are concerned, if the expression considered is multilinear one can resort to the same kind of approach as the one used for the elements of the stiffness matrix. Otherwise one has to resort to quite expensive quadrature formulae, drastically reducing the gain obtained by using an adaptive method.

4.1 The Adaptive Wavelet Collocation Method

A possible way of overcoming the difficulties inherently presented by the wavelet Galerkin method is to avoid on the one hand the computation of integrals by resorting to a collocation method [6], and on the other hand to use interpolating wavelets, where coefficients and point values are related by a simple relation which allows to efficiently treat nonlinearities.

4.1.1 Wavelet Collocation on Uniform Grids

To fix the ideas let us consider a partial differential equation of the form

$$\mathscr{A} u = f \quad \text{in } \Omega, \qquad u = g \quad \text{on } \partial\Omega$$

with $\Omega = (0,1)^2$ and with

$$\mathscr{A} = \sum_{|\alpha|\leq 2} c_\alpha \frac{\partial^{|\alpha|}}{\partial x^\alpha} + \mathscr{F}(x, u(x), \nabla u(x)). \tag{68}$$

We want to compute an approximate solution to (68) by resorting to the interpolating wavelet basis described in Sect. 2.4.1 in the framework of an adaptive scheme. In order to deal with the non linearity in the simplest way, we choose a collocation approach, imposing the equation at a suitable set of collocation points rather than testing the equation against test functions, thus avoiding the necessity of computing integrals.

Let us first describe how the approximate solution is computed for a given nonuniform grid. We start by observing that given any subset Λ_h of Λ we can define a nonuniform grid G_h

$$G_h = \{\zeta_\mu,\ \mu \in K_{j_0}\} \cup \{\xi_\lambda,\ \lambda \in \Lambda_h\},$$

and a corresponding nonuniform approximation space V_h

$$V_h = V_{j_0} \cup span < \Psi_\lambda,\ \lambda \in \Lambda_h >,$$

respectively, including by default the coarse grid $\{\zeta_\mu,\ \mu \in K_{j_0}\}$ and the corresponding coarse space V_{j_0}.

Given the index set $\Lambda_h \in \Lambda$ we can then compute an approximation to the solution u of (68) by looking, by any solution method suitable for the nonlinear equation considered, for $u_h \in V_h$ such that

$$\mathscr{A} u_h(\lambda) = f(\lambda), \quad \forall \lambda \in G_h. \tag{69}$$

The choice of the collocation approach, in the framework of wavelet methods, has numerous advantages over other approaches, like Galerkin or Petrov–Galerkin.

The main advantage is that no integral evaluation is needed and therefore the computational load of both the evaluation of nonlinear terms and of the assembling of the linear system can be kept to the minimum.

Let us consider more in detail how such a scheme can be implemented as opposed to wavelet Galerkin schemes. Typically, four different phases need to be implemented.

Pre-processing: After fixing a finest level j_{max} compute the values of the derivatives of ϑ at the dyadic points:

$$t_k^s = \vartheta^{(s)}\left(\frac{k}{2^{j_{max}-j_0}}\right). \tag{70}$$

This is done recursively by taking advantage of the refinement equation (3). Such a task requires $O(2^{j_{max}-j_0})$ operations, independently of the dimension d of the domain of the problem. Moreover such values can be computed once and for all and then stored (the storage needed is also proportional to $2^{j_{max}-j_0}$, since the function ϑ is compactly supported).

Assembling the collocation matrix: The entries of the collocation matrix relative to the linear part of the operator take the form

$$r_{\lambda,\mu} = \sum_{|\alpha|\leq 2} c^\alpha(\lambda)\frac{\partial^{|\alpha|}\Psi_\mu}{\partial x^\alpha}(\lambda)$$

with possibly Θ_μ in the place of Ψ_μ. Computing each entry of the matrix involves then the evaluation of derivatives of a basis function at a dyadic point, which, in view of (19), is performed in $O(1)$ operations once the fundamental quantities (70) are known.

Evaluating the nonlinear terms: The advantage of using a collocation approach in the context of non uniform wavelet discretization is particularly evident when dealing with nonlinear operators, especially those which are not of multilinear type. In fact, in a straightforward implementation of the Galerkin approach, in order to evaluate $\int \mathscr{F}(x, u_h, \nabla u_h)\Psi_\lambda$ for $u_h \in V_{\Lambda_h}$ one would need to: (1) evaluate u_h and ∇u_h at the nodes of a (fine) quadrature grid, (2) evaluate $\mathscr{F}(x, u, \nabla u)$ at such nodes, and (3) apply a quadrature formula. Resorting to a collocation scheme reduces such a computation to the evaluation of u_h, ∇u_h and $\mathscr{F}$ at the single mesh point λ.

Fast interpolating wavelet transform: This is needed to go back and forth from point values to wavelet coefficients, the former needed for handling nonlinearities, as well as in post-processing, the latter being the actual unknowns of the discrete problem.

4.1.2 The Adaptive Collocation Scheme

We can now present the adaptive version of the collocation scheme. The idea, proposed in [3,42], is to iteratively compute increasingly good approximations to the solution u of our problem, and use the wavelet coefficients of the actual approximate solution at each iteration in order to design a better grid to be used for computing the next approximate solution.

We select the first grid G_h^0 by simply looking at the behavior of the data of the equation. Assume for simplicity that the coefficients and the boundary data are smooth. Letting, for any dyadic grid G_h (V_h denoting the corresponding discrete space) $L_h f$ denote the unique function verifying

$$L_h f \in V_h, \quad L_h f(\lambda) = f(\lambda) \quad \forall \lambda \in G_h,$$

we chose Λ_h^0 is such a way that

$$\|f - L_h^0 f\| \leq \varepsilon.$$

Once the first grid G_h^0 has been selected, we compute a first approximate solution u_h^0 by applying the collocation method in the corresponding space V_h^0, according to the formulation (69).

Following [3], we then analyze the computed solution in order to design a new (better) grid. More precisely, at the n-th step, given a grid G_h^n, and the relative approximate solution

$$u_h^n = \sum_{\mu \in K_{j_0}} u_\mu \Theta_\mu + \sum_{\lambda \in \Lambda_h^n} d_\lambda^n \Psi_\lambda,$$

we compute the next grid G_h^{n+1} by removing the useless points and refining where the approximation is bad. More precisely, for each point ξ_μ, $\mu = (\eta, j, \mathbf{k}) \in \Lambda$, define a set U_μ of neighboring points

$$U_\mu = [0,1]^2 \cap \{\zeta_k^{j+1}, \ k = (2k_1 + \eta_1, \ldots, 2k_d + \eta_d), \quad \eta_i = -1, 0, 1\}.$$

Remark now that if a dyadic point $p = (\mathbf{n})/2^m \in [0,1]^2$ verifies $p \notin \{\zeta_{j_0,\mathbf{k}}, \mathbf{k} \in K_{j_0}\}$, then there exists a unique $\lambda(p) \in \Lambda$ such that $p = \xi_{\lambda(p)}$. Then we can define a neighboring index set $\mathscr{U}_\mu$ corresponding to the set U_μ of neighboring points as

$$\mathscr{U}_\mu = \{\lambda(p), \ p \in U_\mu\} \subset \Lambda.$$

As in [38], the new index set is constructed by looking at the size of the coefficients of the current approximate solution u_h^n. In order to make the procedure more robust [3] proposes however the use of two different tolerances δ_r and δ_c for refining and coarsening.

Define then a set $\Lambda_h^{n,ref}$ of indexes marked for refinement as

$$\Lambda_h^{n,ref} = \{\lambda \in \Lambda_h : |d_\lambda^n| > \delta_r / 2^j\}.$$

We then introduce an updated index set

$$\Lambda_h^{n+1} = \{\lambda : |d_\lambda^n| > \delta_c/(2^j j^2)\} \cup_{\mu \in \Lambda_h^{n,ref}} \mathcal{U}_\mu,$$

and we denote by G_h^{n+1} and V_h^{n+1} the corresponding grid and approximation space. The $(n+1)$-th approximate solution is then computed by solving the collocation Problem (69) on the space V_h^{n+1}, with collocation grid G_h^{n+1}.

Clearly, by adopting an approach similar to the one underlying the Maday, Perrier, Ravel heuristic algorithm for Burger's equation described at the beginning of this section, the present adaptive wavelet collocation method can be also applied to evolution equations: the coefficients of the approximate solution at time step t_n can be used in order to design the grid for the numerical solution at time step t_{n+1} [2].

Though such method lacks the rigorous theoretical justification that is available for the new generation of adaptive wavelet methods that we will describe in the next section, the adaptive wavelet collocation method, which has a strong analogy to the finite difference method with incremental unknowns [13], presents several advantages in terms of simplicity of implementation and of applicability to problems of different nature (in particular, non linear ones). It has been successfully tested on a wide class of problems, both linear and non-linear and both steady-state and evolutionary. In particular we recall applications in the fields of elasticity [4], fluid structure interaction[36], semiconductors [7], geophysical flows [43].

5 The New Generation of Adaptive Wavelet Methods

Before going into details on how more sophisticated adaptive wavelet schemes are defined and before giving at least some idea on how such methods can be rigorously analyzed, let us present an equivalent formulation for the problems which we are going to consider in this section. Such formulation will not only greatly simplify the presentation but it will be at the basis of some of the adaptive wavelet methods that we are going to consider. Throughout this section let us, for $s > 0$, employ the notation $H^{-s}(\Omega) := (H^s(\Omega))'$. For the sake of simplicity let us focus on a simple linear model problem of the following form: find $u \in H^t(\Omega)$, $t \geq 0$, such that

$$Au = f, \tag{71}$$

with $A : H^t(\Omega) \to H^{-t}(\Omega)$ a linear symmetric operator verifying the classical assumptions:

$$\langle Au, u\rangle \gtrsim \|u\|^2_{H^t(\Omega)}, \qquad \langle Au, v\rangle \lesssim \|u\|_{H^t(\Omega)}\|v\|_{H^t(\Omega)}. \tag{72}$$

Assume also that for positive s the operator A is bounded from $H^{t+s}(\Omega)$ to $H^{-t+s}(\Omega)$. Moreover assume that A is local: $\mathrm{supp}(Au) \subseteq \mathrm{supp}(u)$.

Let us re-write the problem as an infinite dimensional "discrete" problem as follows. Select a couple of biorthogonal wavelet bases $\{\psi_\lambda, \lambda \in \Lambda\}$ and $\{\tilde{\psi}_\lambda, \lambda \in \Lambda\}$

for $L^2(\Omega)$ in such a way that the assumptions of Theorem 8 hold for $R = t + s^*$ with $s^* > d/2$; for $\lambda \in \nabla_j$ let $\check{\psi}_\lambda = 2^{-jt}\psi_\lambda$ so that for $v = \sum_{\lambda\in\Lambda} v_\lambda \check{\psi}_\lambda \in L^2(\Omega)$ we have the following simple form for the norm equivalences (64) and (65):

$$\|v\|_{H^t(\Omega)} \simeq \|\mathbf{v}\|_{\ell^2}, \quad \|v\|_{B_\tau^{t+s,\tau}(\Omega)} \simeq \|\mathbf{v}\|_{\ell^\tau} \quad (\tau = (s/d + 1/2)^{-1}), \tag{73}$$

with $\mathbf{v} = (v_\lambda)_\lambda$ denoting the renormalized wavelet coefficient vector. If we expand the unknown solution of our problem as $u = \sum_\lambda u_\lambda \check{\psi}_\lambda$, and we test the equation against the infinite set of test functions $v = \check{\psi}_\mu$ we obtain the following equivalent form of equation (72):

$$\mathscr{R}\mathbf{u} = \mathbf{f}, \tag{74}$$

where $\mathscr{R} = (r_{\lambda,\mu})_{\lambda,\mu}$ is the bi-infinite stiffness matrix and $\mathbf{f} = (f_\lambda)_\lambda$ the bi-infinite right hand side vector, with

$$r_{\lambda,\mu} = \langle A\check{\psi}_\mu, \check{\psi}_\lambda\rangle, \quad f_\lambda = \langle f, \check{\psi}_\lambda\rangle.$$

Clearly the properties of the operator A translate into properties of the bi-infinite matrix $\mathscr{R}$. More precisely we have that both $\|\mathscr{R}\|_{\ell^2\to\ell^2}$ and $\|\mathscr{R}^{-1}\|_{\ell^2\to\ell^2}$ are finite (they depend on the constants appearing in the norm equivalence (73) as well as on the continuity and coercivity constants of the operator A); we have continuity

$$\|\mathscr{R}\mathbf{v}\|_{\ell^2} \le \|\mathscr{R}\|_{\ell^2\to\ell^2}\|\mathbf{v}\|_{\ell^2} \tag{75}$$

and coercivity

$$\mathbf{v}^T\mathscr{R}\mathbf{v} \ge \|\mathscr{R}^{-1}\|^{-1}_{\ell^2\to\ell^2}\|\mathbf{v}\|^2_{\ell^2} \tag{76}$$

of the discrete operator $\mathscr{R}$.

Remark 13. It is not difficult to realize that we have (with $w = \sum_\lambda w_\lambda \check{\psi}_\lambda$)

$$\|\mathbf{f}\|_{\ell^2} \simeq \|f\|_{H^{-t}(\Omega)}, \quad \|\mathscr{R}\mathbf{w}\|_{\ell^2} \simeq \|Aw\|_{H^{-t}(\Omega)}.$$

5.1 A Posteriori Error Estimates

The first class of methods that we are going to consider is based on the use of suitable *a posteriori* error indicators. Clearly the classical error estimators used for instance in the finite element method cannot be used directly on wavelet solutions. On the other hand norm equivalences for the Sobolev spaces with negative index of the form (64) can be exploited for designing rigorous *a posteriori* error estimators [1, 22].

Assume that we have computed an approximation $w_h = \sum_{\lambda\in\Lambda_h} w_\lambda\check{\psi}_\lambda$, to the solution u of (72), with $\Lambda_h \subset \Lambda$ finite index set. It is well known that under our assumptions the operator A is an isomorphism between $H^t(\Omega)$ and its dual $H^{-t}(\Omega)$. Then we can bound

$$\|u - w_h\|_{H^t(\Omega)} \simeq \|A(u - w_h)\|_{H^{-t}(\Omega)} = \|f - Aw_h\|_{H^{-t}(\Omega)},$$

a bound which is the starting point of many error indicators for problems of the type considered here. The norm equivalence in terms of wavelet coefficients provides us a practical way of computing the (equivalent) $H^{-t}(\Omega)$ norm on the right hand side. More precisely, the norm equivalence (64) imply the validity of the bound

$$\|u - w_h\|^2_{H^t(\Omega)} \simeq \sum_{\lambda \in \Lambda} |\langle f - Au_h, \psi_\lambda\rangle|^2 = \|\mathbf{f} - \mathscr{R}\mathbf{w}\|_{\ell^2}. \tag{77}$$

Letting $\mathbf{e} = (e_\lambda)_\lambda = \mathbf{f} - \mathscr{R}\mathbf{w}$, the term $|e_\lambda| = |f_\lambda - \sum_\mu \mathscr{R}_{\lambda,\mu} w_\mu|$ plays then, for $\lambda \in \nabla_j$, the role of an ideal error indicator, that gives an information on the local error on $\mathrm{supp}(\psi_\lambda)$ at frequency $\sim 2^j$.

We will assume from now on that the vectors $\mathbf{f}$ and $\mathbf{w}$ have a finite number of nonzero entries (we will say then that they are *finitely supported*). Even so applying the bound (77) requires unfortunately the evaluation of a matrix vector multiplication involving a bi-infinite matrix and the computation of the resulting infinite vector. If we want the above *a posteriori* error estimate to be applicable in practice we will need to find a way of truncating it to a finite sum while still keeping the validity of an estimate of the form (77). In Sect. 5.4 we will give an idea on how this can be achieved. Let us now see instead how it is possible to use such an error estimate in the design of a convergent adaptive wavelet scheme.

Following [22] we consider, in the framework of an adaptive wavelet Galerkin method, a refinement strategy based on the error indicators. Let $u \in V_h = span < \check{\psi}_\lambda, \lambda \in \Lambda_h >$ be the Galerkin projection of the solution u to (71), that is the unique element of V_h satisfying

$$\langle Au_h, v_h\rangle = \langle f, v_h\rangle \ \forall v_h \in V_h.$$

Letting $\vartheta^* \in (0, 1)$ be given, define the refined index set $\tilde{\Lambda}_h \subset \Lambda$ in such a way that

$$\Lambda_h \subseteq \tilde{\Lambda}_h, \quad \sum_{\lambda \in \tilde{\Lambda}_h} |e_\lambda|^2 \geq \vartheta^* \|\mathbf{e}\|^2_{\ell^2}. \tag{78}$$

In other words the index set $\tilde{\Lambda}_h$ must capture at least a fixed fraction of the error (as estimated by (77)). This refinement strategy turns out to be error reducing, provided the error is measured in the norm induced by the operator A. In order to show this let $\tilde{u}_h \in \tilde{V}_h = span < \check{\psi}_\lambda, \lambda \in \tilde{\Lambda}_h >$ denote the Galerkin projection of u in the refined space $\tilde{V}_h$. We observe that, if $\lambda \in \tilde{\Lambda}_h$ we can write

$$e_\lambda = \langle f - Au_h, \check{\psi}_\lambda\rangle = \langle A(\tilde{u}_h - u_h), \check{\psi}_\lambda\rangle.$$

Then we have

$$\sum_{\lambda \in \tilde{\Lambda}_h} |e_\lambda|^2 = \sum_{\lambda \in \tilde{\Lambda}_h} |\langle A(\tilde{u}_h - u_h), \check{\psi}_\lambda \rangle|^2$$
$$\lesssim \sum_{\lambda \in \Lambda} |\langle A(\tilde{u}_h - u_h), \check{\psi}_\lambda \rangle|^2 \lesssim \|A(\tilde{u}_h - u_h)\|^2_{H^{-t}(\Omega)},$$

from which, thanks to the continuity of A, we obtain

$$\sum_{\lambda \in \tilde{\Lambda}_h} |e_\lambda|^2 \lesssim \|\tilde{u}_h - u_h\|^2_{H^t(\Omega)}.$$

In view of the refinement strategy adopted this implies that

$$\|\tilde{u}_h - u_h\|^2_{H^t(\Omega)} \gtrsim \sum_{\lambda \in \tilde{\Lambda}_h} |e_\lambda|^2 \geq \vartheta^* \|\mathbf{e}\|^2_{\ell^2} \gtrsim \|u - u_h\|^2_{H^t(\Omega)}.$$

If we now introduce the scalar product $(\cdot,\cdot)_A = \langle A\cdot,\cdot\rangle$ and the corresponding norm $\|\cdot\|_A$ we can on the one hand observe that $\|\cdot\|_A \simeq \|\cdot\|_{H^t(\Omega)}$; it is then easy to see that there exists a positive constant $\tilde{\xi}$ that, without loss of generality, we can assume to satisfy $\tilde{\xi} < 1$, such that

$$\|\tilde{u}_h - u_h\|^2_A \geq \tilde{\xi} \|u - u_h\|^2_A.$$

On the other hand, with respect to the scalar product $(\cdot,\cdot)_A$, Galerkin orthogonality implies that $\tilde{u}_h - u_h \in \tilde{V}_h$ is orthogonal to $u - \tilde{u}_h$. Then we can write

$$\|u - u_h\|^2_A = \|(u - \tilde{u}_h) + (\tilde{u}_h - u_h)\|^2_A = \|u - \tilde{u}_h\|^2_A + \|\tilde{u}_h - u_h\|^2_A,$$

whence, for $\xi = 1 - \tilde{\xi}$

$$\|u - \tilde{u}_h\|^2_A = \|u - u_h\|^2_A - \|u_h - \tilde{u}_h\|^2_A \leq \xi \|u - u_h\|^2_A. \tag{79}$$

Remark 14. We will need the norm $\|\cdot\|_A$ also later on and, by abuse of notation, we will also denote by $\|\cdot\|_A$ the equivalent ℓ^2 norm defined by

$$\|\mathbf{w}\|^2_A = \mathbf{w}^T \mathscr{R} \mathbf{w}$$

(the abuse of notation is justified since for $w = \sum_\lambda w_\lambda \check{\psi}_\lambda$ we have $\|w\|_A = \|\mathbf{w}\|_A$).

Thanks to (79), the refinement strategy (78) guarantees then convergence of an adaptive procedure of the following form.

- initial index set $\Lambda^0 = \emptyset$; initial guess $\mathbf{u}^0 = 0$;
- $\Lambda^i, \mathbf{u}^i \to \Lambda^{i+1}, \mathbf{u}^{i+1}$
  - ⋆ compute $\mathbf{e}^i = \mathbf{f} - \mathscr{R}\mathbf{u}^i$
  - ⋆ compute $\Lambda^{(i+1)} :=$ smallest index set such that (78) holds
  - ⋆ compute $\mathbf{u}^{(i+1)}$:coefficients of the Galerkin solution in *span* $< \check{\psi}_\lambda,\ \lambda \in \Lambda^{(i+1)} >$
- iterate until $\|\mathbf{e}^i\|_{\ell^2} \leq \varepsilon$

The bound (79) guarantees convergence of the algorithm. Moreover, provided ϑ^* is small enough, it is possible to estimate a priori the cardinality of the output index set at convergence [34]. The key ingredient is the following lemma.

Lemma 2. *Let $w \in$ span $< \psi_\lambda,\ \lambda \in \Lambda_h >$ and let $\tilde{\Lambda}_h \supset \Lambda_h$ be the smallest possible index set such that (78) holds. Then, if $\vartheta^* < \kappa(\mathscr{R})^{-1/2}$, for $0 < s < s^*$ and $\tau = (s/d + 1/2)^{-1}$ we have that*

$$\#(\tilde{\Lambda}_h \setminus \Lambda_h) \lesssim \|f - Aw\|_{H^{-t}(\Omega)}^{-d/s} \|u\|_{B_\tau^{s,\tau}}^{d/s}$$

($\kappa(\mathscr{R}) = \|\mathscr{R}\|_{\ell^2 \to \ell^2} \|\mathscr{R}^{-1}\|_{\ell^2 \to \ell^2}$ denoting the condition number *of the infinite matrix $\mathscr{R}$).*

Proof. Let $\lambda > 0$ be a constant such that

$$\vartheta^* \leq \kappa(\mathscr{R})^{-1/2}(1 - \kappa(\mathscr{R})\lambda^2)^{1/2}.$$

Let N be the smallest integer such that $\|\mathbf{u} - P_N\mathbf{u}\|_{\ell^2} \leq \lambda \|\mathbf{u} - \mathbf{w}\|_A$, P_N denoting the nonlinear projection operator introduced in Sect. 3.1.4. By applying Theorem 11, and the equivalence $\|\mathbf{u} - \mathbf{w}\|_A \simeq \|\mathbf{u} - \mathbf{w}\|_{\ell^2} \simeq \|\mathscr{R}(\mathbf{u} - \mathbf{w})\|_{\ell^2}$ we can bound [34]

$$N \lesssim \|\mathbf{f} - \mathscr{R}\mathbf{w}\|_{\ell^2}^{-d/s} \|\mathbf{u}\|_{\ell^\tau}^{d/s}.$$

The definition of the $\|\cdot\|_A$ norm and of the constant λ yield

$$\|\mathbf{u} - P_N\mathbf{u}\|_A \leq \|\mathscr{R}\|_{\ell^2 \to \ell^2}^{1/2} \|\mathbf{u} - P_N\mathbf{u}\|_{\ell^2} \leq \lambda \|\mathscr{R}\|_{\ell^2 \to \ell^2}^{1/2} \|\mathbf{u} - \mathbf{w}\|_A.$$

Let now $\hat{\Lambda} = \Lambda_h \cup \operatorname{supp}(P_N(\mathbf{u}))$ denote the set of indexes λ for which the λ-th coefficient of either $\mathbf{w}$ or $P_N(\mathbf{u})$ does not vanish, and let $\hat{u}_h = \sum_{\lambda \in \hat{\Lambda}} \hat{u}_\lambda \check{\psi}_\lambda$ denote the Galerkin projection of u in *span* $< \check{\psi}_\lambda,\ \lambda \in \hat{\Lambda} >$, $\hat{\mathbf{u}} \in \ell^2(\Lambda)$ denoting the corresponding wavelet coefficient vector (whose component vanish outside $\hat{\Lambda}$). The optimality of the Galerkin projection in the $\|\cdot\|_A$ norm implies

$$\|\mathbf{u} - \hat{\mathbf{u}}\|_A \leq \|\mathbf{u} - P_N\mathbf{u}\|_A \leq \lambda \|\mathscr{R}\|_{\ell^2 \to \ell^2}^{1/2} \|\mathbf{u} - \mathbf{w}\|_A.$$

By Galerkin orthogonality we then have

$$\|\mathbf{w} - \hat{\mathbf{u}}\|_A^2 \geq (1 - \|\mathscr{R}\|_{\ell^2 \to \ell^2} \lambda^2) \|\mathbf{u} - \mathbf{w}\|_A^2,$$

using which together with $\hat{u}_\lambda = w_\lambda = 0$ for $\lambda \notin \hat{\Lambda}$ we can write

$$\begin{aligned}
\vartheta^* \|\mathbf{f} - \mathscr{R}\mathbf{w}\|_{\ell^2}^2 &\leq \kappa(\mathscr{R})^{-1}(1 - \kappa(\mathscr{R})\lambda^2)\|\mathbf{f} - \mathscr{R}\mathbf{w}\|_{\ell^2}^2 \\
&= \kappa(\mathscr{R})^{-1}(1 - \kappa(\mathscr{R})\lambda^2)\|\mathscr{R}\mathbf{u} - \mathscr{R}\mathbf{w}\|_{\ell^2}^2 \\
&\leq \|\mathscr{R}^{-1}\|_{\ell^2 \to \ell^2}^{-1}(1 - \kappa(\mathscr{R})\lambda^2)\|\mathbf{u} - \mathbf{w}\|_A^2 \leq \|\mathscr{R}^{-1}\|_{\ell^2 \to \ell^2}^{-1}\|\hat{\mathbf{u}} - \mathbf{w}\|_A^2,
\end{aligned}$$

where we used the bound $\|\mathscr{R}\mathbf{x}\|_{\ell^2} \leq \|\mathscr{R}\|_{\ell^2 \to \ell^2}^{1/2} \|\mathbf{x}\|_A$. Finally the matrix $\mathscr{R}$ being symmetric positive definite allows us to write [34]

$$\|\mathscr{R}^{-1}\|_{\ell^2 \to \ell^2}^{-1} \|\hat{\mathbf{u}} - \mathbf{w}\|_A^2 \leq \sum_{\lambda \in \hat{\Lambda}} |e_\lambda|^2.$$

Since $\Lambda_h \subset \hat{\Lambda}$, by definition of $\tilde{\Lambda}_h$ we conclude

$$\#(\tilde{\Lambda}_h \setminus \Lambda_h) \leq \#(\hat{\Lambda}_h \setminus \Lambda_h) \leq N \lesssim \|\mathbf{f} - \mathscr{R}\mathbf{w}\|_{\ell^2}^{-d/s} \|\mathbf{u}\|_{\ell^\tau}^{d/s}.$$

The thesis follows thanks to Remark 13. □

Provided the solution u belongs to $B_\tau^{t+s,\tau}(\Omega)$ with $\tau = (1/2 + s/d)^{-1}$, we can then use Lemma 2 to estimate the cardinality of the index set that the adaptive procedure selects upon convergence [34]. Let in fact K be such that $\|r^K\|_{\ell^2} > \varepsilon \geq \|r^{K+1}\|_{\ell^2}$. We have that

$$\#(\Lambda^{K+1}) = \sum_{i=0}^{K} \#(\Lambda^{i+1} \setminus \Lambda^i).$$

Now, by Lemma 2 we have that

$$\#(\Lambda^{i+1} \setminus \Lambda^i) \lesssim \|f - Au^i\|_{H^{-t}(\Omega)}^{-d/s} \|u\|_{B_\tau^{t+s,\tau}(\Omega)}^{d/s} \lesssim \|u - u^i\|_A^{-d/s} \|u\|_{B_\tau^{t+s,\tau}(\Omega)}^{d/s}$$

which, combined with

$$\|u - u^K\|_A \leq \xi^{K-i} \|u - u_i\|_A$$

yields

$$\#(\Lambda^{K+1}) \lesssim \|u\|^{d/s} \|u - u^K\|_A^{-d/s} \sum_{i=0}^{K} \xi^{d(K-i)/s} \lesssim \|u\|_{B_\tau^{t+s,\tau}(\Omega)}^{d/s} \|u - u^K\|_A^{-d/s}.$$

Since, by the definition of K we have that $\|u - u^K\|_A \gtrsim \varepsilon$ we immediately obtain

$$\#(\Lambda^{K+1}) \lesssim \varepsilon^{-d/s} \|u\|^{d/s}_{B^{t+s,\tau}_\tau(\Omega)}.$$

5.2 *Nonlinear Wavelet Methods for the Solution of PDE's*

The adaptive procedure presented in the previous section exploits the properties of wavelets to design an adaptive wavelet algorithm of a classical type:

- Given an approximation space it computes an approximation to the solution of the problem within the given space.
- It looks at the computed approximation in order to design a new approximation space using the a posteriori error estimate (77).
- It iterates until the computed solution is satisfactory.

In view of (75) and (76) the discrete form (74) suggests a different approach. The fact that $\|\mathscr{R}\|_{\ell^2 \to \ell^2}$ and $\|\mathscr{R}^{-1}\|_{\ell^2 \to \ell^2}$ are both finite easily implies the following Proposition:

Proposition 4. *There exist a constant ϑ_0, such that $\forall \vartheta$ with $0 < \vartheta < \vartheta_0$ it holds that*

$$\|I - \vartheta \mathscr{R}\|_{\ell^2 \to \ell^2} \leq \sigma < 1. \tag{80}$$

This suggests us to formally write down an iterative solution scheme for the bi-infinite linear system (74). To fix the ideas let us consider a simple Richardson scheme:

5.2.1 The Richardson Scheme for the Continuous Problem

- `initial guess` $\mathbf{u}^0 = 0$
- $\mathbf{u}^n \longrightarrow \mathbf{u}^{n+1}$
  - ⋆ `compute` $\mathbf{r}^n = \mathbf{f} - \mathscr{R}\mathbf{u}^n$
  - ⋆ $\mathbf{u}^{n+1} = \mathbf{u}^n + \vartheta \mathbf{r}^n$
- `iterate until error` $\leq$ `tolerance.`

Thanks to Proposition 4 it is not difficult to prove that this algorithm converges to the solution $\mathbf{u}$ of (74), provided $\vartheta < \vartheta_0$. This formal (since it acts on infinite matrix and vectors) converging scheme is the basis of two nonlinear algorithms which we will consider in the following sections. Before presenting the nonlinear algorithm let us recall that we have the following lemma [21], which improves Proposition 4.

Lemma 3. *There exist two constants $\tau^* < 2$ and ϑ_0, such that $\forall \vartheta$ with $0 < \vartheta < \vartheta_0$ and $\forall \tau$, with $\tau^* < \tau \leq 2$,*

$$\|I - \vartheta \mathscr{R}\|_{\ell^\tau \to \ell^\tau} \leq \rho < 1. \tag{81}$$

Let us start by considering a very simple scheme. We aim at computing an approximation to u in the nonlinear space Σ_N. The idea [11] is to modify the Richardson scheme for the (74) by forcing $u^n = \sum_\lambda u_\lambda^n \check{\psi}_\lambda$ to belong to the nonlinear space Σ_N. This reduces to forcing the iterates $\mathbf{u}^n$ to have at most N non zero entries. The simplest way of doing this is to project u^n onto Σ_N by the nonlinear projector P_N defined by (66). The result is the following scheme:

- `initial guess` $\mathbf{u}^0 = 0$
- $\mathbf{u}^n \to \mathbf{u}^{n+1}$
  - ⋆ `compute` $\mathbf{r}^n = \mathbf{f} - \mathscr{R}\mathbf{u}^n$
  - ⋆ $\mathbf{u}^{n+1} = P_N(\mathbf{u}^n + \vartheta \mathbf{r}^n) \qquad (u^{n+1} = \sum_\lambda u_\lambda^{n+1} \psi_\lambda \in \Sigma_N)$
- `iterate until error` $\leq$ `tolerance.`

The above procedure is once again not practically computable (since it involves the exact computation of the residual, which is obtained by multiplying a finite vector times an infinite matrix). Nevertheless it is interesting to analyze it and prove stability and some form of convergence. This is the object of the following theorem:

Theorem 12. *$\exists \vartheta_0$ s.t. if $u \in B_\tau^{t+s,\tau}(\Omega)$ with $0 < s < \min\{s^*, d/\tau^* - d/2\}$ and $\tau = (s/d + 1/2)^{-1}$ then for $0 < \vartheta < \vartheta_0$ it holds:*

- *Stability: we have $\|\mathbf{u}^n\|_{\ell^2} \lesssim \|\mathbf{f}\|_{\ell^2} + \|\mathbf{u}^0\|_{\ell^2}, \quad \forall n \in \mathbb{N}$*
- *Convergence: for $\mathbf{e}^n = \mathbf{u}^n - \mathbf{u}$ it holds:*

$$\|\mathbf{e}^n\|_{\ell^2} \leq \rho^n \|\mathbf{e}^0\|_{\ell^2} + \frac{C}{1-\rho} N^{-s/d},$$

where C is a constant depending only on the initial data

Proof. **Stability**. We have, using the ℓ^2 boundedness of P_N as well as (81)

$$\|\mathbf{u}^n\|_{\ell^2} = \|P_N(\mathbf{u}^{n-1} + \vartheta(\mathbf{f} - \mathscr{R}\mathbf{u}^{n-1}))\|_{\ell^2} \leq \|(1 - \vartheta\mathscr{R})\mathbf{u}^{n-1} + \vartheta \mathbf{f}\|_{\ell^2}$$

$$\leq \|\vartheta \mathbf{f}\|_{\ell^2} + \rho \|\mathbf{u}^{n-1}\|_{\ell^2}.$$

Iterating this bound for n decreasing to 0 we obtain

$$\|\mathbf{u}^n\|_{\ell^2} \leq \left(\sum_{i=0}^{n-1} \rho^i\right) \|\vartheta \mathbf{f}\|_{\ell^2} + \rho^n \|\mathbf{u}^0\|_{\ell^2},$$

which gives us the stability bound. In the same way we can prove that

$$\|\mathbf{u}^n\|_{\ell^\tau} \lesssim \|\mathbf{f}\|_{\ell^\tau} + \|\mathbf{u}^0\|_{\ell^\tau}, \quad \forall n \in \mathbb{N}. \tag{82}$$

Convergence. We write down an error equation

$$\mathbf{e}^{n+1} = \mathbf{e}^n - \vartheta \mathcal{R} \mathbf{e}^n + \varepsilon^n,$$

with

$$\varepsilon^n = P_N(\mathbf{u}^n + \vartheta(\mathbf{f} - \mathcal{R}\mathbf{u}^n)) - (\mathbf{u}^n + \vartheta(\mathbf{f} - \mathcal{R}\mathbf{u}^n)).$$

We take the ℓ^2 norm, and, using (81) once again we obtain

$$\|\mathbf{e}^{n+1}\|_{\ell^2} \leq \rho\|\mathbf{e}^n\|_{\ell^2} + \|\varepsilon^n\|_{\ell^2} \leq \sum_{i=0}^{n} \rho^{n-i}\|\varepsilon_i\|_{\ell^2} + \rho^{n+1}\|e_0\|_{\ell^2}$$

$$\leq \left(\max_{0\leq k\leq n} \|\varepsilon^k\|_{\ell^2}\right) \sum_{k=1}^{n} \rho^k + \rho^{n+1}\|e_0\|_{\ell^2}.$$

Let us bound ε^k. Using (82) we have

$$\varepsilon^k \leq N^{-s/d}\|\mathbf{u}^k + \vartheta(\mathbf{f} - \mathcal{R}\mathbf{u}^k)\|_{\ell^\tau} \leq N^{-s/d}\|(I - \vartheta\mathcal{R})\mathbf{u}^k + \vartheta\mathbf{f}\|_{\ell^\tau} \lesssim N^{-s/d}.$$

□

The above scheme shows the idea underlying the new generation of nonlinear wavelet schemes. It has however several limitations. First of all, as already observed it is not computable. The iterates are indeed (by construction) finitely supported vectors but the algorithm involves the computation of the residual which requires once again the multiplication of a finite dimensional vector with the infinite matrix $\mathcal{R}$. This limitation can be overcome by replacing the residual $\mathbf{r}$ by an approximate residual (see Sect. 5.4). In order for the algorithm to converge correctly it is moreover necessary to choose the tolerance for the stopping criterion in a suitable way. In fact, it is not difficult to realize, even without a rigorous analysis, that if the tolerance is to big, the algorithm will stop too soon, and will not produce a satisfactory solution. If the tolerance is too small, the algorithm will not converge, since the error generated by the nonlinear projection step at each iteration may be such that the stopping criterion is never met. Moreover we would like our adaptive algorithm to be optimal not only in terms of convergence rate but also in terms of the workload that we need to face in order to obtain the approximate solution.

5.3 *The CDD2 Algorithm*

In [17] Cohen et al. presented a class of non linear wavelet algorithms based on the Richardson scheme for the full discrete system (74), for which they were able to provide a full analysis including stability, convergence with optimal rate, as well as an a priori estimate on the total number of operations required to converge within a tolerance ε.

Their approach started from the following consideration: if we want a feasible nonlinear algorithm we need to be able to approximate the computation of the residual by procedures involving a finite number of operations. We need, in particular, to be able to approximate the right hand side $\mathbf{f}$ with a finite dimensional vector and to compute, for any finite dimensional vector $\mathbf{v}$ an approximation $\mathbf{w}$ to $\mathscr{R}\mathbf{v}$. Moreover, we want to be able to control the effect that the error resulting from such approximate computations has on the overall algorithm. Finally we want to control the number of operations needed to converge within a given tolerance of the solution.

For simplicity let us assume that the righthand side $\mathbf{f}$ is finitely supported and computed exactly, and assume only that a procedure is available performing the following task:

`applyR` given a finitely supported vector $\mathbf{v}$ and a tolerance η returns a finitely supported vector $\mathbf{w}$ (the support of $\mathbf{w}$ not necessarily coinciding with the one of the input vector $\mathbf{v}$) satisfying

$$\|\mathscr{R}\mathbf{v} - \mathbf{w}\|_{\ell^2} \leq \eta.$$

As in the nonlinear Richardson algorithm described in Sect. 5.2, in order to avoid that the dimension of the vectors involved in the computation becomes too big, we need a nonlinear projection procedure. The CDD2 algorithm assumes in particular the availability of a procedure `coarsen` allowing to meet a certain tolerance with a finite vector of the smallest possible support:

`coarsen` given a vector $\mathbf{v}$ and a tolerance η returns the vector $\mathbf{w}$ with smallest support such that

$$\|\mathbf{w} - \mathbf{v}\|_{\ell^2} \leq \eta. \tag{83}$$

If such operations are available we can implement a solution algorithm with the following form.

5.3.1 The CDD2 Algorithm

- `fix tolerance` ε
- `initial tolerance` $\varepsilon_0 = \|\mathscr{R}^{-1}\|_{\ell^2\to\ell^2}\|\mathbf{f}\|_{\ell^2}$
- `initial index set` $\Lambda^0 = \emptyset$; `initial guess` $\mathbf{u}^0 = 0$
- $(\Lambda^j, \mathbf{u}^j, \varepsilon_j) \to (\Lambda^{j+1}, \mathbf{u}^{j+1}, \varepsilon_{j+1})$
  - ⋆ `initialize inner loop:` $\mathbf{v}^0 = \mathbf{u}^j$
  - ⋆ `for` $k = 1, \ldots, K$ $\mathbf{v}^{k-1} \to \mathbf{v}^k$
    - `set` $\eta^k = \rho^k \varepsilon_j$
    - `applyR`$[\mathbf{v}^{k-1}, \eta^k] \to \mathbf{w}^k$
    - $\mathbf{v}^{k+1} = \mathbf{v}^k + \vartheta(\mathbf{f}^k - \mathbf{w}^k)$

  - ⋆ `projection: coarsen`$[\mathbf{v}^K, 0.4\,\varepsilon_j] \to \mathbf{u}^{j+1}$
  - ⋆ `update tolerance:` $\varepsilon_{j+1} = \varepsilon_j/2$
- `iterate until` $\varepsilon_j \le \varepsilon$.

Lemma 4. *If we choose*

$$K = \min\{k : \rho^{k-1}(\vartheta k + \rho) \le 0.1\},$$

then the iterates of the above algorithm satisfy

$$\|\mathbf{u} - \mathbf{u}^j\|_{\ell^2} \lesssim \varepsilon_j = 2^{-j}\varepsilon_0. \tag{84}$$

Proof. We prove (84) by induction on j. For $j = 0$ we have

$$\|\mathbf{u} - \mathbf{u}^0\|_{\ell^2} \le \|\mathbf{u}\|_{\ell^2} \le \|\mathscr{R}^{-1}\|_{\ell^2 \to \ell^2}\|\mathbf{f}\|_{\ell^2} = \varepsilon_0.$$

Let us now assume that (84) holds for the $(j-1)$-th iterate. Let $\tilde{\mathbf{v}}^k$, $k = 1, \dots, K$, denote the sequence obtained by applying a Richardson scheme with the exact multiplication by the infinite matrix $\mathscr{R}$:

$$\tilde{\mathbf{v}}^0 = \mathbf{u}^{j-1}, \quad \tilde{\mathbf{v}}^{k+1} = \tilde{\mathbf{v}}^k + \vartheta(\mathbf{f} - \mathscr{R}\mathbf{v}^k).$$

We have

$$\mathbf{v}^k - \tilde{\mathbf{v}}^k = (1 - \vartheta\mathscr{R})(\mathbf{v}^{k-1} - \tilde{\mathbf{v}}^{k-1}) + \vartheta(\mathbf{w}^k - \mathscr{R}\mathbf{v}^k),$$

whence

$$\begin{aligned}\|\mathbf{v}^k - \tilde{\mathbf{v}}^k\|_{\ell^2} &\le \|1 - \vartheta\mathscr{R}\|_{\ell^2\to\ell^2}\|\mathbf{v}^{k-1} - \tilde{\mathbf{v}}^{k-1}\|_{\ell^2} + \eta_k \le \rho\|\mathbf{v}^{k-1} - \tilde{\mathbf{v}}^{k-1}\|_{\ell^2} + \eta_k \\ &\le \rho(\rho\|\mathbf{v}^{k-2} - \tilde{\mathbf{v}}^{k-2}\| + \eta_{k-1}) + \eta_k \le \dots \le \rho^j\|\mathbf{v}^0 - \tilde{\mathbf{v}}^0\|_{\ell^2} + \sum_{n=0}^{k-1}\rho^k\eta_{k-n}\end{aligned}$$

from which, since $\mathbf{v}^0 = \tilde{\mathbf{v}}^0$ and $\eta^k = \rho^k\varepsilon_{j-1}$

$$\|\mathbf{v}^K - \tilde{\mathbf{v}}^K\|_{\ell^2} \le \vartheta K\rho^{K-1}\varepsilon_{j-1}.$$

On the other hand, standard estimates for Richardson iterations together with the induction assumption yield

$$\|\tilde{\mathbf{v}}^K - \mathbf{u}\|_{\ell^2} \le \rho^K\|\mathbf{u}^{j-1} - \mathbf{u}\|_{\ell^2} \le \rho^K\varepsilon_{j-1}.$$

Then, by using a triangular inequality we can write

$$\|\mathbf{v}^K - \mathbf{u}\|_{\ell^2} \le \|\mathbf{v}^K - \tilde{\mathbf{v}}^K\|_{\ell^2} + \|\tilde{\mathbf{v}}^K - \mathbf{u}\|_{\ell^2} \le \rho^{K-1}(\vartheta K + \rho)\varepsilon_{j-1} \le 0.1\,\varepsilon_{j-1}.$$

Finally, recall that $\mathbf{u}^j$ is obtained by applying `coarsen` to $\mathbf{v}^K$. Using (83) we can then write

$$\|\mathbf{u}^j - \mathbf{u}\| \le \|\mathbf{u}^j - \mathbf{v}^K\|_{\ell^2} + \|\mathbf{v}^K - \mathbf{u}\|_{\ell^2} \le (0.4 + 0.1)\varepsilon_{j-1} = \varepsilon_j.$$

□

The algorithm is then guaranteed to converge. Clearly if we want to have control on the number of operations (for which we need to be able to control the size of the support of the iterates) we need on the one hand to exploit some smoothness information on the solution of the continuous equation, and on the other hand to assume that we are able to optimally estimate the dependence on the input data (tolerance and support of the input vector) of the number of operations needed by the two procedures `applyR` and `coarsen`.

Following [17], we will then assume that there exists an $\hat{s}$ such that, given any tolerance η and any finitely supported vector $\mathbf{v}$, the output $\mathbf{w}$ of `applyR` verifies for any $0 < s < \hat{s}$, with $\tau = (s/d + 1/2)^{-1}$,

$$\#(supp\mathbf{w}) = \#\{\lambda : w_\lambda \neq 0\} \lesssim \|v\|_{\ell^\tau}^{d/s}\eta^{-d/s}, \qquad \|\mathbf{w}\|_{\ell^\tau} \lesssim \|\mathbf{v}\|_{\ell^\tau}.$$

Moreover we assume that the number of arithmetic operations and of sorts needed to compute $\mathbf{w}$ are, respectively, bounded by $C(\eta^{-d/s}\|\mathbf{v}\|_{\ell^\tau} + \#(\operatorname{supp}\mathbf{v}))$ and by $C\#(\operatorname{supp}\mathbf{v})\log(\#(\operatorname{supp}\mathbf{v}))$. Under such an assumption it is possible to improve the convergence result by proving that if $\mathbf{u} \in \ell^\tau$ then the output $\bar{\mathbf{u}}$ of the nonlinear algorithm with target accuracy ε verifies

$$\#(\operatorname{supp}\bar{\mathbf{u}}) \lesssim \|\mathbf{u}\|_{\ell^\tau}^{d/s}\varepsilon^{-d/s}, \qquad \|\bar{\mathbf{u}}\|_{\ell^\tau} \lesssim \|\mathbf{u}\|_{\ell^\tau}.$$

Moreover the total number of operations behaves asymptotically for $\varepsilon \to 0$ as $\varepsilon^{-d/s}$ arithmetic operations and $\varepsilon^{-d/s}|\log(\varepsilon)|$ sorts.

5.4 Operations on Infinite Matrices and Vectors

As already observed several times all the algorithms presented above are only theoretical if we do not provide a way of handling approximately the infinite matrices and vectors involved as well as the operations on such objects, in particular the matrix-vector multiplication $\mathbf{w} = \mathscr{R}\mathbf{v}$

$$w_\lambda = \sum_\mu \mathscr{R}_{\lambda,\mu}v_\mu = \sum_\mu \langle A\check{\psi}_\mu, \check{\psi}_\lambda\rangle v_\mu. \tag{85}$$

The idea is to replace w_λ by an approximation obtained by neglecting, in the sum on the right hand side of (85), the contribution of those indexes μ which are sufficiently “far” (either in space or in frequency) from λ. In fact, we observe that for $\lambda \in \nabla_j$ and $\mu \in \nabla_m$ we have

$$|\langle A\check{\psi}_\lambda, \check{\psi}_\mu\rangle| \lesssim 2^{-(\check{R}+t)|j-m|} i(\lambda,\mu), \tag{86}$$

with $\check{R} < \min\{R-2t, \tilde{R}\}$ and $i(\lambda,\mu) = 1$ if $\operatorname{supp}\check{\psi}_\lambda \cap \operatorname{supp}\check{\psi}_\mu \neq \emptyset$ and vanishing otherwise. If $j \leq m$ this is easily proven by observing that we have

$$|\langle A\check{\psi}_\lambda, \check{\psi}_\mu\rangle| \lesssim \|A\check{\psi}_\lambda\|_{H^{\check{R}}(\Omega)} \|\check{\psi}_\mu\|_{H^{-\check{R}}(\Omega)}, \tag{87}$$

$$\lesssim \|\check{\psi}_\lambda\|_{H^{\check{R}+2t}(\Omega)} \|\check{\psi}_\mu\|_{H^{-\check{R}}(\Omega)} \lesssim 2^{-jt} 2^{j(\check{R}+2t)} 2^{-mt} 2^{-m\check{R}} \tag{88}$$

(recall that $\check{\psi}_\lambda$ and $\check{\psi}_\mu$ are normalized in such a ay that $\|\check{\psi}_\lambda\|_{H^t(\Omega)} \simeq \|\check{\psi}_\mu\|_{H^t(\Omega)} \simeq 1$, which results in the two factors 2^{-jt} and 2^{-mt} on the righthand side of (88)). If $j > m$ we observe that $\langle A\check{\psi}_\lambda, \check{\psi}_\mu\rangle = \langle \check{\psi}_\lambda, A\check{\psi}_\mu\rangle$ and proceed analogously.

The bound (86) suggests to introduce a function $v : \Lambda \times \Lambda \to \mathbb{R}$ defined by

$$\text{For } \lambda \in \nabla_m,\ \mu \in \nabla_j, \quad v(\lambda,\mu) = 2^{-\sigma|j-m|} i(\lambda,\mu),$$

where $0 < \sigma \leq \check{R}+t$ is a parameter to be chosen, and to construct an approximation to $\mathbf{w}$ by only considering the contribution of those basis functions ψ_μ with $v(\mu,\lambda)$ sufficiently big. More precisely, given a tolerance ε we can define

$$\check{w}_\lambda = \sum_{\mu\in\Lambda: v(\lambda,\mu)>\varepsilon} \langle A\check{\psi}_\mu, \check{\psi}_\lambda\rangle v_\mu. \tag{89}$$

The key in proving that $\check{\mathbf{w}}$ is a good approximation to $\mathscr{R}\mathbf{v}$ is based on the observation that for $\mathbf{a} = (a_\lambda)_\lambda \in \ell^2(\Lambda)$ letting $\mathbf{b} = (b_\lambda)_\lambda$ be defined, for $\lambda \in \nabla_j$ by

$$b_\lambda = \sum_m 2^{-(\alpha+d/2)|j-m|} \sum_{\mu\in\nabla_m} i(\lambda,\mu) a_\mu,$$

with $\alpha > 0$, then it holds that $\mathbf{b} = (b_\lambda)_\lambda \in \ell^2(\Lambda)$ and

$$\|\mathbf{b}\|_{\ell^2} \lesssim \|\mathbf{a}\|_{\ell^2}. \tag{90}$$

This is not difficult to see: in fact, due to the particular structure of the index set Λ, we have that, for $\lambda \in \nabla_j$,

$$b_\lambda = \sum_m 2^{-(\alpha+d/2)|j-m|} b_\lambda^m, \qquad b_\lambda^m = \sum_{\mu\in\nabla_m} i(\lambda,\mu) a_\mu,$$

whence

$$\|b\|_{\ell^2} \lesssim \sum_m 2^{-(\alpha+d/2)|j-m|}\|b^m\|_{\ell^2}.$$

Observing that $\#(\{\mu \in \nabla_m : i(\mu,\lambda) \neq 0\}) \lesssim \max\{1, 2^{d(m-j)}\} \lesssim 2^{d|m-j|}$ the result follows easily thanks to the ℓ^2 boundedness of the discrete convolution product with a ℓ^1 function.

In view of (90) it is now not difficult to prove the following Lemma [1, 22].

Lemma 5. *If $\sigma < \check{R} + t - d/2$ then we have*

$$\|\mathbf{w} - \check{\mathbf{w}}\|_{\ell^2} \lesssim \varepsilon \|\mathbf{w}\|_{\ell^2}.$$

Proof. Let $\alpha = \check{R} + t - d/2 - \sigma$ and for $\lambda \in \nabla_j$ and $\mu \in \nabla_m$ let

$$\tilde{v}(\lambda,\mu) = i(\lambda,\mu)2^{-\alpha|j-m|}.$$

It is not difficult to verify that

$$|w_\lambda - \check{w}_\lambda| = |\sum_{\mu\in\Lambda : v(\lambda,\mu)\leq\varepsilon} u_\mu \langle A\check{\psi}_\mu, \check{\psi}_\lambda\rangle| \lesssim \varepsilon \sum_{\mu\in\Lambda : v(\lambda,\mu)\leq\varepsilon} \tilde{v}(\lambda,\mu)|u_\mu|.$$

Applying (90) yields the thesis. □

This kind of approximated matrix vector multiplication algorithms can replace the exact matrix vector multiplication algorithm in the adaptive schemes presented in this section. For instance, they allow to design a feasible refinement strategy analogous to (78) which results in an adaptive scheme converging to a neighborhood of the solution of the problem. In fact, the following corollary is not difficult to prove.

Corollary 4. *Setting $\check{\mathbf{e}} = (\check{e}_\lambda)_\lambda$*

$$\check{e}_\lambda = f_\lambda - \sum_{\mu\in\Lambda_h : v(\lambda,\mu)>\varepsilon} \check{\mathscr{R}}_{\lambda,\mu} u_\mu,$$

we have

$$\|u - u_h\|^2_{H^t(\Omega)} \lesssim \|\check{\mathbf{e}}\|_{\ell^2} + \varepsilon\|f\|^2_{H^{-t}(\Omega)}$$

and

$$\|\check{\mathbf{e}}\|_{\ell^2} \lesssim \|u - u_h\|^2_{H^t(\Omega)} + \varepsilon\|f\|^2_{H^{-t}(\Omega)}.$$

Then, given $\vartheta^* \in (0,1)$ and any tolerance $\varepsilon_0 > 0$, it is possible to prove that there exists a constant μ such that, provided $\varepsilon \leq \mu\varepsilon_0$ if we define $\tilde{\Lambda}_h$ to be the smallest index set such that

$$\sum_{\lambda\in\tilde{\Lambda}_h} |\check{e}_\lambda|^2 \geq \vartheta^* \sum_{\lambda\in\tilde{\Lambda}} 2^{-2j}|\check{e}_\lambda|^2, \quad (91)$$

and we let $\tilde{u}_h \in \tilde{V}_h = span < \psi_\lambda,\ \lambda \in \tilde{\Lambda}_h >$ denote the Galerkin projection of u, then either

$$\|u - \tilde{u}_h\|_A \leq \xi \|u - u_h\|_A$$

with $\xi < 1$ or

$$\sum_{\lambda \in \tilde{\Lambda}} |\check{e}_\lambda|^2 \leq \varepsilon_0$$

(see [22] for a proof).

Remark 15. In all the algorithms presented in this section we assumed that we always treat the vector **f** exactly, so that algorithms, in the form described here, are only feasible if such a vector is finite. We remark that suitable approximations for **f** are possible and their effect can be incorporated in all the algorithm proposed, without substantially changing the kind of results that can be obtained [17].

Remark 16. The simple approximate matrix vector multiplication (89) can be improved to a more sophisticated algorithm where instead of only taking advantage of the decrease properties of the matrix $\mathscr{R}$ (see (86)) one also exploits the decreasing properties (if any) of the vector itself. More in detail it is possible to prove [17] that the decreasing properties of $\mathscr{R}$ imply the existence of a positive summable sequence $(\alpha_j)_{j\geq 0}$ and of matrices $\mathscr{R}_j$ such that $\mathscr{R}_j$ has at most $2^j \alpha_j$ nonzero entries per row and per column and

$$\|\mathscr{R} - \mathscr{R}_j\|_{\ell^2 \to \ell^2} \leq \alpha_j 2^{-j}.$$

Given the a vector $\mathbf{v} \in \Sigma_N$ we can compute an approximation to $\mathscr{R}\mathbf{v}$ as follows:

$$\mathscr{R}\mathbf{v} \sim \mathbf{w}_j = \mathscr{R}_j P_1 \mathbf{v} + \mathscr{R}_{j-1}[P_2\mathbf{v} - P_1\mathbf{v}] + \cdots + \mathscr{R}_0[P_{2^j}\mathbf{v} - P_{2^{j-1}}\mathbf{v}].$$

It is not difficult to prove that here exists τ^* such that for $\tau^* < \tau \leq 2$ the following bound holds with $s = d/\tau - d/2$:

$$\|\mathscr{R}\mathbf{v} - \mathbf{w}_j\|_{\ell^2} \lesssim 2^{-\frac{s}{d}j} \|\mathbf{v}\|_{\ell^\tau}.$$

References

1. S. Bertoluzza, A posteriori error estimates for the wavelet Galerkin method. Appl. Math. Lett. **8**, 1–6 (1995)
2. S. Bertoluzza, Adaptive wavelet collocation method for the solution of Burgers equation. Transport Theor. Stat. Phys. **25**, 339–352 (1996)
3. S. Bertoluzza, An adaptive wavelet collocation method based on interpolating wavelets, in *Multiscale Wavelet Methods for Partial Differential Equations*, ed. by W. Dahmen, A. Kurdila, P. Oswald (Academic Press, New York, 1997), pp. 109–135
4. S. Bertoluzza, L. Castro, Adaptive wavelet collocation for plane elasticity problems. I.M.A.T.I.-C.N.R. Report 26PV08/23/0 (2008)
5. S. Bertoluzza, G.Naldi, Some remarks on wavelet interpolation. Comput. Appl. Math. **13**(1), 13–32 (1994)

6. S. Bertoluzza, G. Naldi, A wavelet collocation method for the numerical solution of partial differential equations. Appl. Comput. Harmon. Anal. **3**, 1–9 (1996)
7. S. Bertoluzza, P. Pietra, *Adaptive Wavelet Collocation for Nonlinear BVPs. Lecture Notes in Control and Information Science* (Springer, London, 1996), pp. 168–174
8. S. Bertoluzza, M. Verani, Convergence of a non-linear wavelet algorithm for the solution of PDE's. Appl. Math. Lett. **16**(1), 113–118 (2003)
9. S. Bertoluzza, Y. Maday, J.C. Ravel, A dynamically adaptive wavelet method for solving partial differential equations. Comput. Meth. Appl. Mech. Eng. **116**(14), 293299 (1994)
10. S. Bertoluzza, C. Canuto, K. Urban, On the adaptive computation of integrals of wavelets. Appl. Numer. Math. **34**, 13–38 (2000)
11. S. Bertoluzza, S. Mazet, M. Verani, A nonlinear Richardson algorithm for the solution of elliptic PDE's. M3AS **13**, 143–158 (2003)
12. C. Canuto, A. Tabacco, K. Urban, The wavelet element method. II. Realization and additional features in 2D and 3D. Appl. Comput. Harmon. Anal. **8**, 123–165 (2000)
13. M. Chen, R. Temam, Incremental unknowns for solving partial differential equations. Numer. Math. **59**, 255–271 (1991)
14. O. Christensen, *An Introduction to Frames and Riesz Bases. Applied and Numerical Harmonic Analysis Series* (Birkhuser, Basel, 2003)
15. A. Cohen, *Numerical Analysis of Wavelet Methods* (Elsevier, Amsterdam, 2003)
16. A. Cohen, R. Masson, Wavelet adaptive method for second order elliptic problems: boundary conditions and domain decomposition. Numer. Math. **86**(2), 193–238 (2000)
17. A. Cohen, W. Dahmen, R. DeVore, Adaptive wavelet methods II: beyond the elliptic case. Found. Comput. Math. **2** (2002), 203–245 (1992)
18. A. Cohen, I. Daubechies, J. Feauveau, Biorthogonal Bases of compactly supported Wavelets. Comm. Pure Appl. Math. **45**, 485–560, (1992)
19. A. Cohen, I. Daubechies, P. Vial, Wavelets on the interval and fast wavelet transforms. Appl. Comput. Harmon. Anal. **1**, 54–81 (1993)
20. A. Cohen, W. Dahmen, R. DeVore, Multiscale decomposition on bounded domains. Trans. Am. Math. Soc. **352**(8), 3651–3685 (2000)
21. A. Cohen, W. Dahmen, R. DeVore, Adaptive wavelet methods for elliption operator equations—convergence rates. Math. Comput. **70**, 27–75 (2000)
22. S. Dahlke, W. Dahmen, R. Hochmuth, R. Schneider, Stable multiscale bases and local error estimation for elliptic problems, Appl. Numer. Math. **23**, 21–47 (1997), MR 98a:65075
23. W. Dahmen, Stability of multiscale transformations. J. Fourier Anal. Appl. **2**(4), 341–361 (1996)
24. W. Dahmen, C.A. Michelli, Using the refinement equation for evaluating integrals of wavelets. SIAM J. Numer. Anal. **30**, 507–537 (1993)
25. W. Dahmen, R. Schneider, Wavelets with complementary boundary conditions—function spaces on the cube. Results Math. **34**, 255–293 (1998)
26. W. Dahmen, R. Schneider, Composite wavelet bases for operator equations. Math. Comput. **68**(228), 1533–1567 (1999)
27. W. Dahmen, R. Stevenson, Element-by-element construction of wavelets satisfying stability and moment monditions. SIAM J. Numer. Anal. **37**(1), 319–352 (1999)
28. W. Dahmen, A. Kunoth, R. Schneider, Wavelet least square methods for boundary value problems. SIAM J. Numer. Anal. **39**, 1985–2013, (2002)
29. I. Daubechies, Orthonormal bases of compactly supported wavelets. Comm. Pure Appl. Math. **41**, 909–996 (1988)
30. I. Daubechies, *Ten Lectures on Wavelets* (Society for Industrial and Applied Mathematics (SIAM), Philadelphia, 1992)
31. G. Deslaurier, S. Dubuc, Symmetric iterative interpolation processes. Constr. Approx. **5**, 46–68 (1989)
32. R.A. DeVore, Nonlinear approximation. Acta Numer. **7**, 51–150, (1998)
33. D. Donoho, Interpolating wavelet transforms. Department of Statistics, Stanford University (1992) http://www-stat.stanford.edu/ donoho/Reports/1992/interpol.pdf

34. T. Gantumur, H. Harbrecht, R. Stevenson, An optimal adaptive wavelet method without coarsening of the iterands. Math. Comput. **76**(258), 615–629 (2007)
35. N. Hall, B. Jawerth, L. Andersson, G. Peters, Wavelets on closed subsets of the real line. Technical Report (1993)
36. N.K.-R. Kevlahan, O.V. Vasilyev, An adaptive wavelet collocation method for fluid–structure interaction. SIAM J. Sci. Comput. **26**(6), 1894–1915 (2005)
37. P.G. Lemarié, Fonctions à support compact dans les analyses multi-résolutions. Rev. Mat. Iberoamericana **7**(2), 157–182 (1991)
38. Y. Maday, V. Perrier, J.C. Ravel, Adaptivité dynamique sur bases d'ondelettes pour l'approximation d'équations aux derivées partielles. C. R. Acad. Sci Paris **312**(Série I), 405–410 (1991)
39. Y. Meyer, Wavelets and operators, in *Different Perspectives on Wavelets*, vol. 47, ed. by I. Daubechies. Proceedings of Symposia in Applied Mathematics, (American Mathematical Society, Providence, RI, 1993), pp. 35–58. From an American Math. Soc. short course, Jan. 11–12, 1993, San Antonio, TX
40. N. Saito, G. Beylkin, Multiresolution representation using the autocorrelation functions of compactly supported wavelets. IEEE Trans. Signal Process. **41**, 3584–3590 (1993)
41. H. Triebel, *Interpolation Theory, Function Spaces, Differential Operators* (North Holland, Amsterdam, 1978)
42. O.V. Vasilyev, S.Paolucci, A dynamically adaptive multilevel wavelet collocation method for solving partial differential equations in a finite domain. J. Comput. Phys. **125**, 498–512 (1996)
43. O.V. Vasilyev, Y.Y. Podladchikov, D.A. Yuen, Modeling of compaction driven flow in poro-viscoelastic medium using adaptive wavelet collocation method. Geophys. Res. Lett. **25**(17), 3239–3242 (1998)

Heterogeneous Mathematical Models in Fluid Dynamics and Associated Solution Algorithms

Marco Discacciati, Paola Gervasio, and Alfio Quarteroni

Abstract Mathematical models of complex physical problems can be based on heterogeneous differential equations, i.e. on boundary-value problems of different kind in different subregions of the computational domain. In this presentation we will introduce a few representative examples, we will illustrate the way the coupling conditions between the different models can be devised, then we will address several solution algorithms and discuss their properties of convergence as well as their robustness with respect to the variation of the physical parameters that characterize the submodels.

1 Introduction and Motivation

For the description and simulation of complex physical phenomena, combination of hierarchical mathematical models can be set up with the aim of reducing the computational complexity. This gives rise to a system of heterogeneous problems, where different kind of differential problems are set up in subdomains (either disjoint or overlapping) of the original computational domain. When facing this

M. Discacciati (✉)
Laboratori de Càlcul Numèric (LaCàN), Universitat Politècnica de Catalunya (UPC BarcelonaTech), Campus Nord UPC - C2, E-08034 Barcelona, Spain
e-mail: marco.discacciati@upc.edu

P. Gervasio
Department of Mathematics, University of Brescia, Brescia, Italy
e-mail: paola.gervasio@unibs.it

A. Quarteroni
MOX, Department of Mathematics, Politecnico di Milano, Milano, Italy

CMCS-EPFL, CH-1015 Lausanne, Switzerland
e-mail: alfio.quarteroni@epfl.ch

S. Bertoluzza et al., *Multiscale and Adaptivity: Modeling, Numerics and Applications*, Lecture Notes in Mathematics 2040, DOI 10.1007/978-3-642-24079-9_2,

kind of coupled problems, two natural issues arise. The former is concerned with the way interface coupling conditions can be devised, the latter with the construction of suitable solution algorithms that can take advantage of the intrinsic splitting nature of the problem at hand. This work will focus on both issues, in the context of heterogeneous boundary-value problems that can be used for fluid dynamics applications.

The outline of this presentation is as follows. After giving the motivation for this investigation, we will present two different approaches for the derivation and analysis of the interface coupling conditions: the one based on the variational formulation, the other on virtual controls. For the former we will consider at first advection–diffusion problems. After carrying out their variational analysis we propose domain decomposition algorithms for their solution, in particular those based on Dirichlet–Neumann, adaptive Robin–Neumann, or Steklov–Poincaré iterations. Then, we will focus on Navier–Stokes/Darcy or Stokes/potential coupled problem presenting their asymptotic analysis together with possible solution techniques.

For the virtual control approach, we will study the case of non-overlapping subdomains for advection–diffusion problems considering in particular possible techniques to solve the optimality system and we will present some numerical results. Then, we will consider the case of domain decomposition with overlap, namely Schwarz methods with Dirichlet/Robin interface conditions. We will investigate the virtual control approach with overlap for the advection–diffusion equations including the case of three virtual controls and we will present some numerical results. Finally, we will illustrate this framework for the case of the Stokes–Darcy coupled problem, and for the coupling of incompressible flows.

In order to motivate our investigation, we begin to analyze the advection–diffusion problem.

Let us consider a bounded domain $\Omega \subset \mathbb{R}^d$ $(d = 1,2,3)$ with Lipschitz boundary and the advection–diffusion equation

$$\begin{cases} Au \equiv \operatorname{div}(-\nu\nabla u + \mathbf{b}u) + b_0 u = f & \text{in } \Omega \\ u = g & \text{on } \partial\Omega, \end{cases} \tag{1}$$

where $\nu > 0$ is a characteristic parameter of the problem, $\mathbf{b} = \mathbf{b}(\mathbf{x})$ a d-dimensional vector valued function, $b_0 = b_0(\mathbf{x})$ and $f = f(\mathbf{x})$ scalar functions, all assigned in Ω, while $g = g(\mathbf{x})$ is assigned on $\partial\Omega$.

The characteristic parameter ν can either represent the thermal diffusivity in heat transfer problems, or the inverse of the Reynolds number in incompressible fluid-dynamics, or another suitable parameter.

Denoting by

$$\mathbb{P}\mathrm{e}_g(\mathbf{x}) = \frac{|\mathbf{b}(\mathbf{x})|}{2\nu} \tag{2}$$

the *global Péclet number*, we call (1) an *advection-dominated* problem when $\mathbb{P}\mathrm{e}_g(\mathbf{x}) \gg 1$.

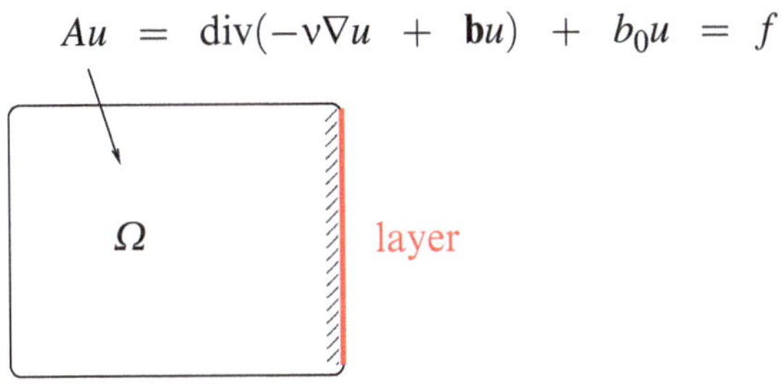

Fig. 1 A simple computational domain and the localization of the boundary layer

We are interested in treating advection dominated problems with boundary layers (see, e.g., Fig. 1), that arise when boundary data are incompatible with the limit (as $\nu \to 0$) of the advection–diffusion equation. As an example, let us consider the one-dimensional advection–diffusion equation

$$\begin{cases} -\nu u''(x) + bu'(x) = 0, \ 0 < x < 1, \\ u(0) = 0, \ u(1) = 1, \end{cases} \tag{3}$$

with $\nu > 0$ and $b > 0$. Problem (3) can be solved exactly and its solution reads

$$u(x) = \frac{e^{bx/\nu} - 1}{e^{b/\nu} - 1}.$$

Such solution exhibits a boundary layer of width $O(\nu/b)$ near to $x = 1$ when the ratio ν/b is small enough, that is when

$$\mathbb{P}\mathrm{e}_g(\mathbf{x}) \gg 1. \tag{4}$$

In Fig. 2 we show the one-dimensional solution $u(x)$ of (3) for two different values of the Péclet number: $\mathbb{P}\mathrm{e}_g(\mathbf{x}) = 0.5$ at left and $\mathbb{P}\mathrm{e}_g(\mathbf{x}) = 100$ at right. Only in the latter case a boundary layer occurs.

When (4) holds, the diffusive term is relevant only in a small part of the domain near to the boundary layer, while it can formally be neglected in the rest of the domain, where the advection phenomenon prevails.

The idea is then: to split the domain in two non-overlapping subdomains Ω_1 and Ω_2 where we denote by $\Gamma = \partial\Omega_1 \cap \partial\Omega_2$ the interface between subdomains, and then to solve a reduced problem as follows (see Fig. 3):

$$\begin{cases} A_1 u_1 \equiv \text{div}(\mathbf{b}u_1) + b_0 u_1 = f & \text{in } \Omega_1 \\ A_2 u_2 \equiv \text{div}(-\nu\nabla u_2 + \mathbf{b}u_2) + b_0 u_2 = f & \text{in } \Omega_2 \\ \text{Boundary conditions} & \text{on } \partial\Omega. \end{cases} \tag{5}$$

The main question that follows is: *how to couple the subproblems*?

To answer this question one should:

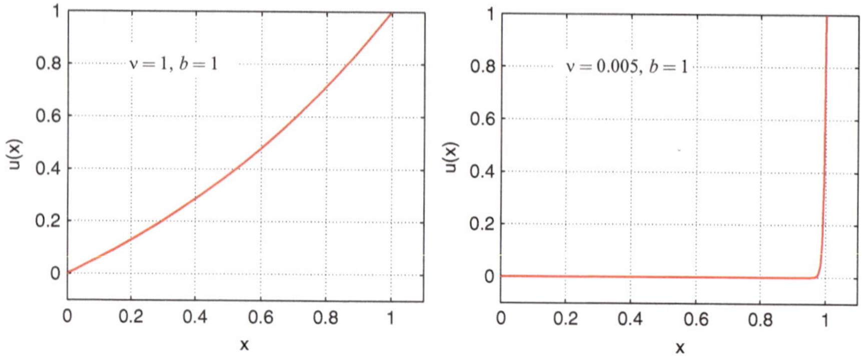

Fig. 2 The exact solution of problem (3). The solution at right exhibits a boundary layer in $x = 1$

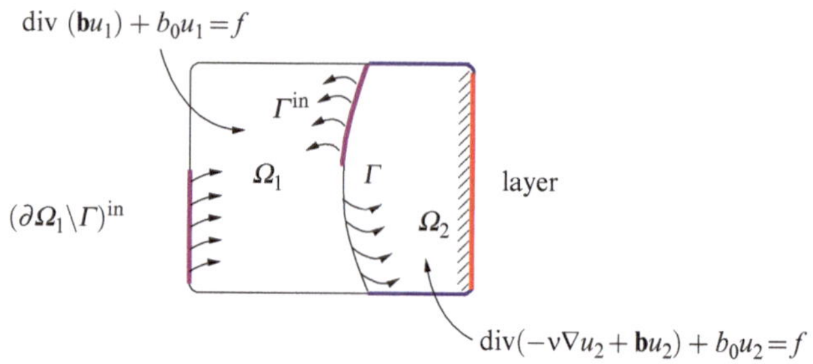

Fig. 3 The reduced problem on the computational domain $\Omega \subset \mathbb{R}^2$

1. Find *interface conditions* on Γ so that the new reduced problem is well posed and its solution is "close to" the original one; then
2. Set up *efficient solution algorithms* to solve the reduced problem.

By a singular perturbation analysis, Gastaldi et al. [30] proposed the following set of interface conditions:

$$\begin{cases} u_1 = u_2 & \text{on } \Gamma^{\text{in}} \\ \mathbf{b}\cdot\mathbf{n}_\Gamma u_1 + \nu\dfrac{\partial u_2}{\partial n_\Gamma} - \mathbf{b}\cdot\mathbf{n}_\Gamma u_2 = 0 & \text{on } \Gamma, \end{cases} \tag{6}$$

where $\mathbf{n}_\Gamma$ is the normal versor to Γ oriented from Ω_1 to Ω_2 and $\Gamma^{\text{in}} = \{\mathbf{x} \in \Gamma : \mathbf{b}(\mathbf{x})\cdot\mathbf{n}_\Gamma(\mathbf{x}) < 0\}$ is the inflow interface for Ω_1.

The coupled formulation (5) and (6) allows the independent solution of a sequence of hyperbolic problems in Ω_1 and elliptic problems in Ω_2, in the framework of iterative processes between subdomains. The different possible treatments of the interface relations is what distinguishes one iterative method from another. In this respect, a very natural approach is defined as follows. Given a suitable initial guess $\lambda^{(0)}$ on Γ^{in} and a suitable relaxation parameter $\vartheta > 0$, it iterates between Ω_1 and Ω_2 until convergence as follows: for $k \geq 0$ do

$$\text{Solve} \quad \begin{cases} A_1 u_1^{(k+1)} = f & \text{in } \Omega_1 \\ u_1^{(k+1)} = g & \text{on } (\partial\Omega_1 \setminus \Gamma)^{\text{in}} \\ u_1^{(k+1)} = \lambda^{(k)} & \text{on } \Gamma^{\text{in}}, \end{cases}$$

$$\text{Solve} \quad \begin{cases} A_2 u_2^{(k+1)} = f & \text{in } \Omega_2 \\ u_2^{(k+1)} = g & \text{on } \partial\Omega_2 \setminus \Gamma \\ -\nu \dfrac{\partial u_2^{(k+1)}}{\partial n_\Gamma} + \mathbf{b} \cdot \mathbf{n}_\Gamma u_2^{(k+1)} = \mathbf{b} \cdot \mathbf{n}_\Gamma u_1^{(k+1)} & \text{on } \Gamma, \end{cases} \tag{7}$$

$$\text{Compute } \lambda^{(k+1)} = (1 - \vartheta)\lambda^{(k)} + \vartheta u_2^{(k+1)}|_{\Gamma^{\text{in}}}.$$

The coupled advection/advection–diffusion problem has been studied in [30] and alternative interface conditions have been proposed in [21, 23, 24]. In [26] the problem has been solved in the context of virtual control approach. We refer to Sects. 2.2, 2.3, 3.1 for a more detailed analysis and solution of this problem.

Another problem which deserves our attention is the generalized Stokes equation (see [51, Sect. 8.2.1]).

Let us refer to an idealised geometrical situation as depicted in Fig. 4, left.

The bounded domain $\Omega \subset \mathbb{R}^d$, $d = 2, 3$, is external to a body whose boundary is Γ_b and we set $\Gamma_\infty := \partial\Omega \setminus \Gamma_b$. The problem we are considering reads: find the vector field $\mathbf{u}$ and the scalar field p such that

$$\begin{cases} \alpha\mathbf{u} - \nu\Delta\mathbf{u} + \nabla p = \mathbf{f}, \quad \text{div}\mathbf{u} = 0 & \text{in } \Omega \\ \mathbf{u} = \mathbf{0} & \text{on } \Gamma_b \\ B\mathbf{u} = \boldsymbol{\varphi}_\infty & \text{on } \Gamma_\infty, \end{cases} \tag{8}$$

where $\mathbf{f}$ and $\boldsymbol{\varphi}_\infty$ are given functions, B denotes the boundary operator on Γ_∞, while $\alpha \geq 0$ is a given parameter. To take $\alpha = 0$ corresponds to solve the Stokes problem. Nevertheless, this problem may arise in the process of solving the full Navier–Stokes

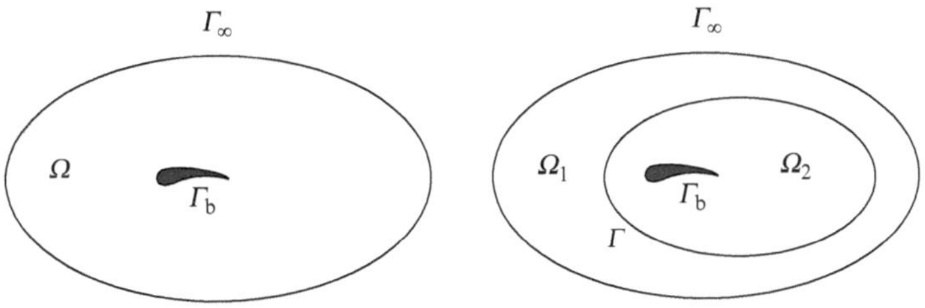

Fig. 4 The geometrical configuration for an external problem (*left*) and a possible non overlapping decomposition of the computational domain (*right*)

equations, when the discretisation of the time derivative is performed by means of a scheme that is explicit in the non-linear convective term. In this case, the parameter $\alpha > 0$ represents the inverse of the time-step and the function $\mathbf{f}$, in fact, depends on the solution at the previous step, i.e. $\mathbf{f} = \mathbf{f}(\mathbf{u}^{(n)})$.

The boundary conditions on Γ_∞ have to be prescribed in a suitable way for assuring well-posedness. In this respect, on a portion $\Gamma_\infty^{\mathrm{in}}$ of Γ_∞ an onset flow $\mathbf{u} = \mathbf{u}_\infty^{\mathrm{in}}$ is given. However, assigning conditions on the outflow section $\Gamma_\infty^{\mathrm{out}}$ may not be simple. It is also clear that all interesting flow features occur in the vicinity of the body due to the role of viscosity in this area.

For this reason, Schenk and Hebeker [56] have proposed the replacement of problem (8) with a reduced one far from the obstacle.

The computational domain Ω is partitioned into a subdomain Ω_2, next to the body, and a far field subdomain Ω_1; the interface between Ω_1 and Ω_2 is denoted by Γ, $\mathbf{n}_\Gamma$ is the unit normal vector on Γ directed from Ω_1 to Ω_2, and $\mathbf{n}$ the unit outward normal vector on $\partial\Omega$. The global Stokes equation (8) is replaced with the following coupled problem, where the viscosity ν is set to 0 in Ω_1:

$$\begin{cases} \alpha \mathbf{u}_1 + \nabla p_1 = \mathbf{f}, \quad \mathrm{div}\mathbf{u}_1 = 0 & \text{in } \Omega_1 \\ \mathbf{u}_1 = \mathbf{u}_\infty^{\mathrm{in}} & \text{on } \Gamma_\infty^{\mathrm{in}} \\ p_1 = 0 & \text{on } \Gamma_\infty^{\mathrm{out}} \\ \alpha \mathbf{u}_2 - \nu \Delta \mathbf{u}_2 + \nabla p_2 = \mathbf{f}, \quad \mathrm{div}\mathbf{u}_2 = 0 & \text{in } \Omega_2 \\ \mathbf{u}_2 = \mathbf{0} & \text{on } \Gamma_{\mathrm{b}}, \end{cases} \tag{9}$$

or equivalently, by applying the divergence operator to $(9)_1$:

$$\begin{cases} \Delta p_1 = \mathrm{div}\mathbf{f} & \text{in } \Omega_1 \\ \dfrac{\partial p_1}{\partial n} = (\mathbf{f} - \alpha \mathbf{u}_\infty^{\mathrm{in}}) \cdot \mathbf{n} & \text{on } \Gamma_\infty^{\mathrm{in}} \\ p_1 = 0 & \text{on } \Gamma_\infty^{\mathrm{out}} \\ \alpha \mathbf{u}_2 - \nu \Delta \mathbf{u}_2 + \nabla p_2 = \mathbf{f}, \quad \mathrm{div}\mathbf{u}_2 = 0 & \text{in } \Omega_2 \\ \mathbf{u}_2 = \mathbf{0} & \text{on } \Gamma_{\mathrm{b}}. \end{cases} \tag{10}$$

Either problem (9) and (10) are incomplete, because the matching conditions that have to be fulfilled on Γ are missing.

In [56] these conditions are recovered through a singular perturbation analysis similar to that carried out for the advection–diffusion problem in [30] and they read:

$$\begin{cases} \dfrac{\partial p_1}{\partial n_\Gamma} = (\mathbf{f} - \alpha \mathbf{u}_2) \cdot \mathbf{n}_\Gamma & \text{on } \Gamma \\ p_1 \mathbf{n}_\Gamma = -\nu (\mathbf{n}_\Gamma \cdot \nabla) \mathbf{u}_2 + p_2 \mathbf{n}_\Gamma & \text{on } \Gamma. \end{cases} \tag{11}$$

Fig. 5 The domain decomposition configuration for an internal problem

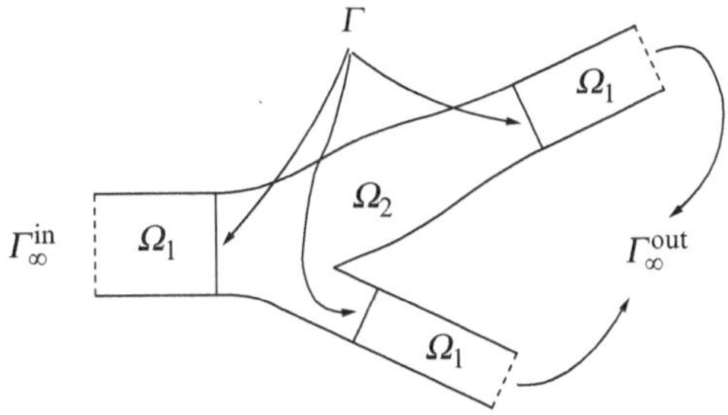

The coupled problem (10) and (11) can be used also for the simulation of the fluid motion inside a bounded domain, as depicted in Fig. 5. In this case the domain Ω_1, in which the reduced problem is solved, is non-connected and separates the interior domain from both inflow and outflow interfaces.

We observe that the system (10) and (11) models two possible different coupled problems. The first one, when $\alpha = 0$, is a Stokes/potential coupling, the vector field $\mathbf{f}$ is independent of the velocity $\mathbf{u}$ and the pressure p_1 is indipendent of the solution $(\mathbf{u}_2, p_2)$. Such coupling can be used to model external flows.

The second one, when $\alpha > 0$, corresponds to the single step of a time-dependent Navier–Stokes/potential coupling where, as said above, the vector field $\mathbf{f}$ depends on the solution at the previous step. This is the case of the simulation of either the flow inside a channel (or the blood flow in the carotid) or a far field condition.

As in the case of the advection–diffusion problem, the interface conditions (11) could be used to set-up an iterative algorithm by subdomains as follows.

Assume that $\widehat{\boldsymbol{\lambda}}^{(0)}$ is given and satisfies $\int_\Gamma \widehat{\boldsymbol{\lambda}}^{(0)} \cdot \mathbf{n}_\Gamma = 0$; for any $k \geq 0$ solve

$$\begin{cases} \Delta p_1^{(k+1)} = \mathrm{div}\mathbf{f} & \text{in } \Omega_1 \\ \dfrac{\partial p_1^{(k+1)}}{\partial n} = (\mathbf{f} - \alpha \mathbf{u}_\infty^{\mathrm{in}}) \cdot \mathbf{n} & \text{on } \Gamma_\infty^{\mathrm{in}} \\ p_1^{(k+1)} = 0 & \text{on } \Gamma_\infty^{\mathrm{out}} \\ \dfrac{\partial p_1^{(k+1)}}{\partial n_\Gamma} = (\mathbf{f} - \alpha \widehat{\boldsymbol{\lambda}}^{(k)}) \cdot \mathbf{n}_\Gamma & \text{on } \Gamma, \end{cases} \tag{12}$$

then solve

$$\begin{cases} \alpha \mathbf{u}_2^{(k+1)} - \nu \Delta \mathbf{u}_2^{(k+1)} + \nabla p_2^{(k+1)} = \mathbf{f}, \qquad \mathrm{div}\mathbf{u}_2^{(k+1)} = 0 & \text{in } \Omega_2 \\ \mathbf{u}_2^{(k+1)} = \mathbf{0} & \text{on } \Gamma_\mathrm{b} \\ \nu(\mathbf{n}_\Gamma \cdot \nabla)\mathbf{u}_2^{(k+1)} - p_2^{(k+1)}\mathbf{n}_\Gamma = -p_1^{(k+1)}\mathbf{n}_\Gamma & \text{on } \Gamma \end{cases} \tag{13}$$

and finally set

$$\widehat{\boldsymbol{\lambda}}^{(k+1)} = (1-\vartheta)\widehat{\boldsymbol{\lambda}}^{(k)} + \vartheta \mathbf{u}_{2|\Gamma}^{(k+1)}, \tag{14}$$

where $\vartheta > 0$ is a relaxation parameter.

Since $\text{div}\mathbf{u}_2^{(k+1)} = 0$ in Ω_2, the trace $\mathbf{u}_{2|\Gamma}^{(k+1)}$ satisfies

$$\int_\Gamma \mathbf{u}_{2|\Gamma}^{(k+1)} \cdot \mathbf{n}_\Gamma = 0,$$

whence $\int_\Gamma \widehat{\boldsymbol{\lambda}}^{(k)} \cdot \mathbf{n}_\Gamma = 0$ for each $k \geq 0$.

The analysis of the coupled problem (10) and (11) and the proof of convergence of the above iterative process (12)–(14) are reported in [56]. The analysis can be performed also by writing the problem in terms of the associated Steklov–Poincaré operators, and then proving convergence by applying an abstract result (see [51, Thm 4.2.2]).

Finally, we introduce a coupled free/porous-media flow problem.

The computational domain is a region naturally split into two parts: one occupied by the fluid, the other by the porous media. More precisely, let $\Omega \subset \mathbb{R}^d$ $(d = 2, 3)$ be a bounded domain, partitioned into two non intersecting subdomains Ω_f and Ω_p separated by an interface Γ, i.e. $\bar{\Omega} = \bar{\Omega}_f \cup \bar{\Omega}_p$, $\Omega_f \cap \Omega_p = \emptyset$ and $\bar{\Omega}_f \cap \bar{\Omega}_p = \Gamma$ (Fig. 6). We suppose the boundaries $\partial\Omega_f$ and $\partial\Omega_p$ to be Lipschitz continuous. From the physical point of view, Γ is a surface separating the domain Ω_f filled by a fluid, from a domain Ω_p formed by a porous medium. We assume that Ω_f has a fixed surface, i.e., we neglect here the case of free-surface flows. The fluid in Ω_f can filtrate through the adjacent porous medium.

The Navier–Stokes equations describe the motion of the fluid in Ω_f: $\forall t > 0$,

$$\begin{cases} \partial_t \mathbf{u}_f - \mathbf{div}\,\mathsf{T}(\mathbf{u}_f, p_f) + (\mathbf{u}_f \cdot \nabla)\mathbf{u}_f = \mathbf{f} & \text{in } \Omega_f \\ \text{div}\,\mathbf{u}_f = 0 & \text{in } \Omega_f, \end{cases} \tag{15}$$

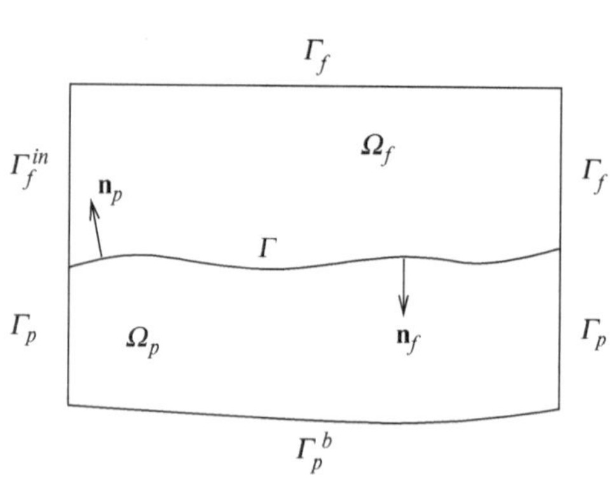

Fig. 6 Representation of a 2D section of a possible computational domain for the Stokes/Darcy coupling

where $\mathsf{T}(\mathbf{u}_f, p_f) = \nu(\nabla \mathbf{u}_f + \nabla^T \mathbf{u}_f) - p_f \mathsf{I}$ is the Cauchy stress tensor, I being the identity tensor. $\nu > 0$ is the kinematic viscosity of the fluid, $\mathbf{f}$ a given volumetric force, while $\mathbf{u}_f$ and p_f are the fluid velocity and pressure, respectively.

The filtration of an incompressible fluid through porous media is often described by Darcy's law. The latter provides the simplest linear relation between velocity and pressure in porous media under the physically reasonable assumption that fluid flows are usually very slow and all the inertial (non-linear) terms may be neglected. Darcy's law introduces a fictitious flow velocity, the *Darcy velocity* or *specific discharge* $\mathbf{q}$ through a given cross section of the porous medium, rather than the true velocity $\mathbf{u}_p$ with respect to the porous matrix:

$$\mathbf{u}_p = \frac{\mathbf{q}}{n}, \tag{16}$$

with n being the *volumetric porosity*, defined as the ratio between the volume of void space and the total volume of the porous medium.

To introduce Darcy's law, we define a scalar quantity φ called *piezometric head* which essentially represents the fluid pressure in Ω_p:

$$\varphi = z + \frac{p_p}{g}, \tag{17}$$

where z is the elevation from a reference level, accounting for the potential energy per unit weight of fluid, p_p is the ratio between the fluid pressure in Ω_p and its viscosity ρ_f, and g is the gravity acceleration.

Then, Darcy's law can be written as

$$\mathbf{q} = -K \nabla \varphi, \tag{18}$$

where K is a symmetric positive definite diagonal tensor $K = (K_{ij})_{i,j=1,\dots,d}$, $K_{ij} \in L^\infty(\Omega_p)$, $K_{ij} > 0$, $K_{ij} = K_{ji}$, called *hydraulic conductivity tensor*, which depends on the properties of the fluid as well as on the characteristics of the porous medium. Let us denote $\mathsf{K} = K/n$.

In conclusion, the motion of an incompressible fluid through a saturated porous medium is described by the following equations:

$$\begin{cases} \mathbf{u}_p = -\mathsf{K}\nabla\varphi & \text{in } \Omega_p \\ \operatorname{div} \mathbf{u}_p = 0 & \text{in } \Omega_p. \end{cases} \tag{19}$$

Finally, to represent the filtration of the free fluid through the porous medium, we have to introduce suitable coupling conditions between the Navier–Stokes and Darcy equations across the common interface Γ. In particular we consider the following three conditions.

1. Continuity of the normal component of the velocity:

$$\mathbf{u}_f \cdot \mathbf{n} = \mathbf{u}_p \cdot \mathbf{n}, \tag{20}$$

where we have indicated $\mathbf{n} = \mathbf{n}_f = -\mathbf{n}_p$ on Γ. This condition is a consequence of the incompressibility of the fluid.

2. Continuity of the normal stresses across Γ (see, e.g., [36]):

$$-\mathbf{n} \cdot \mathsf{T}(\mathbf{u}_f, p_f) \cdot \mathbf{n} = g\varphi. \tag{21}$$

Remark that pressures may be discontinuous across the interface.

3. Finally, in order to have a completely determined flow in the free-fluid region, we have to specify a further condition on the tangential component of the fluid velocity at the interface. An experimental condition was obtained by Beavers and Joseph stating that the slip velocity at the interface differs from the seepage velocity in the porous domain and it is proportional to the shear rate on Γ [5]:

$$\frac{\nu\alpha_{BJ}}{\sqrt{\mathsf{K}}}(\mathbf{u}_f - \mathbf{u}_p)_\tau - (\mathsf{T}(\mathbf{u}_f, p_f) \cdot \mathbf{n})_\tau = 0. \tag{22}$$

By $(\mathbf{v})_\tau$ we indicate the tangential component to the interface of $\mathbf{v}$:

$$(\mathbf{v})_\tau = \mathbf{v} - (\mathbf{v} \cdot \mathbf{n})\mathbf{n}. \tag{23}$$

Since the seepage velocity $\mathbf{u}_p$ is far smaller than the fluid slip velocity $\mathbf{u}_f$ at the interface, Saffman proposed to use the following simplified condition (the so-called Beavers–Joseph–Saffman condition) [53]:

$$\frac{\nu\alpha_{BJ}}{\sqrt{\mathsf{K}}}(\mathbf{u}_f)_\tau - (\mathsf{T}(\mathbf{u}_f, p_f) \cdot \mathbf{n})_\tau = 0. \tag{24}$$

This condition was later derived mathematically by means of homogenization by Jäger and Mikelić [36–38].

The three coupling conditions described in this section have been extensively studied and analysed also in [17, 19, 46, 49, 52].

In conclusion, the coupled Navier–Stokes/Darcy model reads

$$\begin{cases} \partial_t \mathbf{u}_f - \mathbf{div}\, \mathsf{T}(\mathbf{u}_f, p_f) + (\mathbf{u}_f \cdot \nabla)\mathbf{u}_f = \mathbf{f} & \text{in } \Omega_f \\ \operatorname{div} \mathbf{u}_f = 0 & \text{in } \Omega_f \\ \mathbf{u}_p = -\mathsf{K}\nabla\varphi & \text{in } \Omega_p \\ \operatorname{div} \mathbf{u}_p = 0 & \text{in } \Omega_p \\ \mathbf{u}_f \cdot \mathbf{n} = \mathbf{u}_p \cdot \mathbf{n} & \text{on } \Gamma \\ -\mathbf{n} \cdot \mathsf{T}(\mathbf{u}_f, p_f) \cdot \mathbf{n} = g\varphi & \text{on } \Gamma \\ \dfrac{\nu\alpha_{BJ}}{\sqrt{\mathsf{K}}}(\mathbf{u}_f)_\tau - (\mathsf{T}(\mathbf{u}_f, p_f) \cdot \mathbf{n})_\tau = 0 & \text{on } \Gamma. \end{cases} \tag{25}$$

Using Darcy's law we can rewrite the system (19) as an elliptic equation for the scalar unknown φ:

$$-\nabla \cdot (\mathsf{K}\nabla\varphi) = 0 \quad \text{in } \Omega_p. \tag{26}$$

In this case, the differential formulation of the coupled Navier–Stokes/Darcy problem becomes

$$\begin{cases} \partial_t \mathbf{u}_f - \mathbf{div}\, \mathsf{T}(\mathbf{u}_f, p_f) + (\mathbf{u}_f \cdot \nabla)\mathbf{u}_f = \mathbf{f} & \text{in } \Omega_f \\ \text{div}\, \mathbf{u}_f = 0 & \text{in } \Omega_f \\ -\text{div}\, (\mathsf{K}\nabla\varphi) = 0 & \text{in } \Omega_p, \end{cases} \tag{27}$$

with the interface conditions on Γ:

$$\begin{cases} \mathbf{u}_f \cdot \mathbf{n} = -\mathsf{K}\dfrac{\partial\varphi}{\partial n} \\ -\mathbf{n} \cdot \mathsf{T}(\mathbf{u}_f, p_f) \cdot \mathbf{n} = g\varphi \\ \dfrac{\nu\alpha_{BJ}}{\sqrt{\mathsf{K}}}(\mathbf{u}_f)_\tau - (\mathsf{T}(\mathbf{u}_f, p_f) \cdot \mathbf{n})_\tau = 0. \end{cases} \tag{28}$$

We refer to Sects. 2.6, 2.7, 3.4 for a more exhaustive analysis of the Stokes/Darcy coupling.

2 Variational Formulation Approach

The reduced problems presented above will be analysed in this Section in a variational setting, in order to deduce suitable interface conditions which can be rigorously justified. Moreover, different iterative algorithms to solve the reduced problems will be presented.

2.1 The Advection–Diffusion Problem

We consider an open bounded domain $\Omega \subset \mathbb{R}^d$ $(d = 2, 3)$ with Lipschitz boundary $\partial\Omega$, and we split it into two open subsets Ω_1 and Ω_2 such that

$$\overline{\Omega} = \overline{\Omega}_1 \cup \overline{\Omega}_2, \quad \Omega_1 \cap \Omega_2 = \emptyset. \tag{29}$$

Then, we denote by

$$\Gamma = \partial\Omega_1 \cap \partial\Omega_2 \tag{30}$$

the interface between the subdomains (see Fig. 3) and we assume that Γ is of class $C^{1,1}$; $\overset{\circ}{\Gamma}$ will denote the interior of Γ.

Given two scalar functions f and b_0 defined in Ω, a positive function ν defined in $\Omega_2 \cup \overset{\circ}{\Gamma}$, a d-dimensional vector valued function $\mathbf{b}$ defined in Ω satisfying the following inequalities:

$$\begin{aligned} &\exists \nu_0 \in \mathbb{R} : \nu(\mathbf{x}) \geq \nu_0 > 0 && \forall \mathbf{x} \in \Omega_2 \cup \overset{\circ}{\Gamma}, \\ &\exists \sigma_0 \in \mathbb{R} : b_0(\mathbf{x}) + \frac{1}{2}\mathrm{div}\mathbf{b}(\mathbf{x}) \geq \sigma_0 > 0 && \forall \mathbf{x} \in \Omega, \end{aligned} \tag{31}$$

we are interested in finding two functions u_1 and u_2 (defined in $\overline{\Omega}_1$ and $\overline{\Omega}_2$, respectively) such that u_1 statisfies the advection–reaction equation

$$A_1 u_1 \equiv \mathrm{div}(\mathbf{b}u_1) + b_0 u_1 = f \qquad \text{in } \Omega_1, \tag{32}$$

while u_2 satisfies the advection–diffusion–reaction equation

$$A_2 u_2 \equiv -\mathrm{div}(\nu \nabla u_2) + \mathrm{div}(\mathbf{b}u_2) + b_0 u_2 = f \qquad \text{in } \Omega_2. \tag{33}$$

For each subdomain, we distinguish between the *external* (or physical) boundary $\partial\Omega \cap \partial\Omega_k = \partial\Omega_k \setminus \Gamma$ (for $k = 1, 2$) and the *internal* one, i.e. the interface Γ.

Moreover, for any non-empty subset $S \subseteq \partial\Omega_1$, we define

$$\text{The } \textit{inflow} \text{ part of } S: \quad S^{\mathrm{in}} = \{\mathbf{x} \in S : \mathbf{b}(\mathbf{x}) \cdot \mathbf{n}(\mathbf{x}) < 0\}, \tag{34}$$

where $\mathbf{n}(\mathbf{x})$ is the outward unit normal vector on S,

$$\text{The } \textit{outflow} \text{ part of } S: \quad S^{\mathrm{out}} = \{\mathbf{x} \in S : \mathbf{b}(\mathbf{x}) \cdot \mathbf{n}(\mathbf{x}) \geq 0\}. \tag{35}$$

Boundary conditions for problem (32) must be assigned on $\partial\Omega_1^{\mathrm{in}}$.

For a given suitable function g defined on $\partial\Omega$, we denote by g_1 and g_2 the restriction of g to $(\partial\Omega_1 \setminus \Gamma)^{\mathrm{in}}$ and $\partial\Omega_2 \setminus \Gamma$, respectively, and we set the following Dirichlet boundary conditions on the external boundaries:

$$\begin{aligned} u_1 &= g_1 && \text{on } (\partial\Omega_1 \setminus \Gamma)^{\mathrm{in}}, \\ u_2 &= g_2 && \text{on } \partial\Omega_2 \setminus \Gamma. \end{aligned} \tag{36}$$

Finally, let us denote by $\mathbf{n}_\Gamma$ the normal versor to Γ oriented from Ω_1 to Ω_2, so that $\mathbf{n}_\Gamma(\mathbf{x}) = \mathbf{n}_1(\mathbf{x}) = -\mathbf{n}_2(\mathbf{x})$, $\forall \mathbf{x} \in \Gamma$.

2.2 Variational Analysis for the Advection–Diffusion Equation

The basic steps of the analysis carried out in [30] are summarized here.

1. Given a positive function ν in Ω, we denote by $\mathscr{P}_\Omega(\nu)$ the advection–diffusion problem (1) in Ω. For any $\varepsilon > 0$, we introduce a smooth function ν_ε defined in Ω_2, which is a regularization of ν according with continuity to ε on Γ. Then, ν_ε^* is the globally defined viscosity defined as (see Fig. 7)

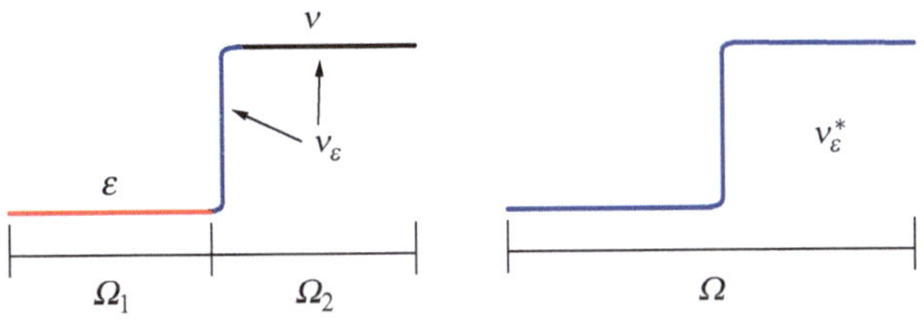

Fig. 7 The viscosity ν_ε^* for the regularized problem. $\nu_\varepsilon|_{\Omega_2} \to \nu$ when $(\varepsilon \to 0)$

$$\nu_\varepsilon^* = \begin{cases} \varepsilon & \text{in } \Omega_1 \\ \nu_\varepsilon & \text{in } \Omega_2 \, . \end{cases}$$

We denote by $\mathscr{P}_\Omega(\nu_\varepsilon^*) \equiv [\mathscr{P}_{\Omega_1}(\varepsilon)/\mathscr{P}_{\Omega_2}(\nu_\varepsilon)]$ the following advection–diffusion problem:

$$\begin{cases} -\varepsilon\Delta u_{1,\varepsilon} + \operatorname{div}(\mathbf{b}u_{1,\varepsilon}) + b_0 u_{1,\varepsilon} = f & \text{in } \Omega_1 \\ \operatorname{div}(-\nu_\varepsilon \nabla u_{2,\varepsilon} + \mathbf{b}u_{2,\varepsilon}) + b_0 u_{2,\varepsilon} = f & \text{in } \Omega_2 \\ \varepsilon\dfrac{\partial u_{1,\varepsilon}}{\partial n_\Gamma} - \mathbf{b}\cdot\mathbf{n}_\Gamma u_{1,\varepsilon} = \nu_\varepsilon \dfrac{\partial u_{2,\varepsilon}}{\partial n_\Gamma} - \mathbf{b}\cdot\mathbf{n}_\Gamma u_{2,\varepsilon} & \text{on } \Gamma \\ u_{1,\varepsilon} = u_{2,\varepsilon} & \text{on } \Gamma \\ u = g & \text{on } \partial\Omega . \end{cases} \tag{37}$$

2. For any $\varepsilon > 0$, let $\mathscr{V}_\Omega(\varepsilon)$ be the variational formulation associated to $\mathscr{P}_\Omega(\varepsilon)$. Solving $\mathscr{V}_\Omega(\varepsilon)$ means to look for the solution $u_\varepsilon \in V$ of

$$a_\varepsilon(u_\varepsilon, w_\varepsilon) = F(w_\varepsilon), \qquad \forall w_\varepsilon \in V. \tag{38}$$

If we take $g \equiv 0$, this means to set $V = H_0^1(\Omega)$ and to solve

$$a_\varepsilon(u_\varepsilon, w_\varepsilon) = \int_\Omega [(\varepsilon\nabla u_\varepsilon - \mathbf{b}u_\varepsilon)\cdot\nabla w_\varepsilon + b_0 u_\varepsilon w_\varepsilon]\,d\mathbf{x}, \qquad F(w_\varepsilon) = \int_\Omega f w_\varepsilon d\mathbf{x} \tag{39}$$

for any $w_\varepsilon \in V$.

Otherwise, if $g \neq 0$ the formulation is the same, however the right hand side has to be modified as follows:

$$F_g(w_\varepsilon) = F(w_\varepsilon) - a_\varepsilon(R_g, w_\varepsilon),$$

where R_g is a suitable lifting of the boundary data g, so that the final solution reads $u_\varepsilon + R_g$ (see [50]).

3. By asymptotic analysis on $\mathscr{V}_{\Omega_1}(\varepsilon)$, recover the reduced problem $\mathscr{P}_{\Omega_1}(0)$, so that

$$\mathscr{P}_{\Omega}(\nu_\varepsilon^*) \to [\mathscr{P}_{\Omega_1}(0)/\mathscr{P}_{\Omega_2}(\nu)] \quad \text{when } \varepsilon \to 0\,.$$

The new coupled problem $[\mathscr{P}_{\Omega_1}(0)/\mathscr{P}_{\Omega_2}(\nu)]$ inherits from the limit process a proper set of interface conditions.

According to the analysis performed in [30], $u_{1,\varepsilon}$ converges weakly in $L^2(\Omega_1)$ and $u_{2,\varepsilon}$ converges weakly in $H^1(\Omega_2)$ when $\varepsilon \to 0$, moreover the limit $(u_1, u_2) \in L^2(\Omega_1) \times H^1(\Omega_2)$ satisfies the following reduced coupled problem:

$$\begin{cases} \operatorname{div}(\mathbf{b}u_1) + b_0 u_1 = f & \text{in } \Omega_1 \\ \operatorname{div}(-\nu\nabla u_2 + \mathbf{b}u_2) + b_0 u_2 = f & \text{in } \Omega_2 \\ -\mathbf{b}\cdot\mathbf{n}_\Gamma u_1 = \nu\dfrac{\partial u_2}{\partial n_\Gamma} - \mathbf{b}\cdot\mathbf{n}_\Gamma u_2 & \text{on } \Gamma \\ u_1 = u_2 & \text{on } \Gamma^{\text{in}} \\ u_1 = g_1 & \text{on } (\partial\Omega_1 \setminus \Gamma)^{\text{in}} \\ u_2 = g_2 & \text{on } \partial\Omega_2 \setminus \Gamma. \end{cases} \tag{40}$$

The *interface conditions* $(40)_{3,4}$ express the continuity of the flux across the whole interface Γ and the continuity of the solution across the inflow interface Γ^{in}, respectively. No continuity condition is imposed on Γ^{out}, as a matter of fact, u_1 and u_2 exhibit a jump across Γ^{out} which is proportional to $\nu_{|\Gamma}$.

Note that the interface conditions $(40)_{3,4}$ can be equivalently expressed as

$$\begin{aligned} & u_1 = u_2 && \text{on } \Gamma^{\text{in}}, \\ & \mathbf{b}\cdot\mathbf{n}_\Gamma u_1 + \nu\frac{\partial u_2}{\partial n_\Gamma} - \mathbf{b}\cdot\mathbf{n}_\Gamma u_2 = 0 && \text{on } \Gamma^{\text{out}} \\ & \nu\frac{\partial u_2}{\partial n_\Gamma} = 0 && \text{on } \Gamma^{\text{in}}. \end{aligned} \tag{41}$$

In order to proceed with the analysis of the coupled problem, we introduce the following notations. Let A be an open bounded subset in $\mathbb{R}^d$, with Lipschitz continuous boundary. For any open subset $\Gamma \subset \partial A$, we define the weighted L^2-space

$$L^2_{\mathbf{b}}(\Gamma) = \{\varphi : \Gamma \to \mathbb{R} :\ \sqrt{|\mathbf{b}\cdot\mathbf{n}_\Gamma|}\varphi \in L^2(\Gamma)\}, \tag{42}$$

and the trace space

$$H^{1/2}_{00}(\Gamma) = \{\varphi \in L^2(\Gamma) :\ \exists\tilde{\varphi} \in H^{1/2}(\partial A) :\ \tilde{\varphi}|_\Gamma = \varphi,\ \tilde{\varphi}|_{\partial A\setminus\Gamma} = 0\}. \tag{43}$$

The space $L^2_{\mathbf{b}}(\Gamma)$ endowed with the norm

$$\|\varphi\|_{L^2_{\mathbf{b}}(\Gamma)} = \left(\int_\Gamma |\mathbf{b}\cdot\mathbf{n}_\Gamma|\varphi^2 \mathrm{d}\Gamma\right)^{1/2}$$

is a Hilbert space.

The following result has been proved in [30]:

Theorem 1. *Assume the following regularity properties on the data: $\partial\Omega_1$ and $\partial\Omega_2$ are Lipschitz continuous, piecewise $C^{1,1}$; Γ is of class $C^{1,1}$;*

$$\begin{aligned} &\nu \in L^\infty(\Omega_2), \quad \mathbf{b} \in \left[W^{1,\infty}(\Omega)\right]^2, \quad b_0 \in L^\infty(\Omega), \quad f \in L^2(\Omega), \\ &g \in H^{-1/2}(\partial\Omega) : \; g_1 \in L^2_{\mathbf{b}}((\partial\Omega_1 \setminus \Gamma)^{\mathrm{in}}), \quad g_2 \in H^{1/2}(\partial\Omega_2 \setminus \Gamma). \end{aligned} \tag{44}$$

Finally assume (31).

Then there is a unique pair $(u_1, u_2) \in L^2(\Omega_1) \times H^1(\Omega_2)$ which solves (40), where: $(40)_1$ and $(40)_2$ hold in the sense of distributions in Ω_1 and Ω_2, respectively; interface condition $(41)_1$ holds a.e. on Γ^{in}, interface condition $(41)_2$ holds in $(H^{1/2}_{00}(\Gamma^{\mathrm{out}}))'$; interface condition $(41)_3$ holds in $(H^{1/2}_{00}(\Gamma^{\mathrm{in}}))'$. Finally, problem (40) is limit of a family of globally elliptic variational problems.

From now on, the solution (u_1, u_2) of the heterogeneous problem (40) will be named *heterogeneous solution*.

Other interface conditions have been proposed in the literature to close system (32), (33), (36). For instance, the conditions

$$\begin{aligned} &-\mathbf{b}\cdot\mathbf{n}_\Gamma u_1 = \nu\frac{\partial u_2}{\partial \mathbf{n}_\Gamma} - \mathbf{b}\cdot\mathbf{n}_\Gamma u_2 \;\; \text{on } \Gamma^{\mathrm{out}} \\ &u_1 = u_2, \qquad \frac{\partial u_1}{\partial \mathbf{n}_\Gamma} = \frac{\partial u_2}{\partial \mathbf{n}_\Gamma} \qquad \text{on } \Gamma^{\mathrm{in}}, \end{aligned} \tag{45}$$

have been proposed in [21] and are based on absorbing boundary condition theory. The following set (see [23, 24]):

$$\begin{aligned} &u_1 = u_2 \qquad \text{on } \Gamma \\ &\frac{\partial u_1}{\partial \mathbf{n}_\Gamma} = \frac{\partial u_2}{\partial \mathbf{n}_\Gamma} \quad \text{on } \Gamma^{\mathrm{in}} \end{aligned} \tag{46}$$

takes into account the requirement of glueing the solutions across the interface with high regularity.

However, the coupled problem with either one of these set of conditions (45) and (46) cannot be regarded as a limit of the original complete variational problem as the viscosity ε tends to zero in Ω_1.

Another possible approach to set suitable interface conditions was proposed in [25] for the one-dimensional case with constant coefficients and it is based on the factorization of the differential operator. To briefly explain it, let us take $\Omega = (x_1, x_2)$ and let $x_0 \in \Omega$ denote the position of the interface between Ω_1 and Ω_2, i.e. $\Omega_1 = (x_1, x_0)$ and $\Omega_2 = (x_0, x_2)$. The method consists in the following steps:

- Factorize the differential operator $A_2 \cdot = -\nu\partial_x^2 \cdot + b\partial_x \cdot + b_0 \cdot$ as

$$A_2 = (b\partial_x - b\lambda^+)\left(-\frac{\nu}{b}\partial_x + \frac{\nu}{b}\lambda^-\right),$$

  where $\lambda^{\pm} = (b \pm \sqrt{b^2 + 4\nu b_0})/(2\nu)$, with $\lambda^+ > 0$ and $\lambda^- < 0$.
- Compute the function $\tilde{u}_1(x) = \tilde{u}_1(x_1)e^{\lambda^+(x-x_1)} + \frac{1}{b}\int_{x_1}^{x} f(t)e^{\lambda^+(x-t)}dt$, which is the solution of the modified advection–reaction equation $\tilde{A}_1\tilde{u}_1 = b\tilde{u}_1' - b\lambda^+\tilde{u}_1' = f$ in Ω_1 with a suitable boundary condition at $x = x_1$.
- Solve the advection diffusion problem $A_2 u_2 = f$ in Ω_2 with the following interface condition at $x = x_0$:

$$-\frac{\nu}{b}u_2'(x_0) + \frac{\nu}{b}\lambda^- u_2(x_0) = \left(-\frac{\nu}{b}u_1'(x_1) + \frac{\nu}{b}\lambda^- u_1(x_1) - \tilde{u}_1(x_1)\right)e^{-\lambda^+ x_1} + \tilde{u}_1(x_0).$$

- Solve the advection reaction problem $A_1 u_1 = bu_1' + b_0 u_1 = f$ in Ω_1 with either $u_1(x_0) = u_2(x_0)$ if $b < 0$, or a suitable boundary condition at $x = x_1$ if $b > 0$.

It is shown in [25] that the L^2–norm error between the heterogeneous solution and the global elliptic one behaves like ν (for $\nu \to 0$) in the domain Ω_1, while in Ω_2 it exponentially decreases with ν when $b < 0$ and it behaves like ν^m ($m = 1, 2, \ldots$) when $b > 0$. The integer m depends on the accuarcy of the boundary condition imposed at $x = x_1$.

2.3 *Domain Decomposition Algorithms for the Solution of the Reduced Advection–Diffusion Problem*

In this Section we will present two iterative domain decomposition methods to solve the coupled problem (40), starting from the interface conditions $(40)_{3,4}$. Moreover we will reformulate the heterogeneous problem in terms of the Steklov–Poincaré equation at the interface.

2.3.1 Dirichlet–Neumann algorithm

The interface conditions $(40)_3$ and $(40)_4$ provide, respectively, Dirichlet or Neumann data at the interface Γ. Then we can use the condition $(40)_3$ as an inflow (Dirichlet) condition for the advection problem in Ω_1 and the condition $(40)_4$ as a Neumann condition for the elliptic problem in Ω_2. The algorithm, named *Dirichlet–Neumann*

(DN) method, produces two sequences of functions $\{u_1^{(k)}\}$ and $\{u_2^{(k)}\}$ converging to the solutions u_1 and u_2, respectively, of the heterogeneous problem as follows.

Given $\lambda^{(0)} \in L^2_{\mathbf{b}}(\Gamma^{\text{in}})$, for $k \geq 0$ do:

$$\text{Solve} \quad \begin{cases} A_1 u_1^{(k+1)} = f & \text{in } \Omega_1 \\ u_1^{(k+1)} = g & \text{on } (\partial\Omega_1 \setminus \Gamma)^{\text{in}} \\ u_1^{(k+1)} = \lambda^{(k)} & \text{on } \Gamma^{\text{in}}, \end{cases}$$

$$\text{Solve} \quad \begin{cases} A_2 u_2^{(k+1)} = f & \text{in } \Omega_2 \\ u_2^{(k+1)} = g & \text{on } \partial\Omega_2 \setminus \Gamma \\ -\nu \dfrac{\partial u_2^{(k+1)}}{\partial n_\Gamma} + \mathbf{b} \cdot \mathbf{n}_\Gamma u_2^{(k+1)} = \mathbf{b} \cdot \mathbf{n}_\Gamma u_1^{(k+1)} & \text{on } \Gamma, \end{cases} \tag{47}$$

$$\text{Compute } \lambda^{(k+1)} = (1-\vartheta)\lambda^{(k)} + \vartheta u_2^{(k+1)}|_{\Gamma^{\text{in}}},$$

where $\vartheta > 0$ is a suitable relaxation parameter.

The convergence properties of this method are analysed in [30], while several numerical results can be found in [22]. The convergence of DN method is guaranteed by the following theorem [30].

Theorem 2. *Let us consider the assumptions of Theorem 1. There exists $\delta > 0$ such that, if $\lambda^{(0)} \in L^2_{\mathbf{b}}(\Gamma^{\text{in}})$ and $\vartheta \in (0, 1+\delta)$, then the sequence $(u_1^{(k)}, u_2^{(k)})$ converges to a limit pair (u_1, u_2) in the following sense:*

$$u_1^{(k)} \to u_1 \text{ in } L^2(\Omega_1), \qquad u_2^{(k)} \to u_2 \text{ in } H^1(\Omega_2).$$

The limit pair provides the unique solution to the coupled problem (40).

Other research papers connected with this approach are [2, 9, 29, 55].

We note that, when $\Gamma^{\text{out}} = \Gamma$, the DN algorithm (47) converges in one iteration, since the solution in Ω_1 is independent of the solution in Ω_2 and, once u_1 is known, the solution in Ω_2 is obtained by a single "Neumann step".

On the contrary, when $\Gamma^{\text{in}} = \Gamma$, the coupled problem (40) can be solved without iterations. As a matter of fact, by re-writing the interface condition $(47)_6$ as in (41), we note that the solution in Ω_2 is uniquely determined, independently of a trace function λ on Γ. Consequently, the solution in Ω_1 is uniquely defined by the interface condition $(41)_1$.

2.3.2 Adaptive Robin Neumann Algorithm

Another iterative algorithm, that can be invoked to solve the reduced advection–diffusion problem (40) reads as follows. Given the functions $\lambda^{(0)} \in L^2_{\mathbf{b}}(\Gamma^{\text{in}})$, $\mu^{(0)} \in L^2_{\mathbf{b}}(\Gamma^{\text{out}})$ and $u_2^{(0)} \in H^1(\Omega_2)$, for $k \geq 0$ do:

$$
\text{Solve} \quad \begin{cases}
\operatorname{div}(\mathbf{b}u_1^{(k+1)}) + b_0 u_1^{(k+1)} = f & \text{in } \Omega_1 \\
u_1^{(k+1)} = g & \text{on } (\partial\Omega_1 \setminus \Gamma)^{\text{in}} \\
-\mathbf{b}\cdot\mathbf{n}_\Gamma u_1^{(k+1)} = \nu \dfrac{\partial u_2^{(k)}}{\partial n_\Gamma} - \mathbf{b}\cdot\mathbf{n}_\Gamma \lambda^{(k)} & \text{on } \Gamma^{\text{in}},
\end{cases}
$$

$$
\text{Solve} \quad \begin{cases}
\operatorname{div}(-\nu\nabla u_2^{(k+1)} + \mathbf{b}u_2^{(k+1)}) + b_0 u_2^{(k+1)} = f & \text{in } \Omega_2 \\
u_2^{(k+1)} = g & \text{on } \partial\Omega_2 \setminus \Gamma \\
\nu \dfrac{\partial u_2^{(k+1)}}{\partial n_\Gamma} - \mathbf{b}\cdot\mathbf{n}_\Gamma u_2^{(k+1)} = -\mathbf{b}\cdot\mathbf{n}_\Gamma \mu^{(k)} & \text{on } \Gamma^{\text{out}} \\
\nu \dfrac{\partial u_2^{(k+1)}}{\partial n_\Gamma} = 0 & \text{on } \Gamma^{\text{in}},
\end{cases}
\tag{48}
$$

$$
\text{Compute} \quad \begin{cases}
\lambda^{(k+1)} = (1-\vartheta)\lambda^{(k)} + \vartheta u_2^{(k+1)} \text{ on } \Gamma^{\text{in}} \\
\mu^{(k+1)} = (1-\vartheta)\mu^{(k)} + \vartheta u_1^{(k+1)} \text{ on } \Gamma^{\text{out}}.
\end{cases}
$$

The algorithm (48) is obtained as the limit, when $\varepsilon \to 0$, of the Adaptive-Robin–Neumann (ARN) method proposed in [10] for the homogeneous global elliptic problem (37). In its original form, ARN method reads given $\lambda^{(0)}$, $\mu^{(0)}$ and $u_2^{(0)}$, for $k \geq 0$ do

$$
\text{Solve} \quad \begin{cases}
-\varepsilon\Delta u_{1,\varepsilon}^{(k+1)} + \operatorname{div}(\mathbf{b}u_{1,\varepsilon}^{(k+1)}) + b_0 u_{1,\varepsilon}^{(k+1)} = f & \text{in } \Omega_1 \\
u_{1,\varepsilon}^{(k+1)} = g & \text{on } (\partial\Omega_1 \setminus \Gamma)^{\text{in}} \\
\varepsilon \dfrac{\partial u_{1,\varepsilon}^{(k+1)}}{\partial n_\Gamma} - \mathbf{b}\cdot\mathbf{n}_\Gamma u_{1,\varepsilon}^{(k+1)} = \nu_\varepsilon \dfrac{\partial u_{2,\varepsilon}^{(k)}}{\partial n_\Gamma} - \mathbf{b}\cdot\mathbf{n}_\Gamma \lambda^{(k)} & \text{on } \Gamma_1^{\text{in}} = \Gamma^{\text{in}} \\
\varepsilon \dfrac{\partial u_{1,\varepsilon}^{(k+1)}}{\partial n_\Gamma} = \nu_\varepsilon \dfrac{\partial u_{2,\varepsilon}^{(k)}}{\partial n_\Gamma} & \text{on } \Gamma_1^{\text{out}} = \Gamma^{\text{out}},
\end{cases}
$$

$$
\text{Solve} \quad \begin{cases}
\operatorname{div}(-\nu_\varepsilon\nabla u_{2,\varepsilon}^{(k+1)} + \mathbf{b}u_{2,\varepsilon}^{(k+1)}) + b_0 u_{2,\varepsilon}^{(k+1)} = f & \text{in } \Omega_2 \\
u_{2,\varepsilon}^{(k+1)} = g & \text{on } \partial\Omega_2 \setminus \Gamma \\
\nu_\varepsilon \dfrac{\partial u_{2,\varepsilon}^{(k+1)}}{\partial n_\Gamma} - \mathbf{b}\cdot\mathbf{n}_\Gamma u_{2,\varepsilon}^{(k+1)} = \varepsilon \dfrac{\partial u_{1,\varepsilon}^{(k+1)}}{\partial n_\Gamma} - \mathbf{b}\cdot\mathbf{n}_\Gamma \mu^{(k)} & \text{on } \Gamma_2^{\text{in}} = \Gamma^{\text{out}} \\
\nu_\varepsilon \dfrac{\partial u_{2,\varepsilon}^{(k+1)}}{\partial n_\Gamma} = \varepsilon \dfrac{\partial u_{1,\varepsilon}^{(k+1)}}{\partial n_\Gamma} & \text{on } \Gamma_2^{\text{out}} = \Gamma^{\text{in}},
\end{cases}
\tag{49}
$$

$$
\text{Compute} \quad \begin{cases}
\lambda^{(k+1)} = (1-\vartheta)\lambda^{(k)} + \vartheta u_{2,\varepsilon}^{(k+1)} \text{ on } \Gamma^{\text{in}} \\
\mu^{(k+1)} = (1-\vartheta)\mu^{(k)} + \vartheta u_{1,\varepsilon}^{(k+1)} \text{ on } \Gamma^{\text{out}}.
\end{cases}
$$

The idea of this method is to impose a Robin interface condition on the local (i.e. referred to that subdomain) inflow interface Γ_i^{in} $(i = 1, 2)$ and a Neumann interface condition on the local outflow interface Γ_i^{out} $(i = 1, 2)$.

Coming back to the heterogeneous coupling, it is straightforward to prove that, if the choice of ϑ guarantees the convergence of ARN method, then the limit solution of ARN (48) coincides with the solution of the heterogeneous problem (40). Moreover, if $u_2^{(0)}$ is chosen with null normal derivative on the interface Γ and $\vartheta = 1$, then ARN (48) and DN (47) methods coincide.

When either $\Gamma^{\rm in} = \Gamma$ or $\Gamma^{\rm out} = \Gamma$ we can conclude that no iterations are need for ARN method, as for DN.

Remark 1. We want to remark here that in the Dirichlet/Neumann method, the Neumann condition $(47)_6$ is in fact a conormal derivative associated to the differential operator A_2. On the contrary, in the ARN method the Neumann condition (as $(48)_7$) is a pure normal derivative on the interface, while the conormal derivative $(48)_6$ is called Robin condition, in agreement with the classical definition of Robin boundary condition. Following the latter notation, actually the Dirichlet/Neumann method should be a Dirichlet/Robin method.

2.3.3 Steklov–Poincaré Based Solution Algorithms

Let us consider the heterogeneous problem (40) with homogeneous Dirichlet conditions on $\partial\Omega$, i.e., $g \equiv 0$. Let $\lambda \in H_{00}^{1/2}(\Gamma)$ denote the unknown trace of the solution u_2 on Γ. Thanks to the interface condition $(40)_4$, the solution (u_1, u_2) of (40) can be written as

$$u_1 = u_1^\lambda + w_1, \quad u_2 = u_2^\lambda + w_2,$$

where w_1 and w_2 depend on the assigned function f and are the solution of

$$\begin{cases} A_1 w_1 = f & \text{in } \Omega_1 \\ w_1 = 0 & \text{on } \partial\Omega_1^{\rm in}, \end{cases} \qquad \begin{cases} A_2 w_2 = f & \text{in } \Omega_2 \\ w_2 = 0 & \text{on } \partial\Omega_2, \end{cases} \tag{50}$$

while u_1^λ and u_2^λ are the solutions of

$$\begin{cases} A_1 u_1^\lambda = 0 & \text{in } \Omega_1 \\ u_1^\lambda = 0 & \text{on } (\partial\Omega_1 \setminus \Gamma)^{\rm in} \\ u_1^\lambda = \lambda_{|\Gamma^{\rm in}} & \text{on } \Gamma^{\rm in}, \end{cases} \qquad \begin{cases} A_2 u_2^\lambda = 0 & \text{in } \Omega_2 \\ u_2 = 0 & \text{on } \partial\Omega_2 \setminus \Gamma \\ u_2^\lambda = \lambda & \text{on } \Gamma. \end{cases} \tag{51}$$

Given $\lambda \in H_{00}^{1/2}(\Gamma)$, we define the Steklov–Poincaré operators S_1 and S_2 such that

$$S_1\lambda = \begin{cases} \mathbf{b}\cdot\mathbf{n}_\Gamma u_1^\lambda & \text{on } \Gamma^{\rm out} \\ 0 & \text{on } \Gamma^{\rm in} \end{cases} \tag{52}$$

and

$$S_2\lambda = \begin{cases} \nu \dfrac{\partial u_2^\lambda}{\partial n_\Gamma} - \mathbf{b}\cdot\mathbf{n}_\Gamma u_2^\lambda & \text{on } \Gamma^{\text{out}} \\ \nu \dfrac{\partial u_2^\lambda}{\partial n_\Gamma} & \text{on } \Gamma^{\text{in}}. \end{cases} \tag{53}$$

Actually, $S_1\lambda$ depends only on the values of λ on Γ^{in}.

Then the interface conditions $(40)_3$ can be equivalently expressed in terms of Steklov–Poincaré operators as

$$S\lambda \equiv S_1\lambda + S_2\lambda = \chi, \tag{54}$$

where

$$\chi = \begin{cases} -\mathbf{b}\cdot\mathbf{n}_\Gamma w_1 - \nu \dfrac{\partial w_2}{\partial n_\Gamma} + \mathbf{b}\cdot\mathbf{n}_\Gamma w_2 & \text{on } \Gamma^{\text{out}} \\ -\nu \dfrac{\partial w_2}{\partial n_\Gamma} & \text{on } \Gamma^{\text{in}}. \end{cases} \tag{55}$$

The operator $S : H_{00}^{1/2}(\Gamma) \to (H_{00}^{1/2}(\Gamma))'$ is the so-called Steklov–Poincaré operator and the (54) is the Steklov–Poincaré equation associated to the heterogeneous problem (40). The solution of (40) can be reached by sequentially solving the problems (50), (54) and (51).

Several methods may be invoked to solve the Steklov–Poincaré equation (54). To start, let us consider the preconditioned Richardson method

$$\begin{cases} \lambda^{(0)} \text{ given} \\ P(\lambda^{(k+1)} - \lambda^{(k)}) = \vartheta(\chi - S\lambda^{(k)}), \ \text{for } k \geq 0, \end{cases} \tag{56}$$

where P is the preconditioner and $\vartheta > 0$ an acceleration parameter.
Thanks to the well-posedness of the ellitpic problem in Ω_2, the operator S_2 is invertible and we can use it as preconditioner, so that (56) becomes

$$\begin{cases} \lambda^{(0)} \text{ given} \\ \lambda^{(k+1)} = (1-\vartheta)\lambda^{(k)} + \vartheta S_2^{-1}(\chi - S_1\lambda^{(k)}), \qquad \text{for } k \geq 0. \end{cases} \tag{57}$$

By comparing (57) with (47), we recognize that the Dirichlet–Neumann method is equivalent to the Richardson iterative method applied to the Steklov–Poincaré equation (54) with preconditioner S_2, since the identity $u_2^{(k+1)}|_\Gamma = S_2^{-1}(\chi - S_1\lambda^{(k)})$ holds.

After a discretization of the heterogeneous problem (by, e.g., finite elements or spectral methods) it is possible to write the discrete counterpart of both the Steklov–Poincaré equation (54) and the Dirichlet–Neumann algorithm (47).

It can be be proven that the Dirichlet–Neumann algorithm converges, for suitable choices of the relaxation parameter ϑ, independently of the discretization parameter h for finite elements or N for spectral methods (see, e.g., [30] for a proof in the spectral method context). This because the local Steklov–Poincaré operator S_2 is spectrally equivalent to the global Steklov–Poincaré operator S.

Krylov methods are valid alternatives to Richardson iterations to solve the preconditioned Steklov–Poincaré equation

$$S_2^{-1} S \lambda = S_2^{-1} \chi. \tag{58}$$

In the next section we will provide numerical results about the numerical solution of the coupled problem (40) by using either Dirichlet–Neumann method (47), Adaptive Robin-Neumann method (48) and the preconditioned Bi-CGStab [57] on (58).

2.4 Numerical Results for the Advection–Diffusion Problem

In this Section we will provide the numerical solution of a test case in two-dimensional computational domains. The discretization of the differential equation inside each subdomain is performed by quadrilateral conformal Spectral Element Methods (SEM). We refer to [8] for a detailed description of these methods, while here we recall in brief their basic features.

Let $\mathscr{T} = \{T_m\}_{m=1}^M$ be a partition of the computational domain $\Omega \subset \mathbb{R}^d$, where each element T_m is obtained by a bijective and differentiable transformation F_m from the reference (or parent) element $\hat{\Omega}^d = (-1,1)^d$. On the reference element we define the finite dimensional space $\hat{\mathbb{Q}}_N = \mathrm{span}\{\hat{x}_1^{j_1} \cdots \hat{x}_d^{j_d} : 0 \le j_1, \ldots, j_d \le N\}$ and, for any $T_m \in \mathscr{T}$: $T_m = \mathbf{F}_m(\hat{\Omega}^d)$, set $h_m = \mathrm{diam}(T_m)$ and

$$V_{N_m}(T_m) = \{v : v = \hat{v} \circ \mathbf{F}_m^{-1} \text{ for some } \hat{v} \in \hat{\mathbb{Q}}_{N_m}\}.$$

The SEM multidimensional space is

$$X_\delta = \{v \in C^0(\overline{\Omega}) : v_{|T_m} \in V_{N_m}(T_m), \ \forall T_m \in \mathscr{T}\},$$

where δ is an abridged notation for "discrete", that accounts for the local geometric sizes $\{h_m\}$ and the local polynomial degrees $\{N_m\}$, for $m = 1, \ldots, M$.

Let us consider the variational formulation (38) and, for simplicity, impose the homogeneous Dirichlet condition on the boundary (i.e. $g \equiv 0$). The SEM approximation of the solution of (38) is the function $u_\delta \in V_\delta = X_\delta \cap H_0^1(\Omega)$, such that

$$\sum_m a_{T_m}(u_\delta, v_\delta) = \sum_m (f, v_\delta)_{T_m} \quad \forall v_\delta \in V_\delta \tag{59}$$

holds, where a_{T_m} and $(f,v)_{T_m}$ denote the restrictions to T_m of the bilinear form and the L_2-inner product (respectively) defined in (39).

Since the high computational cost in evaluating integrals in (59), the bilinear form a_{T_m} and the L_2-inner product $(f,v)_{T_m}$ are often approximated by a discrete bilinear form a_{N_m,T_m} and a discrete inner product $(f,v)_{N_m,T_m}$, respectively, in which exact integrals are replaced by Numerical Integration (NI) based on Legendre–Gauss–Lobatto formulas.

The SEM-NI approximation of the solution of (38) will be the function $u_\delta \in V_\delta$ such that

$$\sum_m a_{N_m,T_m}(u_\delta, v_\delta) = \sum_m (f, v_\delta)_{N_m,T_m} \quad \forall v_\delta \in V_\delta. \tag{60}$$

We consider now a test case and we compare the convergence rate of the iterative methods explained in Sect. 2.3. We will denote by DN the Dirichlet Neumann method (47), by ARN the Adaptive Robin–Neumann method (48) and by BiCGStab-SP the preconditioned BiCGstab method applied to the preconditioned Steklov–Poincaré equation (58). Our aim is twofold. From one hand we will represent the numerical solution of the heterogeneous problem (40), on the other hand we want to investigate and compare the convergence rate of the iterative methods versus the magnitude of the viscosity ν and the discretization size (i.e. the local geometric sizes h_m and the local polynomial degrees N_m).

Test case #1: Let us consider problem (40). The computational domain $\Omega = (-1,1)^2$ is split in $\Omega_1 = (-1,0.8) \times (-1,1)$ and $\Omega_2 = (0.8,1) \times (-1,1)$. The interface is $\Gamma = \{0.8\} \times (-1,1)$. The data of the problem are: $\mathbf{b} = [y,0]^t$, $b_0 = 1$, $f = 1$ and the inflow interface is $\Gamma^{\text{in}} = \{0.8\} \times (-1,0)$. Dirichlet boundary conditions are imposed on the vertical sides of Ω, precisely $g = 1$ on $\{-1\} \times (0,1)$, $g = 0$ on $\{1\} \times (-1,1)$, while homogeneous Neumann conditions are imposed on the horizontal sides of Ω_2. The viscosity will be specified below.

In Fig. 8 the SEM-NI solutions for $\nu = 10^{-2}$ and $\nu = 10^{-3}$ are shown. A non-uniform partition in 3×6 (4×6, resp.) quadrilaterals has been considered in Ω_1 (Ω_2, resp.). The same polynomial degree $N = 8$ has been fixed inside each spectral element. The jump of the solution across Γ^{out} is evident for $\nu = 0.01$, in particular we have obtained $\|u_1 - u_2\|_{L^\infty(\Gamma^{\text{out}})} \simeq 0.237$ when $\nu = 0.01$ and $\|u_1 - u_2\|_{L^\infty(\Gamma^{\text{out}})} \simeq 0.020$ when $\nu = 0.001$.

Now we want to compare DN, ARN and BiCGStab-SP methods for what concerns the convergence rate and the computational efficiency.

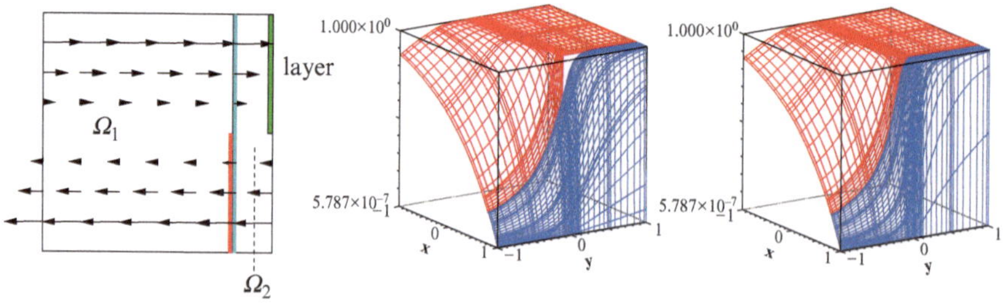

Fig. 8 Test case #1. The data of the test case (*left*) and the heterogeneous solution for $\nu = 0.01$ (*center*) and $\nu = 0.001$ (*right*)

The convergence of both DN and ARN is measured by the stopping test on the difference between two iterates, i.e.

$$\begin{aligned} &\|\lambda^{(k+1)} - \lambda^{(k)}\| \leq \varepsilon && \text{for DN} \\ &\max\{\|\lambda^{(k+1)} - \lambda^{(k)}\|, \|\mu^{(k+1)} - \mu^{(k)}\|\} \leq \varepsilon && \text{for ARN,} \end{aligned} \tag{61}$$

while the convergence of BiCGStab-SP is measured by the stopping test on the residual $r^{(k+1)} = \chi - S\lambda^{(k+1)}$, i.e.

$$\frac{\|r^{(k+1)}\|}{\|r^{(0)}\|} \leq \varepsilon. \tag{62}$$

The convergence of both DN and ARN methods depends on the choice of the relaxation parameter ϑ, on the contrary, the BiCGStab-SP algorithm does not require to set any acceleration parameter.

In Fig. 9 we report the number of iterations of both DN and ARN methods in order to converge up to a tolerance of 10^{-6} for $\nu = 0.01$ and we conclude that, for this test case, the optimal value of ϑ is $\vartheta_{opt} = 1$. Analogous results are obtained for smaller values of the viscosity.

In Table 1 we report the number of iterations needed by every iterative scheme (DN, ARN, BiCGstab-SP) to converge up to a tolerance of 10^{-6}, versus the polynomial degree N. For both DN and ARN method we set $\vartheta = 1$. The partition of Ω is not uniform and it coincides with that used to represent the numerical solutions in Fig. 8. The discretization we have used is fine enough to guarantee the absence of spurious oscillations due to large Péclet number.

As we can see, the convergence rate of all methods is independent of both polynomial degree N and viscosity ν.

The BiCGStab-SP method requires the smallest number of iterations, nevertheless each Bi-CGStab iteration costs about two and a half iterations of either DN or ARN. As a matter of fact, each iteration of DN (or equivalenlty ARN) requires the solution of an advection problem in Ω_1 plus the solution of an elliptic problem

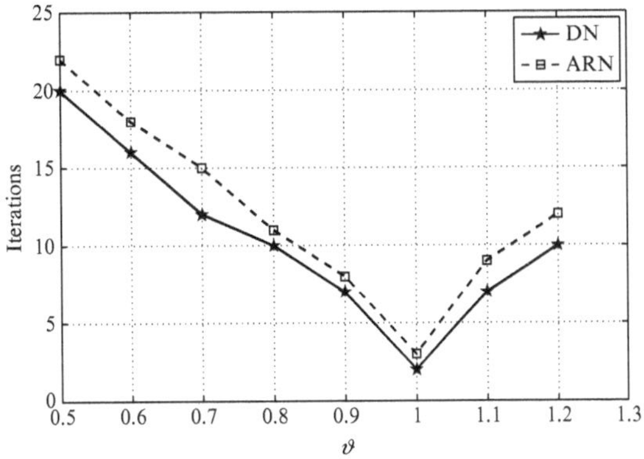

Fig. 9 Test case #1 with $\nu = 0.01$. DN and ARN iterations to satisfy the stopping test (61) versus the relaxation parameter ϑ

Table 1 Test case #1. Number of iterations to satisfy stopping test with $\varepsilon = 10^{-6}$

N	$\nu = 0.1$			$\nu = 0.01$			$\nu = 0.001$		
	DN	ARN	SP	DN	ARN	SP	DN	ARN	SP
4	2	3	1	2	3	1	2	3	1
6	2	3	1	2	3	1	2	3	1
8	2	3	1	2	3	1	2	3	1
10	2	3	1	2	3	1	2	3	1
12	2	3	1	2	3	1	2	3	1
14	2	3	1	2	3	1	2	3	1
16	2	3	1	2	3	1	2	3	1

The relaxation parameter is $\vartheta = 1$ in both DN and ARN. SP is an abridged notation for BiCGStab-SP method.

in Ω_2. On the contrary, each iteration of BiCGstab-SP requires two matrix vector products to compute the residual $r^{(k)} = \chi - S\lambda^{(k)}$ plus the solution of two linear systems on the preconditioner $S_2 z^{(k)} = r^{(k)}$, meaning that we have to solve two advection problems in Ω_1 plus three elliptic problems in Ω_2 at each iteration.

For this test case, we conclude that all three methods are very efficient and their computational costs are comparable. Nevertheless, both DN and ARN methods require a priori knowledge of the optimal relaxation parameter ϑ.

2.5 Navier–Stokes/Potential Coupled Problem

Models similar to the (Navier–)Stokes/Darcy problem introduced in Sect. 1 can be used in external aerodynamics to describe the motion of an incompressible fluid around a body such as, for example, a ship, a boat or a submerged body in a water basin. In fact, such problems can be studied by decomposing the computational domain into two parts: a region Ω_2 close to the body where, due to the viscosity effects, all the interesting features of the flow occur, and an outer region Ω_1 far away from the body where one can neglect the viscosity effects. See, e.g., Fig. 10.

Therefore, suitable heterogeneous differential models comprising Navier–Stokes equations, Euler equations, potential flows and other models from fluid dynamics could be envisaged (see, e.g., [3, 35]).

Here, we present a simple model where in Ω_2 we consider the full Navier–Stokes equations, while in Ω_1 we adopt a Laplace equation for the velocity potential.

A coupled heterogeneous model of this kind has been studied in [56] considering a computational domain as in Fig. 11 and the following generalized Stokes problem:

$$\begin{cases} \alpha \mathbf{u}_\varepsilon - \nu_\varepsilon \Delta \mathbf{u}_\varepsilon + \nabla p_\varepsilon = \mathbf{f} & \text{in } \Omega \\ \nabla \cdot \mathbf{u}_\varepsilon = 0 & \text{in } \Omega \\ \mathbf{u}_\varepsilon = \mathbf{0} & \text{on } \Gamma_b, \end{cases} \tag{63}$$

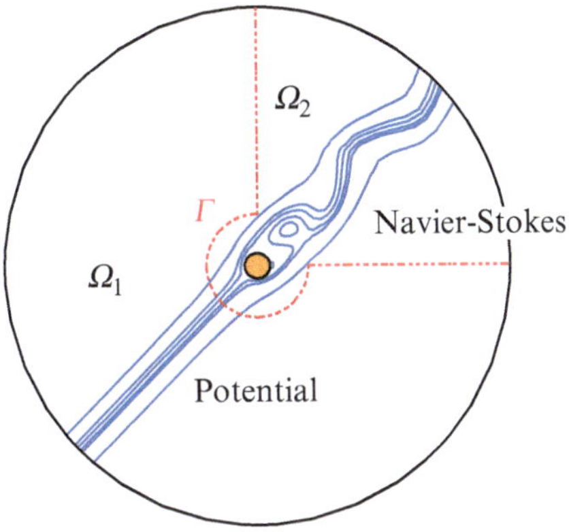

Fig. 10 Flow around a cylinder computed using a Navier–Stokes/potential coupled problem

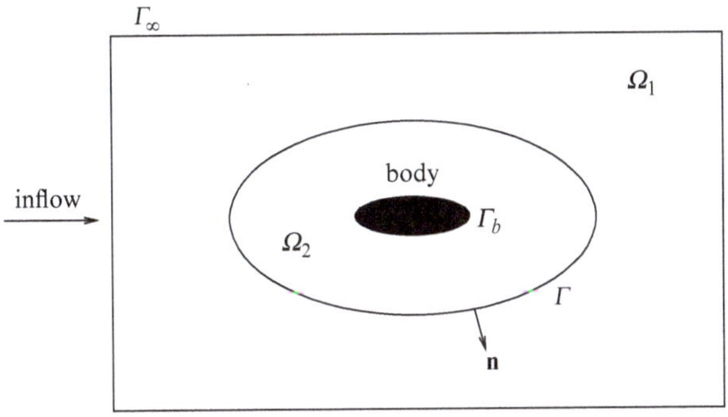

Fig. 11 Representation of the computational domain for an external aerodynamics problem

with suitable boundary conditions on the outer boundary Γ_∞. The viscosity is $\nu_\varepsilon = \nu$ in Ω_2, while $\nu_\varepsilon = \varepsilon$ in Ω_1.

In [56] a vanishing viscosity argument is used letting $\varepsilon \to 0$ in Ω_1 in order to set up a suitable global model and to define the correct interface conditions across Γ. Precisely, the following limit coupled problem was characterized:

$$\begin{cases} \alpha\mathbf{u} - \nu\Delta\mathbf{u} + \nabla p = \mathbf{f} & \text{in } \Omega_2 \\ \nabla \cdot \mathbf{u} = 0 & \text{in } \Omega_2 \\ \Delta q = \nabla \cdot \mathbf{f} & \text{in } \Omega_1 \end{cases} \tag{64}$$

with suitable boundary conditions and the coupling conditions across the interface Γ

$$\begin{aligned} \frac{\partial q}{\partial \mathbf{n}_\Gamma} &= (\mathbf{f} - \alpha\mathbf{u}) \cdot \mathbf{n}_\Gamma && \text{on } \Gamma \\ -\nu\frac{\partial \mathbf{u}}{\partial \mathbf{n}_\Gamma} + p\mathbf{n}_\Gamma &= q\mathbf{n}_\Gamma && \text{on } \Gamma. \end{aligned} \tag{65}$$

$\mathbf{n}_\Gamma$ denotes the unit normal vector on Γ directed from Ω_2 to Ω_1. We remark that, apart from the physical meaning of the variables, the coupling conditions (65) are similar in their structure to those used for the Navier–Stokes/Darcy coupling (28). In fact, $(65)_1$ corresponds to $(28)_1$, and in $(65)_2$ the pressure is still discontinuous across the interface, even if there is no distinction between the normal and the tangential components of the stress tensor as in $(28)_2$ and $(28)_3$.

Because of these similarities, the analysis that we shall develop in Sect. 2.6 for the Navier–Stokes/Darcy problem could be accommodated to account also for the heterogenous coupling (64) and (65).

However, one has to keep in mind that the physical meaning of the two coupled problems is very different. In the Navier–Stokes/Darcy case we have two viscous models where Darcy equation and the coupling conditions can be obtained by homogenization in the limit $\varepsilon \to 0$ in Ω_p, where ε represents the size of the pores in the porous medium. On the other hand, the Navier–Stokes/potential model couples viscous and inviscid equations, the latter being obtained in the limit $\nu \to 0$ like also the corresponding coupling conditions.

2.6 *Asymptotic Analysis of the Coupled Navier–Stokes/Darcy Problem*

We focus now on the coupled Navier–Stokes/Darcy problem (27) and (28), however we confine ourselves to the steady problem by dropping the time-derivative in the momentum equation $(27)_1$:

$$-\mathbf{div}\,\mathsf{T}(\mathbf{u}_f, p_f) + (\mathbf{u}_f \cdot \nabla)\mathbf{u}_f = \mathbf{f} \quad \text{in } \Omega_f. \tag{66}$$

Even when considering the time-dependent problem, a similar kind of "steady" problem can be found when using an implicit finite difference time-advancing scheme. In that case, however, an extra reaction term $\alpha \mathbf{u}_f$ would show up on the left-hand side of (66), where the positive coefficient α plays the role of inverse of the time-step. This reaction term would not affect our forthcoming analysis, though.

To discuss possible boundary conditions on the external boundary of Ω_f and Ω_p, let us split the boundaries $\partial\Omega_f$ and $\partial\Omega_p$ as $\partial\Omega_f = \Gamma \cup \Gamma_f^{in}$ and $\partial\Omega_p = \Gamma \cup \Gamma_p \cup \Gamma_p^b$, as shown in Fig. 6, left.

For the Darcy equation we assign the piezometric head $\varphi = \varphi_p$ on Γ_p; moreover, we require that the normal component of the velocity vanishes on the bottom surface, that is, $\mathbf{u}_p \cdot \mathbf{n}_p = 0$ on Γ_p^b.

For the Navier–Stokes problem, several combinations of boundary conditions are possible, representing different kinds of flow problems. Here, we assign a non-null inflow $\mathbf{u}_f = \mathbf{u}_{in}$ on Γ_f^{in} and a no-slip condition $\mathbf{u}_f = \mathbf{0}$ on the remaining boundary Γ_f.

To summarize, the coupled problem (66)–(28) is supplemented with the boundary conditions:

$$\mathbf{u}_f = \mathbf{u}_{in} \text{ on } \Gamma_f^{in}, \quad \mathbf{u}_f = \mathbf{0} \text{ on } \Gamma_f,$$
$$\varphi = \varphi_p \text{ on } \Gamma_p, \quad \mathsf{K}\frac{\partial \varphi}{\partial n} = 0 \text{ on } \Gamma_p^b. \tag{67}$$

We introduce the following functional spaces:

$$\begin{aligned} H_f &= \{\mathbf{v} \in (H^1(\Omega_f))^d \ : \ \mathbf{v} = \mathbf{0} \text{ on } \Gamma_f \cup \Gamma_f^{in}\}, \\ \widetilde{H}_f &= \{\mathbf{v} \in (H^1(\Omega_f))^d \ : \ \mathbf{v} = \mathbf{0} \text{ on } \Gamma_f \cup \Gamma\}, \\ Q &= L^2(\Omega_f), \quad H_p = \{\psi \in H^1(\Omega_p) \ : \ \psi = 0 \text{ on } \Gamma_p\}. \end{aligned} \tag{68}$$

We denote by $|\cdot|_1$ and $\|\cdot\|_1$ the H^1-seminorm and norm, respectively, and by $\|\cdot\|_0$ the L^2-norm; it will always be clear form the context whether we are referring to spaces on Ω_f or Ω_p.

The space $W = H_f \times H_p$ is a Hilbert space with norm

$$\|\underline{w}\|_W = \left(\|\mathbf{w}\|_1^2 + \|\psi\|_1^2\right)^{1/2} \qquad \forall \underline{w} = (\mathbf{w}, \psi) \in W.$$

Finally, we consider on Γ the trace space $\Lambda = H_{00}^{1/2}(\Gamma)$ and denote its norm by $\|\cdot\|_\Lambda$ (see [42]).

We introduce a continuous extension operator

$$E_f : (H^{1/2}(\Gamma_f^{in}))^d \to \widetilde{H}_f\,. \tag{69}$$

Then $\forall \mathbf{u}_{in} \in (H_{00}^{1/2}(\Gamma_f^{in}))^d$ we can construct a vector function $E_f \mathbf{u}_{in} \in \widetilde{H}_f$ such that $E_f \mathbf{u}_{in|\Gamma_f^{in}} = \mathbf{u}_{in}$.

We introduce another continuous extension operator:

$$E_p : H^{1/2}(\Gamma_p^b) \to H^1(\Omega_p) \quad \text{such that } E_p \varphi_p = 0 \text{ on } \Gamma. \tag{70}$$

Then, for all $\varphi \in H^1(\Omega_p)$ we define the function $\varphi_0 = \varphi - E_p \varphi_p$.

Finally, we define the following bilinear forms:

$$\begin{aligned} a_f(\mathbf{v}, \mathbf{w}) &= \int_{\Omega_f} \frac{\nu}{2}(\nabla \mathbf{v} + \nabla^T \mathbf{v}) \cdot (\nabla \mathbf{w} + \nabla^T \mathbf{w}) \qquad \forall \mathbf{v}, \mathbf{w} \in (H^1(\Omega_f))^d, \\ b_f(\mathbf{v}, q) &= -\int_{\Omega_f} q \operatorname{div} \mathbf{v} \qquad \forall \mathbf{v} \in (H^1(\Omega_f))^d, \quad \forall q \in Q, \\ a_p(\varphi, \psi) &= \int_{\Omega_p} \nabla \psi \cdot \mathsf{K} \nabla \varphi \quad \forall \varphi, \psi \in H^1(\Omega_p), \end{aligned} \tag{71}$$

and, for all $\mathbf{v}, \mathbf{w}, \mathbf{z} \in (H^1(\Omega_f))^d$, the trilinear form

$$c_f(\mathbf{w}; \mathbf{z}, \mathbf{v}) = \int_{\Omega_f} [(\mathbf{w} \cdot \nabla)\mathbf{z}] \cdot \mathbf{v} = \sum_{i,j=1}^{d} \int_{\Omega_f} w_j \frac{\partial z_i}{\partial x_j} v_i\,. \tag{72}$$

Now, if we multiply (66) by $\mathbf{v} \in H_f$ and integrate by parts we obtain

$$a_f(\mathbf{u}_f, \mathbf{v}) + c_f(\mathbf{u}_f; \mathbf{u}_f, \mathbf{v}) + b_f(\mathbf{v}, p_f) - \int_\Gamma \mathbf{n} \cdot \mathsf{T}(\mathbf{u}_f, p_f)\,\mathbf{v} = \int_{\Omega_f} \mathbf{f} \cdot \mathbf{v}\,.$$

Notice that we can write

$$-\int_\Gamma \mathbf{n} \cdot \mathsf{T}(\mathbf{u}_f, p_f)\,\mathbf{v} = -\int_\Gamma [\mathbf{n} \cdot \mathsf{T}(\mathbf{u}_f, p_f) \cdot \mathbf{n}]\mathbf{v} \cdot \mathbf{n} - \int_\Gamma (\mathsf{T}(\mathbf{u}_f, p_f) \cdot \mathbf{n})_\tau \cdot (\mathbf{v})_\tau,$$

so that we can incorporate in weak form the interface conditions $(28)_2$ and $(28)_3$ as follows:

$$-\int_\Gamma \mathbf{n} \cdot \mathsf{T}(\mathbf{u}_f, p_f)\,\mathbf{v} = \int_\Gamma g\varphi(\mathbf{v} \cdot \mathbf{n}) + \int_\Gamma \frac{\nu\alpha_{BJ}}{\sqrt{\mathsf{K}}}(\mathbf{u}_f)_\tau \cdot (\mathbf{v})_\tau\,.$$

Finally, we consider the lifting $E_f\mathbf{u}_{in}$ of the boundary datum and we split $\mathbf{u}_f = \mathbf{u}_f^0 + E_f\mathbf{u}_{in}$ with $\mathbf{u}_f^0 \in H_f$; we recall that $E_f\mathbf{u}_{in} = \mathbf{0}$ on Γ and we get

$$\begin{aligned} &a_f(\mathbf{u}_f^0, \mathbf{v}) + c_f(\mathbf{u}_f^0 + E_f\mathbf{u}_{in}; \mathbf{u}_f^0 + E_f\mathbf{u}_{in}, \mathbf{v}) + b_f(\mathbf{v}, p_f) \\ &\quad + \int_\Gamma g\varphi(\mathbf{v} \cdot \mathbf{n}) + \int_\Gamma \frac{\nu\alpha_{BJ}}{\sqrt{\mathsf{K}}}(\mathbf{u}_f)_\tau \cdot (\mathbf{v})_\tau = \int_{\Omega_f} \mathbf{f} \cdot \mathbf{v} - a_f(E_f\mathbf{u}_{in}, \mathbf{v}). \quad (73) \end{aligned}$$

From $(27)_2$ we find

$$b_f(\mathbf{u}_f^0, q) = -b_f(E_f\mathbf{u}_{in}, q) \qquad \forall q \in Q. \quad (74)$$

On the other hand, if we multiply $(27)_3$ by $\psi \in H_p$ and integrate by parts we get

$$a_p(\varphi, \psi) + \int_\Gamma \mathsf{K}\frac{\partial\varphi}{\partial n}\psi = 0\,.$$

Now we incorporate the interface condition $(28)_1$ in weak form as

$$a_p(\varphi, \psi) - \int_\Gamma (\mathbf{u}_f \cdot \mathbf{n})\psi = 0,$$

and, considering the splitting $\varphi = \varphi_0 + E_p\varphi_p$ we obtain

$$a_p(\varphi_0, \psi) - \int_\Gamma (\mathbf{u}_f \cdot \mathbf{n})\psi = -a_p(E_p\varphi_p, \psi). \quad (75)$$

We multiply (75) by g and sum to (73) and (74); then, we define

$$\begin{aligned} \mathscr{A}(\underline{v}, \underline{w}) &= a_f(\mathbf{v}, \mathbf{w}) + g\,a_p(\varphi, \psi) + \int_\Gamma g\,\varphi(\mathbf{w} \cdot \mathbf{n}) - \int_\Gamma g\,\psi(\mathbf{v} \cdot \mathbf{n}) \\ &\quad + \int_\Gamma \frac{\nu\alpha_{BJ}}{\sqrt{\mathsf{K}}}(\mathbf{w})_\tau \cdot (\mathbf{v})_\tau, \quad (76) \end{aligned}$$

$$\mathscr{C}(\underline{v};\underline{w},\underline{u}) = c_f(\mathbf{v};\mathbf{w},\mathbf{u}),$$
$$\mathscr{B}(\underline{w},q) = b_f(\mathbf{w},q),$$

for all $\underline{v} = (\mathbf{v},\varphi)$, $\underline{w} = (\mathbf{w},\psi)$, $\underline{u} = (\mathbf{u},\xi) \in W$, $q \in Q$. Finally, we define the following linear functionals:

$$\begin{aligned}\langle \mathscr{F},\underline{w}\rangle &= \int_{\Omega_f} \mathbf{f}\cdot\mathbf{w} - a_f(E_f\mathbf{u}_{in},\mathbf{w}) - g\, a_p(E_p\varphi_p,\psi),\\ \langle \mathscr{G},q\rangle &= -b_f(E_f\mathbf{u}_{in},q),\end{aligned} \tag{77}$$

for all $\underline{w} = (\mathbf{w},\psi) \in W$, $q \in Q$.

Adopting these notations, the weak formulation of the coupled Navier–Stokes/Darcy problem reads

find $\underline{u} = (\mathbf{u}_f^0,\varphi_0) \in W$, $p_f \in Q$ such that

$$\begin{cases} \mathscr{A}(\underline{u},\underline{v}) + \mathscr{C}(\underline{u}+\underline{u}^*;\underline{u}+\underline{u}^*,\underline{v}) + \mathscr{B}(\underline{v},p_f) = \langle \mathscr{F},\underline{v}\rangle & \forall \underline{v} = (\mathbf{v},\psi) \in W\\ \mathscr{B}(\underline{u},q) = \langle \mathscr{G},q\rangle & \forall q \in Q, \end{cases} \tag{78}$$

with $\underline{u}^* = (E_f\mathbf{u}_{in},0) \in \widetilde{H}_f \times H^1(\Omega_p)$.

Remark that the interface conditions (28) have been incorporated in the weak formulation as natural conditions on Γ: in particular, $(28)_2$ and $(28)_3$ are natural conditions for the Navier–Stokes problem, while $(28)_1$ becomes a natural condition for Darcy's problem.

The well-posedness of (78) can be proved quite easily in the case of the Stokes/Darcy coupling, i.e. when we neglect the trilinear form $\mathscr{C}(\cdot;\cdot,\cdot)$. Indeed, in this case the existence and uniqueness of the solution follows from the classical theory of Brezzi for saddle-point problems after proving the continuity of $\mathscr{A}(\cdot,\cdot)$, its coerciveness on the kernel of $\mathscr{B}(\cdot,\cdot)$ and that an inf-sup condition holds between the spaces W and Q. For details of this analysis we refer to [18].

The case of the Navier–Stokes/Darcy problem is more involved. In particular, in this case we could prove the well-posedness only under some hypotheses on the data similar to those required for the sole Navier–Stokes equations. Moreover, uniqueness is guaranteed only in the case of small enough filtration velocities $\mathbf{u}_f \cdot \mathbf{n}$ across Γ. The analysis that we have carried out is based on classical results for nonlinear saddle-point problems (see, e.g., [31]). We refer the reader to [4, 19]. Similar results have been proved using a different approach in [32].

2.7 *Solution Techniques for the Navier–Stokes/Darcy Coupling*

A possible approach to solve the Navier–Stokes/Darcy problem is to exploit its naturally decoupled structure keeping separated the fluid and the porous media parts and exchanging information between surface and groundwater flows only through boundary conditions at the interface. From the computational point of view, this strategy is useful at the stage of setting up effective methods to solve the problem numerically.

Therefore, we apply a domain decomposition technique at the differential level to study the Navier–Stokes/Darcy coupled problem. Our aim will be to introduce and analyze a generalized Steklov–Poincaré interface equation (see [51]) associated to our problem, in order to reformulate it solely in terms of interface unknowns. This re-interpretation is crucial to set up iterative procedures between the subdomains Ω_f and Ω_p, that can be used at the discrete level.

Here we illustrate the main ideas behind this approach, and refer to [19] for a complete analysis.

We choose a suitable governing variable on the interface Γ. Considering the interface conditions $(28)_1$ and $(28)_2$, we can foresee two different strategies to select the interface variable:

1. We can set the interface variable λ as the trace of the normal velocity on the interface:
$$\lambda = \mathbf{u}_f \cdot \mathbf{n} = -\mathsf{K}\frac{\partial \varphi}{\partial n}. \tag{79}$$
2. We can define the interface variable σ as the trace of the piezometric head on Γ:
$$\sigma = \varphi = -\frac{1}{g}\mathbf{n} \cdot \mathsf{T}(\mathbf{u}_f, p_f) \cdot \mathbf{n}. \tag{80}$$

Both choices are suitable from the mathematical viewpoint since they guarantee well-posed subproblems in the fluid and the porous medium part.

We discuss here the approach in the case of the Stokes/Darcy coupling considering the choice of the interface variable λ as in (79). We refer the reader to [15] for the second case (80).

For simplicity, from now on we consider the following condition on the interface:

$$(\mathbf{u}_f)_\tau = 0 \quad \text{on } \Gamma \tag{81}$$

instead of $(28)_3$.

Consider the auxiliary problems:

$$\begin{cases} -\mathbf{div}\, \mathsf{T}(\mathbf{u}^*, p^*) = \mathbf{f} & \text{in } \Omega_f \\ \mathrm{div}\, \mathbf{u}^* = 0 & \text{in } \Omega_f \\ \mathbf{u}^* = \mathbf{u}_{in} & \text{on } \Gamma_f^{in} \\ (\mathbf{u}^*)_\tau = 0 & \text{on } \Gamma \\ \mathbf{u}^* \cdot \mathbf{n} = 0 & \text{on } \Gamma, \end{cases} \qquad \begin{cases} -\mathrm{div}\, (\mathsf{K}\nabla\varphi^*) = 0 & \text{in } \Omega_p \\ \varphi^* = \varphi_p & \text{on } \Gamma_p \\ \mathsf{K}\dfrac{\partial \varphi^*}{\partial n} = 0 & \text{on } \Gamma_p^b \\ \mathsf{K}\dfrac{\partial \varphi^*}{\partial n} = 0 & \text{on } \Gamma. \end{cases} \tag{82}$$

Then, assuming to know the value of $\lambda \in \Lambda_0$, with

$$\Lambda_0 = \{\mu \in H_{00}^{1/2}(\Gamma) \,:\, \textstyle\int_\Gamma \mu = 0\},$$

we consider the problems:

$$\begin{cases} -\mathbf{div}\,\mathsf{T}(\mathbf{u}^\lambda, p^\lambda) = \mathbf{0} & \text{in } \Omega_f \\ \text{div}\,\mathbf{u}^\lambda = 0 & \text{in } \Omega_f \\ \mathbf{u}^\lambda = \mathbf{0} & \text{on } \Gamma_f^{in} \\ (\mathbf{u}^\lambda)_\tau = 0 & \text{on } \Gamma \\ \mathbf{u}^\lambda \cdot \mathbf{n} = \lambda & \text{on } \Gamma, \end{cases} \qquad \begin{cases} -\text{div}\,(\mathsf{K}\nabla\varphi^\lambda) = 0 & \text{in } \Omega_p \\ \varphi^\lambda = 0 & \text{on } \Gamma_p \\ \mathsf{K}\dfrac{\partial\varphi^\lambda}{\partial n} = 0 & \text{on } \Gamma_p^b \\ \mathsf{K}\dfrac{\partial\varphi^\lambda}{\partial n} = \lambda & \text{on } \Gamma. \end{cases} \tag{83}$$

We can prove that the solution of the Stokes–Darcy problem can be expressed as: $\mathbf{u}_f = \mathbf{u}^\lambda + \mathbf{u}^*$, $p_f = p^\lambda + p^*$, $\varphi = \varphi^\lambda + \varphi^*$, where $\lambda \in \Lambda_0$ is the solution of the Steklov–Poincaré equation

$$(S_f + S_p)\lambda = \chi \quad \text{on } \Gamma. \tag{84}$$

S_f and S_p are the local Steklov–Poincaré operators formally defined as

$$S_f : \Lambda_0 \to \Lambda_0' \text{ such that } S_f\lambda = \mathbf{n}\cdot\mathsf{T}(\mathbf{u}^\lambda, p^\lambda)\cdot\mathbf{n} \text{ on } \Gamma,$$

while

$$S_p : \Lambda_0 \to \Lambda_0' \text{ such that } S_p\lambda = g\varphi^\lambda \text{ on } \Gamma.$$

Finally,

$$\chi = -\mathbf{n}\cdot\mathsf{T}(\mathbf{u}^*, p^*)\cdot\mathbf{n} - g\varphi^* \text{ on } \Gamma.$$

The analysis of the operators S_f and S_p as well as the study of the well-posedness of the interface equation (84) have been carried out in [18]. In particular, we have proved that the operator S_f is invertible on the trace space Λ_0 and it is spectrally equivalent to $S_f + S_p$, i.e., there exist two positive constants k_1 and k_2 (independent of η) such that

$$k_1\langle S_f\eta, \eta\rangle \;\le\; \langle S\eta, \eta\rangle \;\le\; k_2\langle S_f\eta, \eta\rangle \qquad \forall \eta \in \Lambda_0.$$

The same property holds at the discrete level considering conforming finite element approximations of S_f and S_p with constants k_1 and k_2 that do not depend on the grid size h. This property makes the operator S_f an attractive preconditioner to solve the interface problem (84) via an iterative method like, e.g., Richardson or the Conjugate Gradient, yielding a convergence rate independent of h.

For example, we can consider the following Richardson iterations: given $\lambda^{(0)} \in \Lambda_0$, for $k \ge 0$,

$$\lambda^{(k+1)} = \lambda^{(k)} + \vartheta S_f^{-1}(\chi - (S_f + S_p)\lambda^{(k)}) \quad \text{on } \Gamma, \tag{85}$$

where $0 < \vartheta < 1$ is a suitable relaxation parameter.

This method requires at each step to apply S_p and S_f^{-1}, i.e., recalling the definitions of these operators, to solve a Darcy problem in Ω_p with given flux across Γ and a Stokes problem in Ω_f with assigned normal stress on Γ. More precisely, we can rewrite (85) as: let $\lambda^{(0)} \in \Lambda$ be an initial guess; for $k \ge 0$,

$$\text{Solve} \quad \begin{cases} -\text{div}\,(\mathsf{K}\nabla\varphi^{(k+1)}) = 0 & \text{in } \Omega_p \\ \varphi^{(k+1)} = \varphi_p & \text{on } \Gamma_p \\ \mathsf{K}\dfrac{\partial\varphi^{(k+1)}}{\partial n} = 0 & \text{on } \Gamma_p^b \\ \mathsf{K}\dfrac{\partial\varphi^{(k+1)}}{\partial n} = \lambda^{(k)} & \text{on } \Gamma, \end{cases}$$

$$\text{Solve} \quad \begin{cases} -\mathbf{div}\,\mathsf{T}(\mathbf{u}^{(k+1)}, p^{(k+1)}) = \mathbf{f} & \text{in } \Omega_f \\ \text{div}\,\mathbf{u}^{(k+1)} = 0 & \text{in } \Omega_f \\ \mathbf{u}^{(k+1)} = \mathbf{u}_{in} & \text{on } \Gamma_f^{in} \\ (\mathbf{u}^{(k+1)})_\tau = 0 & \text{on } \Gamma \\ -\mathbf{n}\cdot\mathsf{T}(\mathbf{u}^{(k+1)}, p^{(k+1)})\cdot\mathbf{n} = g\varphi^{(k+1)} & \text{on } \Gamma, \end{cases} \tag{86}$$

$$\text{Compute} \quad \lambda^{(k+1)} = (1-\vartheta)\lambda^{(k)} + \vartheta\mathbf{u}^{(k+1)}\cdot\mathbf{n} \quad \text{on } \Gamma.$$

Remark that this algorithm has the same structure as the Dirichlet–Neumann method in the domain decomposition framework.

Another possible algorithm that we have studied in [20] is a sequential Robin-Robin method which at each iteration requires to solve a Darcy problem in Ω_p followed by a Stokes problem in Ω_f, both with Robin conditions on Γ. Precisely, the algorithm reads as follows.

Having assigned a trace function $\eta^0 \in L^2(\Gamma)$, and two acceleration parameters $\gamma_f \geq 0$ and $\gamma_p > 0$, for each $k \geq 0$:

$$\text{Solve} \quad \begin{cases} -\text{div}\,(\mathsf{K}\nabla\varphi^{(k+1)}) = 0 & \text{in } \Omega_p \\ \varphi^{(k+1)} = \varphi_p & \text{on } \Gamma_p \\ \mathsf{K}\dfrac{\partial\varphi^{(k+1)}}{\partial n} = 0 & \text{on } \Gamma_p^b \\ -\gamma_p\mathsf{K}\dfrac{\partial\varphi^{(k+1)}}{\partial n} + g\varphi_{|\Gamma}^{(k+1)} = \eta^{(k)} & \text{on } \Gamma, \end{cases}$$

$$\text{Solve} \quad \begin{cases} -\mathbf{div}\,\mathsf{T}(\mathbf{u}^{(k+1)}, p^{(k+1)}) = \mathbf{f} & \text{in } \Omega_f \\ \text{div}\,\mathbf{u}^{(k+1)} = 0 & \text{in } \Omega_f \\ \mathbf{u}^{(k+1)} = \mathbf{u}_{in} & \text{on } \Gamma_f^{in} \\ (\mathbf{u}^{(k+1)})_\tau = 0 & \text{on } \Gamma \\ \mathbf{n}\cdot\mathsf{T}(\mathbf{u}_f^{k+1}, p_f^{k+1})\cdot\mathbf{n} + \gamma_f\mathbf{u}_f^{(k+1)}\cdot\mathbf{n} & \\ \qquad = -g\varphi_{|\Gamma}^{(k+1)} - \gamma_f\mathsf{K}\dfrac{\partial\varphi^{(k+1)}}{\partial n} & \text{on } \Gamma, \end{cases} \tag{87}$$

$$\text{Compute} \quad \eta^{(k+1)} = -\mathbf{n}\cdot\mathsf{T}(\mathbf{u}_f^{(k+1)}, p_f^{(k+1)})\cdot\mathbf{n} + \gamma_p\mathbf{u}_f^{(k+1)}\cdot\mathbf{n} \quad \text{on } \Gamma.$$

Both the Stokes problem in Ω_f and the Darcy problem in Ω_p are well-posed and, at convergence, we recover the solution $(\mathbf{u}_f, p_f) \in H_f \times Q$ and $\varphi \in H_p$ of the coupled Stokes/Darcy problem. Indeed, denoting by φ^* the limit of the sequence φ^k in $H^1(\Omega_p)$ and by $(\mathbf{u}_f^*, p_f^*)$ that of $(\mathbf{u}_f^k, p_f^k)$ in $(H^1(\Omega_f))^d \times Q$, we obtain

$$- \gamma_p \mathsf{K} \frac{\partial \varphi^*}{\partial n} + g\varphi^*_{|\Gamma} = -\mathbf{n} \cdot \mathsf{T}(\mathbf{u}_f^*, p_f^*) \cdot \mathbf{n} + \gamma_p \mathbf{u}_f^* \cdot \mathbf{n} \quad \text{on } \Gamma \,, \tag{88}$$

so that we have

$$(\gamma_f + \gamma_p)\mathbf{u}_f^* \cdot \mathbf{n} = -(\gamma_f + \gamma_p)\mathsf{K}\frac{\partial \varphi^*}{\partial n} \quad \text{on } \Gamma \,,$$

yielding, since $\gamma_f + \gamma_p \neq 0$, $\mathbf{u}_f^* \cdot \mathbf{n} = -\mathsf{K}\frac{\partial \varphi^*}{\partial n}$ on Γ, and also, from (88), that $\mathbf{n} \cdot \mathsf{T}(\mathbf{u}_f^*, p_f^*) \cdot \mathbf{n} = -g\varphi^*_{|\Gamma}$ on Γ. Thus, the two interface conditions $(28)_1$ and $(28)_2$ are satisfied, and we can conclude that the limit functions $\varphi^* \in H_p$ and $(\mathbf{u}_f^*, p_f^*) \in H_f \times Q$ are the solutions of the coupled Stokes/Darcy problem.

A proof of convergence is presented in [20] and it follows the guidelines of the theory by Lions [41] for the Robin–Robin method (see also [51, Sect. 4.5]).

A crucial point in the algorithm is the choice of the acceleration parameters γ_f and γ_p. A general strategy is not available, but thanks to a reinterpretation of the Robin–Robin method as an alternating direction scheme à la Peaceman–Rachford (see [48]), we were able to give some hints on how to choose them. We refer to [20].

We will illustrate the numerical behavior of the Dirichlet–Neumann and of the Robin–Robin algorithms in Sect. 2.8.

Finally, we address the case of the Navier–Stokes/Darcy coupling. Also to this nonlinear problem we can asociate an interface equation similar to (84) still involving the operator S_p but a nonlinear operator $\tilde{S}_f$ analogous to S_f. Formally, we can represent $\tilde{S}_f : \Lambda_0 \to \Lambda_0'$ as the operator associated to the Navier–Stokes problem:

$$\begin{cases} -\mathbf{div}\, \mathsf{T}(\mathbf{u}^\lambda, p^\lambda) + (\mathbf{u}^\lambda \cdot \nabla)\mathbf{u}^\lambda = \mathbf{0} & \text{in } \Omega_f \\ \text{div}\, \mathbf{u}^\lambda = 0 & \text{in } \Omega_f \\ \mathbf{u}^\lambda = \mathbf{0} & \text{on } \Gamma_f^{in} \\ (\mathbf{u}^\lambda)_\tau = 0 & \text{on } \Gamma \\ \mathbf{u}^\lambda \cdot \mathbf{n} = \lambda & \text{on } \Gamma, \end{cases} \tag{89}$$

such that $\tilde{S}_f \lambda = \mathbf{n} \cdot \mathsf{T}(\mathbf{u}^\lambda, p^\lambda) \cdot \mathbf{n}$ on Γ.

Then, we can write the interface problem:

$$\text{find } \lambda \in \Lambda_0 : \quad \tilde{S}_f(\lambda) + S_p\lambda = \chi_p \quad \text{on } \Gamma, \tag{90}$$

with $\chi_p =$, and prove its equivalence to the global coupled problem.

A rigorous presentation of this approach can be found in [4].

The set-up of effective iterative methods for the interface problem (90) is not straightforward. In particular, no results are available yet on the characterization of suitable operators spectrally equivalent to $\tilde{S}_f + S_p$. In [4, 19] we have proposed and analyzed two classical schemes, fixed-point or Newton, for (90) showing their equivalence to the following algorithms, respectively.

Fixed-point iterations: Given $\mathbf{u}_f^0 \in H_f$, for $k \geq 1$, find $\mathbf{u}_f^{(k)} \in H_f$, $p_f^{(k)} \in Q$, $\varphi^{(k)} \in H_p$ such that, for all $\mathbf{v} \in H_f, q \in Q, \psi \in H_p$,

$$\begin{cases} a_f(\mathbf{u}_f^{(k)}, \mathbf{v}) + c_f(\mathbf{u}_f^{(k-1)}; \mathbf{u}_f^{(k)}, \mathbf{v}) + b_f(\mathbf{v}, p_f^{(k)}) \\ \qquad + \displaystyle\int_\Gamma g\,\varphi^{(k)}(\mathbf{v} \cdot \mathbf{n}) + \int_\Gamma \frac{\nu \alpha_{BJ}}{\sqrt{\mathsf{K}}} (\mathbf{u}_f^{(k)})_\tau \cdot (\mathbf{v})_\tau = \int_{\Omega_f} \mathbf{f} \cdot \mathbf{v} \\ b_f(\mathbf{u}_f^{(k)}, q) = 0 \\ a_p(\varphi^{(k)}, \psi) = \displaystyle\int_\Gamma \psi(\mathbf{u}_f^{(k)} \cdot \mathbf{n}). \end{cases} \tag{91}$$

Newton-like methods: Let $\mathbf{u}_f^0 \in H_f$ be given; then, for $k \geq 1$, find $\mathbf{u}_f^{(k)} \in H_f$, $p_f^{(k)} \in Q, \varphi^{(k)} \in H_p$ such that, for all $\mathbf{v} \in H_f, q \in Q, \psi \in H_p$,

$$\begin{cases} a_f(\mathbf{u}_f^{(k)}, \mathbf{v}) + c_f(\mathbf{u}_f^{(k)}; \mathbf{u}_f^{(k-1)}, \mathbf{v}) + c_f(\mathbf{u}_f^{(k-1)}; \mathbf{u}_f^{(k)}, \mathbf{v}) + b_f(\mathbf{v}, p_f^{(k)}) \\ \quad + \displaystyle\int_\Gamma g\varphi^n(\mathbf{v} \cdot \mathbf{n}) + \int_\Gamma \frac{\nu \alpha_{BJ}}{\sqrt{\mathsf{K}}} (\mathbf{u}_f^{(k)})_\tau \cdot (\mathbf{v})_\tau \\ \quad = c_f(\mathbf{u}_f^{(k-1)}; \mathbf{u}_f^{(k-1)}, \mathbf{v}) + \displaystyle\int_{\Omega_f} \mathbf{f} \cdot \mathbf{v} \\ b_f(\mathbf{u}_f^{(k)}, q) = 0 \\ a_p(\varphi^{(k)}, \psi) = \displaystyle\int_\Gamma \psi(\mathbf{u}_f^{(k)} \cdot \mathbf{n}). \end{cases} \tag{92}$$

Some numerical results will be presented in Sect. 2.8.

2.8 Numerical Results for the Navier–Stokes/Darcy Problem

We consider a regular triangulation $\mathscr{T}_h$ of the domain $\overline{\Omega}_f \cup \overline{\Omega}_p$, depending on a positive parameter $h > 0$, made up of triangles T. We assume that the triangulations $\mathscr{T}_{fh}$ and $\mathscr{T}_{ph}$ induced on the subdomains Ω_f and Ω_p are compatible on Γ, that is they share the same edges therein. Finally, we suppose the triangulation induced on Γ to be quasi-uniform (see, e.g., [50]).

Several choices of finite element spaces can be made. If we indicate by $\mathbf{W}_h$ and Q_h the finite element spaces which approximate the velocity and pressure fields, respectively, for the Navier–Stokes problem, there must exist a positive constant $\beta^* > 0$, independent of h, such that the classical inf-sup condition is satisfied, i.e., $\forall q_h \in Q_h, \exists \mathbf{v}_h \in \mathbf{W}_h, \mathbf{v}_h \neq \mathbf{0}$, such that

$$\int_{\Omega_f} q_h \operatorname{div} \mathbf{v}_h \geq \beta^* \|\mathbf{v}_h\|_{H^1(\Omega_f)} \|q_h\|_{L^2(\Omega_f)}.$$

No additional compatibility condition is required when coupling with the Darcy equations. Thus, for our tests we use the $\mathbb{P}_2 - \mathbb{P}_1$ Taylor–Hood finite elements for Stokes or Navier–Stokes and $\mathbb{P}_2$ elements for Darcy equation.

We investigate the convergence properties of algorithm (86) (or, equivalently, (85)) and the PCG algorithm for (84) with preconditioner S_f^{-1}. For the moment we set the physical parameters ν, K, g to 1. We consider the computational domain $\Omega \subset \mathbb{R}^2$ with $\Omega_f = (0,1) \times (1,2)$, $\Omega_p = (0,1) \times (0,1)$ and the interface $\Gamma = (0,1) \times \{1\}$. The boundary conditions and the forcing terms are chosen in such a way that the exact solution of the coupled Stokes/Darcy problem is

$$(\mathbf{u}_f)_1 = -\cos\left(\frac{\pi}{2}y\right)\sin\left(\frac{\pi}{2}x\right), \quad (\mathbf{u}_f)_2 = \sin\left(\frac{\pi}{2}y\right)\cos\left(\frac{\pi}{2}x\right) - 1 + x,$$

$$p_f = 1 - x, \quad \varphi = \frac{2}{\pi}\cos\left(\frac{\pi}{2}x\right)\cos\left(\frac{\pi}{2}y\right) - y(x-1),$$

where $(\mathbf{u}_f)_1$ and $(\mathbf{u}_f)_2$ are the components of the velocity field $\mathbf{u}_f$ (see [19]).

Four different regular conforming meshes have been considered whose number of elements in Ω and of nodes on Γ are reported in Table 2, together with the number of iterations to convergence. A tolerance 10^{-10} has been prescribed for the convergence tests based on the relative residues. In the Dirichlet–Neumann-like algorithm (86) we set the relaxation parameter $\vartheta = 0.7$.

Figure 12 shows the computed residues for the adopted iterative methods when using the finest mesh (logarithmic scale on the y-axis).

These numerical tests show that the discrete preconditioner S_f is optimal with respect to the grid parameter h since the corresponding preconditioned methods yield convergence in a number of iterations independent of h.

We consider now the influence of the physical parameters, which govern the coupled problem, on the convergence rate. We use the PCG method as it embeds the choice of dynamic optimal acceleration parameters. We take the same computational

Table 2 Number of iterations obtained on different grids

Number of mesh elements	Number of nodes on Γ	Algorithm (86) ($\vartheta = 0.7$)	PCG for (84) (preconditioner S_f^{-1})
172	13	18	5
688	27	18	5
2,752	55	18	5
11,008	111	18	5

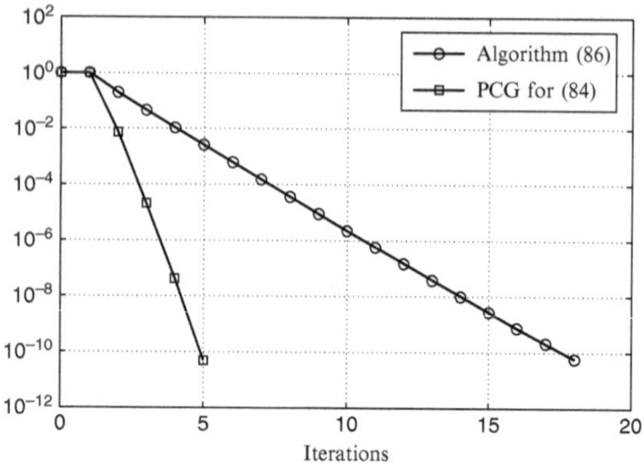

Fig. 12 Computed relative residues for the interface variable λ

Table 3 Iterations using the PCG method (preconditioner S_f^{-1}) with respect to several values of ν and K

ν	K	$h = 1/7$	$h = 1/14$	$h = 1/28$	$h = 1/56$
1	1	5	5	5	5
10^{-1}	10^{-1}	11	11	10	10
10^{-2}	10^{-1}	15	19	18	17
10^{-3}	10^{-2}	20	54	73	56
10^{-4}	10^{-3}	20	59	#	#
10^{-6}	10^{-4}	20	59	148	#

domain, but here the boundary data and the forcing terms are chosen in such a way that the exact solution of the coupled problem is (see [19]):

$$(\mathbf{u}_f)_1 = y^2 - 2y + 1, \quad (\mathbf{u}_f)_2 = x^2 - x, \quad p_f = 2\nu(x + y - 1) + \frac{g}{3\mathsf{K}},$$

$$\varphi = \frac{1}{\mathsf{K}}\left(x(1-x)(y-1) + \frac{y^3}{3} - y^2 + y\right) + \frac{2\nu}{g}x.$$

The most relevant physical quantities for the coupling are the fluid viscosity ν and the hydraulic conductivity K. Therefore, we test our algorithms with respect to different values of ν and K, and set the other physical parameters to 1. We consider a convergence test based on the relative residue with tolerance 10^{-10}.

In Table 3 we report the number of iterations necessary for several choices of ν and K. The symbol # indicates that the method did not converge within 150 iterations.

We can see that the convergence of the algorithm is troublesome when the values of ν and K decrease. In fact, in that case the method converges in a large number of iterations which increases when h decreases, losing its optimality properties. The subdomain iterative method that we have proposed is then effective only when the product $\nu\mathsf{K}$ is sufficiently large, while dealing with small values causes severe difficulties.

Table 4 Number of iterations to solve problem the modified Stokes/Darcy problem using (93) for different values of ν, K and γ

ν	K	γ	Iterations on the mesh with grid size			
			$h = 1/7$	$h = 1/14$	$h = 1/28$	$h = 1/56$
		10	15	24	28	28
10^{-3}	10^{-2}	10^2	12	14	16	14
		10^3	8	9	9	8
		10^3	15	23	28	33
10^{-6}	10^{-4}	10^4	13	14	17	18
		10^5	8	9	9	9

Table 5 Number of iterations using the Robin–Robin method with respect to ν, K and four different grid sizes h; the acceleration parameters are $\gamma_f = 0.3$ and $\gamma_p = 0.1$

ν	K	$h = 1/7$	$h = 1/14$	$h = 1/28$	$h = 1/56$
10^{-4}	10^{-3}	19	19	19	19
10^{-6}	10^{-4}	20	20	20	20
10^{-6}	10^{-7}	20	20	20	20

However, the algorithm still performs well if, instead of the steady Stokes problem, one considers the generalized Stokes momentum equation:

$$\gamma \mathbf{u}_f - \mathbf{div}\, \mathsf{T}(\mathbf{u}_f, p_f) = \tilde{\mathbf{f}} \quad \text{in } \Omega_f, \tag{93}$$

where γ can represent the inverse of a time step within a time discretization using, e.g., the implicit Euler method. Some numerical results are reported in Table 4 (see also [16]).

On the other hand, the Robin–Robin method (87) performs quite well in presence of small values of ν and K. We present hereafter a test considering the same setting as for Table 3. The analogy with the Peaceman–Rachford method has suggested us to set $\gamma_f = 0.3$ and $\gamma_p = 0.1$ (see [15] for more details). In Table 5 we report the number of iterations obtained using the Robin–Robin method for some small values of ν and K and for four different computational grids. A convergence test based on the relative increment of the trace of the discrete normal velocity on the interface $\mathbf{u}^k_{fh} \cdot \mathbf{n}_{|\Gamma}$ has been considered with tolerance 10^{-9}. (See [20].)

Finally, we present some numerical tests for the Navier–Stokes/Darcy coupling using the fixed-point and Newton algorithms of Sect. 2.7. The computational domain and the finite element discretization are the same as in the previous tests. (See also [4].)

In a first test, we set the boundary conditions in such a way that the analytical solution for the coupled problem is $\mathbf{u}_f = (e^{x+y} + y, -e^{x+y} - x)$, $p_f = \cos(\pi x)\cos(\pi y) + x$, $\varphi = e^{x+y} - \cos(\pi x) + xy$. In order to check the behavior of the iterative methods with respect to the grid parameter h, we set the physical parameters (ν, K, g) all equal to 1.

Table 6 Number of iterations for the iterative methods with respect to h

h	Fixed-point	Newton
$h = 1/7$	9	5
$h = 1/14$	9	5
$h = 1/28$	9	5

Table 7 Number of iterations of the fixed-point (FP) and Newton (N) methods with respect to the parameters ν and K

ν	K	$h = 1/7$		$h = 1/14$		$h = 1/28$	
		FP	N	FP	N	FP	N
1	1	7	5	7	5	7	5
1	10^{-4}	5	4	5	4	5	4
10^{-1}	10^{-1}	10	5	10	5	10	5
10^{-2}	10^{-1}	17	6	17	6	17	6
10^{-2}	10^{-3}	14	5	14	5	14	5

The algorithms are stopped as soon as $\|\mathbf{x}^n - \mathbf{x}^{n-1}\|_2 / \|\mathbf{x}^n\|_2 \leq 10^{-10}$, where $\|\cdot\|_2$ is the Euclidean norm and $\mathbf{x}^n$ is the vector of the nodal values of $(\mathbf{u}_f^n, p_f^n, \varphi^n)$. Our initial guess is $\mathbf{u}_f^0 = \mathbf{0}$.

The number of iterations obtained using the fixed-point algorithm (91), and the Newton method (92) are displayed in Table 6. Both methods converge in a number of iterations which does not depend on h.

A second test is carried out in order to assess the influence of the physical parameters on the convergence rate of the algorithms. In this case, the analytical solution is $\mathbf{u}_f = ((y-1)^2 + (y-1) + \sqrt{\mathsf{K}}/\alpha_{BJ}, x(x-1))$, $p_f = 2\nu(x+y-1)$, and $\varphi = \mathsf{K}^{-1}(x(1-x)(y-1) + (y-1)^3/3) + 2\nu x$. We choose several values for the physical parameters ν and K as indicated in Table 7.

3 Virtual Control Approach

The *virtual control* approach represents an alternative approach to the variational asymptotic one, to solve heterogeneous problems.

It is based on the optimal control theory that has been introduced in domain decomposition method with overlapping subdomains to treat both heterogeneous couplings, involving Navier–Stokes and full potential operators [13, 28], and homogeneous problems, either elliptic and parabolic (see [12, 27, 43–45]). In the pioneering papers of Glowinski et al. [12, 27], this method was referred to as a *Least Square* formulation of the multi domain problem.

The basic idea of this approach consists in introducing two "virtual" controls which play the role of unknown Dirichlet data on the interfaces of the decomposition and in minimizing the L^2-norm of the difference between the hyperbolic and the elliptic solutions (defined inside the two subdomains) on the overlap.

The virtual control approach for heterogeneous advection–diffusion operators was introduced and analysed in [26] and there it has been extended to non-overlapping subdomain decompositions (with sharp interfaces). In the latter situation, the virtual controls are defined on the unique interface and the cost functional to be minimized has to be chosen accurately in order to guarantee the well posedness of the optimal control problem.

Finally, in [1] two different formulations of the heterogeneous advection–diffusion problem with either two and three virtual controls have been analysed for overlapping decompositions.

In the following subsection we will give a detailed description of virtual control approach with either overlapping and non-overlapping decompositions for the heterogeneous problems introduced in Sect. 1.

Here we only note that the virtual control approach without overlap is more efficient than the overlapping version, however the former requires a more definite a priori knowledge on structure of interface conditions. On the contrary, the virtual control approach with overlap is more general and it can be regarded as a rigorous translation of a common practice in engineering community based on solving both problems in a common region and using simple "Dirichlet" type conditions at subdomain boundaries.

3.1 *Virtual Control Approach Without Overlap for AD Problems*

The idea of this approach consists in formulating an optimal control problem [39] featuring both control and observation on the interface Γ. We introduce two functions λ_1 and λ_2 defined on the interface Γ and called *virtual controls* (Fig. 13), such that they represent the unknown Dirichlet data on Γ for u_1 and u_2, respectively, i.e.

$$u_1 = \lambda_1 \quad \text{on } \Gamma^{\text{in}}, \qquad u_2 = \lambda_2 \quad \text{on } \Gamma. \tag{94}$$

By collecting differential equations (32) and (33), the external boundary conditions (36) and the interface condition (94), we consider the following problem: given λ_1, λ_2, find $u_1 = u_1(\lambda_1)$ and $u_2 = u_2(\lambda_2)$ such that

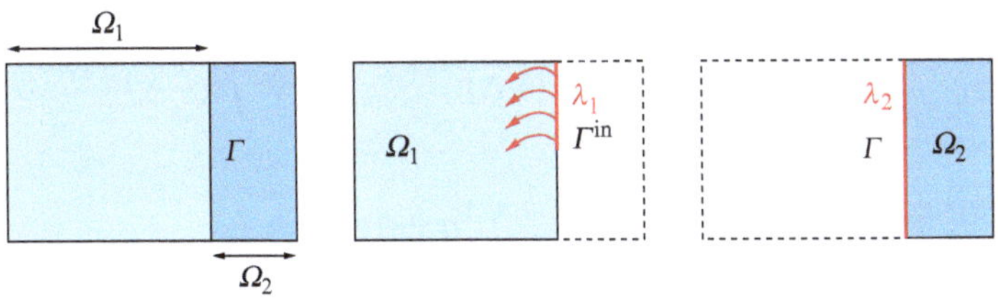

Fig. 13 Virtual control without overlap

$$\begin{cases} A_1 u_1 = \mathrm{div}(\mathbf{b} u_1) + b_0 u_1 = f & \text{in } \Omega_1 \\ u_1 = g_1 & \text{on } (\partial\Omega_1 \setminus \Gamma)^{\text{in}} \\ u_1 = \lambda_1 & \text{on } \Gamma^{\text{in}} \end{cases} \tag{95}$$

and

$$\begin{cases} A_2 u_2 = -\mathrm{div}(\nu \nabla u_2) + \mathrm{div}(\mathbf{b} u_2) + b_0 u_2 = f & \text{in } \Omega_2 \\ u_2 = g_2 & \text{on } \partial\Omega_2 \setminus \Gamma \\ u_2 = \lambda_2 & \text{on } \Gamma. \end{cases} \tag{96}$$

In the case where $\Gamma^{\text{in}} = \emptyset$, no λ_1 is needed since there is no need to prescribe any boundary data on Γ^{in} for problem (95).

The virtual controls λ_1 and λ_2 are determined in such a way that the solutions u_1 and u_2 of (95) and (96) adjust in the best possible way on Γ. More precisely, we look for the solution of the minimization problem

$$\inf_{\lambda_1,\, \lambda_2} J(\lambda_1, \lambda_2), \tag{97}$$

where $J(\lambda_1, \lambda_2)$ is a suitably chosen cost functional.

Various instances have been proposed and analyzed in [26]. Consider, for example,

$$J(\lambda_1, \lambda_2) = \frac{1}{2} \|u_1(\lambda_1) - u_2(\lambda_2)\|^2_{L^2_{\mathbf{b}}(\Gamma^{\text{in}})} + \frac{1}{2} \|\varphi_1(\lambda_1) + \varphi_2(\lambda_2)\|^2_{H^{-1/2}(\Gamma)}, \tag{98}$$

where

$$\varphi_1(\lambda_1) = -\mathbf{b} \cdot \mathbf{n}_\Gamma u_1(\lambda_1), \quad \varphi_2(\lambda_2) = -\nu \frac{\partial u_2(\lambda_2)}{\partial \mathbf{n}_\Gamma} + \mathbf{b} \cdot \mathbf{n}_\Gamma u_2(\lambda_2) \tag{99}$$

are the fluxes on Γ associated to the differential operators A_1 and A_2 (respectively) and $H^{-1/2}(\Gamma)$ is the dual space of $H^{1/2}_{00}(\Gamma)$. Denoting by $-\Delta_\Gamma$ the Laplace Beltrami operator on Γ, for any $\psi, \varphi \in H^{-1/2}(\Gamma)$ we define the following inner product (see, e.g., [39]):

$$(\psi, \varphi)_{H^{-1/2}(\Gamma)} = \int_\Gamma (-\Delta_\Gamma)^{-1/4} \psi \, (-\Delta_\Gamma)^{-1/4} \varphi \, d\Gamma = \int_\Gamma (-\Delta_\Gamma)^{-1/2} \psi \, \varphi \, d\Gamma \tag{100}$$

and the related norm $\|\psi\|_{H^{-1/2}(\Gamma)} = (\psi, \psi)^{1/2}_{H^{-1/2}(\Gamma)}$.

We note that the observation is performed on the whole interface Γ for what concerns the gap on the fluxes, whereas it is restricted to the inflow interface Γ^{in} for that on the velocities.

From now on, by *solution of the virtual control approach* we will mean the solution of the minimization problem (97), with J defined in (98) and with $u_i(\lambda_i)$ (for $i = 1, 2$) the solutions of problems (95) and (96), respectively.

Problems (95) and (96) are well posed. As a matter of fact, the following result holds (see, e.g., [30]):

Theorem 3. *Under assumptions (44), if $g_1 \in L^2_{\mathbf{b}}((\partial\Omega_1 \setminus \Gamma)^{\text{in}})$ and $\lambda_1 \in L^2_{\mathbf{b}}(\Gamma^{\text{in}})$, then the first-order problem (95) admits a unique solution $u_1 = u_1(\lambda_1) \in L^2(\Omega_1)$. Moreover $u_1 \in L^2_{\mathbf{b}}(\partial\Omega_1)$ and $div(\mathbf{b}u_1) \in L^2(\Omega_1)$.*

As of problem (96), if $g_2 \in H^{1/2}(\partial\Omega_2 \setminus \Gamma)$ and $\lambda_2 \in H^{1/2}(\Gamma)$, and moreover there exists a function $\mu \in H^{1/2}(\partial\Omega_2)$ with $g_2 = \mu_{|(\partial\Omega_2\setminus\Gamma)}$ and $\lambda_2 = \mu_{|\Gamma}$, then there exists a unique solution $u_2(\lambda_2)$ of (96) belonging to $H^1(\Omega_2)$. (See, e.g., [30].)

We introduce the following spaces:

$$
\begin{aligned}
&V_1 = \{w \in L^2(\Omega_1) : \text{div}(\mathbf{b}w) \in L^2(\Omega_1),\ \ w_{|\Gamma} \in L^2_{\mathbf{b}}(\Gamma)\}, \quad \Lambda_1 = L^2_{\mathbf{b}}(\Gamma^{\text{in}}),\\
&V_2 = H^1(\Omega_2),\\
&\Lambda_2 = \left\{\lambda_2 \in H^{1/2}(\Gamma) :\ \exists \mu \in H^{1/2}(\partial\Omega_2) \text{ s.t. } \lambda_2 = \mu|_\Gamma \text{ and } g_2 = \mu|_{\partial\Omega_2\setminus\Gamma}\right\},\\
&\mathbf{V} = V_1 \times V_2, \quad \mathbf{\Lambda} = \Lambda_1 \times \Lambda_2.
\end{aligned}
\tag{101}
$$

In order to prove the existence of solution of the minimization problem (97), we define two pairs of auxiliary problems:
find $(w_1^f, w_2^f) \in \mathbf{V}$ such that

$$
\begin{cases}
A_1 w_1^f = f & \text{in } \Omega_1\\
w_1^f = g_1 & \text{on } (\partial\Omega_1 \setminus \Gamma)^{\text{in}}\\
w_1^f = 0 & \text{on } \Gamma^{\text{in}},
\end{cases}
\qquad
\begin{cases}
A_2 w_2^f = f & \text{in } \Omega_2\\
w_2^f = g_2 & \text{on } \partial\Omega_2 \setminus \Gamma\\
w_2^f = 0 & \text{on } \Gamma,
\end{cases}
\tag{102}
$$

and find $(u_1^{\lambda_1}, u_2^{\lambda_2}) \in \mathbf{V}$ such that

$$
\begin{cases}
A_1 u_1^{\lambda_1} = 0 & \text{in } \Omega_1\\
u_1^{\lambda_1} = 0 & \text{on } (\partial\Omega_1 \setminus \Gamma)^{\text{in}}\\
u_1^{\lambda_1} = \lambda_1 & \text{on } \Gamma^{\text{in}},
\end{cases}
\qquad
\begin{cases}
A_2 u_2^{\lambda_2} = 0 & \text{in } \Omega_2\\
u_2^{\lambda_2} = 0 & \text{on } \partial\Omega_2 \setminus \Gamma\\
u_2^{\lambda_2} = \lambda_2 & \text{on } \Gamma.
\end{cases}
\tag{103}
$$

Moreover we define the fluxes on the interface Γ associated to the solutions $u_1^{\lambda_1}$ and $u_2^{\lambda_2}$ as

$$
\varphi_1^{\lambda_1} = -\mathbf{b}\cdot\mathbf{n}_\Gamma u_1^{\lambda_1}, \quad \varphi_2^{\lambda_2} = -\nu\frac{\partial u_2^{\lambda_2}}{\partial \mathbf{n}_\Gamma} + \mathbf{b}\cdot\mathbf{n}_\Gamma u_2^{\lambda_2},
\tag{104}
$$

while those associated to the solutions w_1^f and w_2^f are

$$\chi_1 = -\mathbf{b} \cdot \mathbf{n}_\Gamma w_1^f, \quad \chi_2 = -\nu \frac{\partial w_2^f}{\partial \mathbf{n}_\Gamma} + \mathbf{b} \cdot \mathbf{n}_\Gamma w_2^f. \tag{105}$$

The cost functional J can be split as

$$J(\lambda_1, \lambda_2) = J^0(\lambda_1, \lambda_2) + \mathscr{A}(\lambda_1, \lambda_2), \tag{106}$$

where

$$J^0(\lambda_1, \lambda_2) = \frac{1}{2} \|\lambda_1 - \lambda_2\|^2_{L^2_{\mathbf{b}}(\Gamma^{\mathrm{in}})} + \frac{1}{2} \left\| \varphi_1^{\lambda_1} + \varphi_2^{\lambda_2} \right\|^2_{H^{-1/2}(\Gamma)},$$

while $\mathscr{A}$ is an affine functional which reads

$$\mathscr{A}(\lambda_1, \lambda_2) = \frac{1}{2} \|\chi_1 + \chi_2\|^2_{H^{-1/2}(\Gamma)} + \left(\chi_1 + \chi_2, \varphi_1^{\lambda_1} + \varphi_2^{\lambda_2}\right)_{H^{-1/2}(\Gamma)}.$$

If all data are smooth enough, the existence of $\boldsymbol{\lambda} = (\lambda_1, \lambda_2)$ achieving $\inf J(\lambda_1, \lambda_2)$ in a possibly very large abstract space $\boldsymbol{\Lambda}$, follows from the property of $(J^0(\lambda_1, \lambda_2))^{1/2}$ to be a norm (see [26, Sect. 5]).

3.1.1 The Optimality System

By following standard arguments of optimal control theory for elliptic problems (see [39]), we derive now the *optimality system* corresponding to the minimization problem (97).
Let us write the minimization problem (97) in a variational setting, i.e., we look for the solution $\boldsymbol{\lambda} = (\lambda_1, \lambda_2) \in \boldsymbol{\Lambda}$ such that

$$\langle \nabla J(\boldsymbol{\lambda}), \boldsymbol{\mu} \rangle = 0 \quad \forall \boldsymbol{\mu} \in \boldsymbol{\Lambda}. \tag{107}$$

The partial derivative of J are

$$\begin{aligned} \left\langle \frac{\partial J}{\partial \lambda_1}, \mu_1 \right\rangle &= (\lambda_1 - \lambda_2, \mu_1)_{L^2_{\mathbf{b}}(\Gamma^{\mathrm{in}})} + \left(\varphi_1(\lambda_1) + \varphi_1(\lambda_2), \varphi_1^{\mu_1}\right)_{H^{-1/2}(\Gamma)} \quad \forall \mu_1 \in \Lambda_1, \\ \left\langle \frac{\partial J}{\partial \lambda_2}, \mu_2 \right\rangle &= -(\lambda_1 - \lambda_2, \mu_2)_{L^2_{\mathbf{b}}(\Gamma^{\mathrm{in}})} + \left(\varphi_1(\lambda_1) + \varphi_1(\lambda_2), \varphi_2^{\mu_2}\right)_{H^{-1/2}(\Gamma)} \quad \forall \mu_2 \in \Lambda_2, \end{aligned} \tag{108}$$

where, for any $(\mu_1, \mu_2) \in \boldsymbol{\Lambda}$, $\varphi_1^{\mu_1}$ and $\varphi_2^{\mu_2}$ follow the definition of the fluxes as in (104), while $u_1^{\mu_1}$, $u_2^{\mu_2}$ are defined as in (103).

From the definition (100), for $i = 1, 2$ we obtain

$$\left(\varphi_1(\lambda_1) + \varphi_1(\lambda_2), \varphi_i^{\mu_i}\right)_{H^{-1/2}(\Gamma)} = \int_\Gamma (-\Delta_\Gamma)^{-1/2} (\varphi_1(\lambda_1) + \varphi_1(\lambda_2))\, \varphi_i^{\mu_i}\, \mathrm{d}\Gamma \tag{109}$$

and, in particular for the flux $\varphi_1^{\mu_1}$, it holds

$$\begin{aligned}&\int_{\Gamma}(-\Delta_{\Gamma})^{-1/2}(\varphi_1(\lambda_1)+\varphi_2(\lambda_2))\,\varphi_1^{\mu_1}\mathrm{d}\Gamma\\&=\int_{\Gamma^{\mathrm{in}}}(-\Delta_{\Gamma})^{-1/2}(\varphi_1(\lambda_1)+\varphi_2(\lambda_2))\,(-\mathbf{b}\cdot\mathbf{n}_{\Gamma})\mu_1\mathrm{d}\Gamma\\&\quad+\int_{\Gamma^{\mathrm{out}}}(-\Delta_{\Gamma})^{-1/2}(\varphi_1(\lambda_1)+\varphi_1(\lambda_2))\,(-\mathbf{b}\cdot\mathbf{n}_{\Gamma})u_1^{\mu_1}\mathrm{d}\Gamma.\end{aligned}\tag{110}$$

By defining the adjoint problems

$$\begin{cases}A_1^*p_1\equiv-\mathbf{b}\cdot\nabla p_1+b_0p_1=0 & \text{in } \Omega_1\\ p_1=0 & \text{on } (\partial\Omega_1\setminus\Gamma)^{\mathrm{out}}\\ p_1=(-\Delta_{\Gamma})^{-1/2}(\varphi_1(\lambda_1)+\varphi_2(\lambda_2)) & \text{on } \Gamma^{\mathrm{out}}\end{cases}\tag{111}$$

and

$$\begin{cases}A_2^*p_2\equiv-\mathrm{div}(\nu\nabla p_2)-\mathbf{b}\cdot\nabla p_2+b_0p_2=0 & \text{in } \Omega_2\\ p_2=0 & \text{on } \partial\Omega_2\setminus\Gamma\\ p_2=(-\Delta_{\Gamma})^{-1/2}(\varphi_1(\lambda_1)+\varphi_2(\lambda_2)) & \text{on } \Gamma\end{cases}\tag{112}$$

and, by making use of Green's formula, we have

$$\begin{aligned}\int_{\Gamma^{\mathrm{out}}}(-\Delta_{\Gamma})^{-1/2}(\varphi_1(\lambda_1)+\varphi_1(\lambda_2))\,(-\mathbf{b}\cdot\mathbf{n}_{\Gamma})u_1^{\mu_1}\mathrm{d}\Gamma&=\int_{\Gamma^{\mathrm{out}}}p_1\,(-\mathbf{b}\cdot\mathbf{n}_{\Gamma})u_1^{\mu_1}\mathrm{d}\Gamma\\&=\int_{\Gamma^{\mathrm{in}}}(\mathbf{b}\cdot\mathbf{n}_{\Gamma})p_1\mu_1\mathrm{d}\Gamma\end{aligned}$$

while

$$\begin{aligned}\int_{\Gamma}(-\Delta_{\Gamma})^{-1/2}(\varphi_1(\lambda_1)+\varphi_1(\lambda_2))\,\varphi_2^{\mu_2}\mathrm{d}\Gamma&=\int_{\Gamma}p_2\left(-\nu\frac{\partial u_2^{\mu_2}}{\partial\mathbf{n}_{\Gamma}}+\mathbf{b}\cdot\mathbf{n}_{\Gamma}u_2^{\mu_2}\right)\mathrm{d}\Gamma\\&=-\int_{\Gamma}\nu\frac{\partial p_2}{\partial\mathbf{n}_{\Gamma}}\mu_2,\end{aligned}$$

whence

$$\begin{aligned}\left\langle\frac{\partial J}{\partial\lambda_1},\mu_1\right\rangle&=\int_{\Gamma^{\mathrm{in}}}(-\mathbf{b}\cdot\mathbf{n}_{\Gamma})\,[(\lambda_1-\lambda_2)+(p_2-p_1)]\,\mu_1\mathrm{d}\Gamma,\\\left\langle\frac{\partial J}{\partial\lambda_2},\mu_2\right\rangle&=\int_{\Gamma^{\mathrm{in}}}(\mathbf{b}\cdot\mathbf{n}_{\Gamma})(\lambda_1-\lambda_2)\mu_2\mathrm{d}\Gamma-\int_{\Gamma}\nu\frac{\partial p_2}{\partial\mathbf{n}_{\Gamma}}\mu_2\mathrm{d}\Gamma\end{aligned}\tag{113}$$

for any $\mu_1\in\Lambda_1$ and $\mu_2\in\Lambda_2$. In conclusion, the solution of the minimization problem (97) satisfies the following *optimality system* (in distributional sense):

$$
(OS) \quad \begin{cases} \text{- state equations (95) and (96);} \\ \text{- adjoint equations (111) and (112);} \\ \text{- Euler equations:} \\ (\lambda_1 - \lambda_2) + p_2 - p_1 = 0 & \text{on } \Gamma^{\text{in}} \\ \mathbf{b} \cdot \mathbf{n}_\Gamma (\lambda_1 - \lambda_2) - \nu \dfrac{\partial p_2}{\partial \mathbf{n}_\Gamma} = 0 & \text{on } \Gamma^{\text{in}} \\ -\nu \dfrac{\partial p_2}{\partial \mathbf{n}_\Gamma} = 0 & \text{on } \Gamma^{\text{out}}. \end{cases}
$$

3.1.2 Computation of the Laplace–Beltrami Operator

The computation of the discrete counterpart of $(-\Delta_\Gamma)^{-1/2}(\varphi_1(\lambda_1) + \varphi_2(\lambda_2))$ when $\Omega \subset \mathbb{R}^2$ can be made as follows.

Given a differentiable function u in an open neighbourhood of Γ, the tangential gradient of u is defined by (see, e.g., [11])

$$
\nabla_\Gamma u(\mathbf{x}) = \nabla u(\mathbf{x}) - (\nabla u(\mathbf{x}) \cdot \mathbf{n}_\Gamma(\mathbf{x}))\mathbf{n}_\Gamma(\mathbf{x}), \qquad \forall \mathbf{x} \in \Gamma, \tag{114}
$$

where ∇ denotes the usual gradient in $\mathbb{R}^2$. The *Laplace–Beltrami operator* can be defined through the weak relation:

$$
\int_\Gamma -\Delta_\Gamma u \, w \mathrm{d}\Gamma = \int_\Gamma \nabla_\Gamma u \cdot \nabla_\Gamma w \mathrm{d}\Gamma, \tag{115}
$$

for any function w differentiable in an open neighbourhood of Γ vanishing at the end-points of Γ. In particular, if Γ is a segment parallel to the y-axis, it reduces to

$$
\int_\Gamma -\Delta_\Gamma u \, w \mathrm{d}\Gamma = \int_\Gamma \frac{\partial u}{\partial y} \frac{\partial w}{\partial y} \mathrm{d}\Gamma. \tag{116}
$$

In a finite dimensional context, if $A_{\Gamma,h}$ denotes the symmetric positive definite matrix associated to the discretization of (116), we approximate $(-\Delta_\Gamma)^{1/2}$ by the square root of $A_{\Gamma,h}$, that is the s.p.d. matrix $A_{\Gamma,h}^{1/2}$ defined by

$$
A_{\Gamma,h}^{1/2} = P \Lambda^{1/2} P^T, \tag{117}
$$

where Λ and P are the eigenvalues and eigenvectors matrices, respectively, of $A_{\Gamma,h}$.

Alternatively, the fractional Laplace–Beltrami operator $(-\Delta_\Gamma)^{-1/2}$ can be defined through a Neumann to Dirichlet map defined from $H^{-1/2}(\Gamma)$ to $H^{1/2}(\Gamma)$. Precisely, for any $\varphi \in H^{-1/2}(\Gamma)$ we solve the problem

$$\begin{cases} -\Delta u + u = 0 & \text{in } \Omega_1 \\ \dfrac{\partial u}{\partial \mathbf{n}} = 0 & \text{on } \partial\Omega_1 \setminus \Gamma \\ \dfrac{\partial u}{\partial \mathbf{n}_\Gamma} = \varphi & \text{on } \Gamma \end{cases} \tag{118}$$

and we set $(-\Delta_\Gamma)^{-1/2}\varphi = u|_\Gamma$. The differential problem (118) may be solved in Ω_2 instead of Ω_1.

3.1.3 Recovering the Interface Conditions

In order to recover the interface conditions we are imposing on the interface Γ, we eliminate the adjoint state variables p_1 and p_2 from the optimality system *(OS)*.

Let us set $\widehat{\mathbf{b}} = -\mathbf{b}$, $\overline{b_0} = b_0 - \text{div}\widehat{\mathbf{b}}$ and

$$\Gamma_b^{\text{in}} = \Gamma^{\text{out}}, \quad \Gamma_b^{\text{out}} = \Gamma^{\text{in}}, \quad (\partial\Omega_1 \setminus \Gamma)_b^{\text{in}} = (\partial\Omega_1 \setminus \Gamma)^{\text{out}}.$$

Thanks to (111), (112) and Euler equations in *(OS)*, the functions p_1 and p_2 satisfy the following coupled problem in Ω:

$$\begin{cases} \text{div}(\widehat{\mathbf{b}} p_1) + \overline{b_0} p_1 = 0 & \text{in } \Omega_1 \\ -\text{div}(\nu \nabla p_2) + \text{div}(\widehat{\mathbf{b}} p_2) + \overline{b_0} p_2 = 0 & \text{in } \Omega_2 \\ p_1 = 0 & \text{on } (\partial\Omega_1 \setminus \Gamma)_b^{\text{in}} \\ p_2 = 0 & \text{on } \partial\Omega_2 \setminus \Gamma \\ -\nu \dfrac{\partial p_2}{\partial \mathbf{n}_\Gamma} + (\widehat{\mathbf{b}} \cdot \mathbf{n}_\Gamma) p_2 = (\widehat{\mathbf{b}} \cdot \mathbf{n}_\Gamma) p_1 & \text{on } \Gamma_b^{\text{out}} \\ p_1 = p_2 & \text{on } \Gamma_b^{\text{in}} \\ \nu \dfrac{\partial p_2}{\partial \mathbf{n}_\Gamma} = 0 & \text{on } \Gamma_b^{\text{in}}. \end{cases} \tag{119}$$

By noting that $\overline{b_0} + \frac{1}{2}\text{div}\widehat{\mathbf{b}} = b_0 + \frac{1}{2}\text{div}\mathbf{b} \geq \sigma_0 > 0$ (see (31)) and by applying Theorem 1, problem (119) admits the unique solution $p_1 = 0$ in Ω_1, $p_2 = 0$ in Ω_2. Therefore, $(112)_3$ implies $\varphi_1(\lambda_1) + \varphi_2(\lambda_2) = 0$ on Γ, while the first Euler equation in (OS) implies that $\lambda_1 - \lambda_2 = 0$ on Γ^{in}, i.e. the following conditions hold on the interface:

$$\begin{aligned} &\varphi_1(\lambda_1) + \varphi_2(\lambda_2) = 0 && \text{on } \Gamma \\ &\lambda_1 = \lambda_2 && \text{on } \Gamma^{\text{in}}. \end{aligned} \tag{120}$$

In conclusion the following result holds:

Theorem 4. *If $\boldsymbol{\lambda}$ is the solution of the minimization problem (97) with J defined in (98), then the state solutions u_1 and u_2 of (95) and (96) satisfy the interface conditions (120). Moreover the pair $(u_1(\lambda_1), u_2(\lambda_2))$, obtained by the virtual control approach coincides with the solution of the heterogeneous problem (40).*

Thanks to the interface condition $(120)_2$, the virtual control problem may be reformulated in terms of a unique control function λ defined on Γ and coinciding with λ_2. The control λ_1, previously introduced, will coincide now with the restriction of λ to Γ^{in}.

By this reduction, the virtual control problem (97) becomes:
look for the solution of the minimization problem

$$\inf_{\lambda \in \Lambda_2} J_1(\lambda) \qquad \text{with} \qquad J_1(\lambda) = \frac{1}{2}\|\varphi_1(\lambda) + \varphi_2(\lambda)\|^2_{H^{-1/2}(\Gamma)}, \tag{121}$$

with

$$\varphi_1(\lambda) = -\mathbf{b}\cdot\mathbf{n}_\Gamma u_1(\lambda), \quad \varphi_2(\lambda) = -\nu\frac{\partial u_2(\lambda)}{\partial \mathbf{n}_\Gamma} + \mathbf{b}\cdot\mathbf{n}_\Gamma u_2(\lambda) \tag{122}$$

and $u_1 = u_1(\lambda)$, $u_2 = u_2(\lambda)$ solutions of (95) and (96) with $\lambda_2 = \lambda$, $\lambda_1 = \lambda|_{\Gamma^{\text{in}}}$. By working as done for the two-controls formulation, the derivative of the cost functional J_1 reads

$$\begin{aligned}\langle J_1'(\lambda), \mu\rangle = &\int_{\Gamma^{\text{in}}}(-\mathbf{b}\cdot\mathbf{n}_\Gamma)(p_2 - p_1)\mu \, d\Gamma \\ &- \int_\Gamma \nu\frac{\partial p_2}{\partial \mathbf{n}_\Gamma}\mu \, d\Gamma\end{aligned} \tag{123}$$

for any $\mu \in \Lambda_2$.

The corresponding optimality system $(OS1)$ reads

$$(OS1)\begin{cases} \text{- state equations (95) and (96) with } \lambda_2 = \lambda,\ \lambda_1 = \lambda|_{\Gamma^{\text{in}}} \\ \text{- adjoint equations (111) and (112) with } \varphi_i(\lambda) \text{ instead of} \\ \quad \varphi_i(\lambda_i), \text{ for } i = 1,2; \\ \text{- Euler equations:} \\ (\mathbf{b}\cdot\mathbf{n}_\Gamma)(p_2 - p_1) + \nu\dfrac{\partial p_2}{\partial \mathbf{n}_\Gamma} = 0 \qquad \text{on } \Gamma^{\text{in}} \\ \nu\dfrac{\partial p_2}{\partial \mathbf{n}_\Gamma} = 0 \qquad \text{on } \Gamma^{\text{out}}. \end{cases}$$

Remark 2. Another cost functional proposed in [26] is

$$\tilde{J}(\lambda_1, \lambda_2) = \frac{1}{2}\|u_1(\lambda_1) - u_2(\lambda_2)\|^2_{L^2_{\mathbf{b}}(\Gamma)} + \frac{1}{2}\|\varphi_1(\lambda_1) + \varphi_2(\lambda_2)\|^2_{H^{-1/2}(\Gamma)}. \tag{124}$$

In this case the observation is performed on the whole interface for both fluxes and velocities. The minimization problem (97), in which the functional J is replaced by $\tilde{J}$, admits a unique solution too (see [26]), however it is not guaranteed that $\inf \tilde{J}(\lambda_1, \lambda_2) = 0$, so that no interface conditions are explicitely associated to this minimization problem.

Remark 3. We finally remark that the cost functional to be minimized is set up starting from known interface conditions, it is problem dependent and it requires a priori knowledge of the coupled problem. When the latter are not available, it is more suitable to consider the virtual control approach with overlap, that we will introduce in Sect. 3.3.

3.1.4 How to Solve the Optimality System

A first intuitive way to solve the optimality system $(OS1)$ consists in invoking a Krylov method to seek the solution λ of the Euler equation of $(OS1)$. Let us write the Euler equation, in distributional sense, as

$$J_1'(\lambda) = 0. \tag{125}$$

When we solve it by a Krylov method, like either GMRES or Bi-CGStab, we have to evaluate the action of the functional J_1' on the iterate $\lambda^{(k)}$ at each iteration $k \geq 0$ and this means to perform the steps summarized in the following algorithm.

Algorithm 1

1. solve the primal problems (95) and (96) with $\lambda^{(k)}$ instead of λ_i, for $i = 1, 2$;
2. compute the fluxes $\varphi_1(\lambda^{(k)})$, $\varphi_2(\lambda^{(k)})$ and the function $s^{(k)} = (-\Delta_\Gamma)^{-1/2}(\varphi_1(\lambda^{(k)}) + \varphi_2(\lambda^{(k)}))$
3. solve the dual problems (111), (112) with $s^{(k)}$ instead of $(-\Delta_\Gamma)^{-1/2}(\varphi_1(\lambda_1) + \varphi_2(\lambda_2))$
4. compute $J_1'(\lambda^{(k)})$ by (123), which reads (in distributional sense):

$$J_1'(\lambda) = \begin{cases} (-\mathbf{b} \cdot \mathbf{n}_\Gamma)(p_2 - p_1) - \nu \dfrac{\partial p_2}{\partial \mathbf{n}_\Gamma} & \text{on } \Gamma^{\text{in}} \\ -\nu \dfrac{\partial p_2}{\partial \mathbf{n}_\Gamma} & \text{on } \Gamma^{\text{out}}. \end{cases} \tag{126}$$

The solution of the Euler equation $J_1'(\lambda) = 0$, by a Krylov method with the use of *Algorithm* 1, is an alternative to the solution of the Steklov–Poincaré equation (54).

By properly replacing the definition of both state and adjoint equations and by correctly writing the derivatives of the cost functional, *Algorithm* 1 can be adapted to solve the optimality system associated to the minimization of $\tilde{J}$.

Solving $J'(\lambda) = 0$ is equivalent to solve the Schur complement with respect to the control variable λ derived from the optimal system $(OS1)$.

Table 8 Test case #1. Comparison between the cost functionals (121) and (124)

ν	$J_1(\lambda)$			$\tilde{J}(\lambda_1, \lambda_2)$		
	$\Vert u_1 - u_2 \Vert_{L^\infty(\Gamma^{out})}$	$\inf J_1$	#it	$\Vert u_1 - u_2 \Vert_{L^\infty(\Gamma^{out})}$	$\inf \tilde{J}$	#it
0.1	2.330e−1	1.242e−12	18	1.367e−1	5.239e−6	44
0.05	1.221e−1	3.137e−12	27	5.275e−2	3.286e−7	61
0.01	1.346e−2	6.989e−11	60	7.146e−4	6.234e−10	134
0.005	1.075e−2	2.294e−11	82	5.049e−4	8.749e−10	177

#it stands for the number of Bi-CGStab-VC iterations.

3.1.5 Numerical Results for Decompositions Without Overlap

Let us consider the Test case #1 introduced in Sect. 2.4. First of all we compare the numerical solutions obtained by the virtual control approach by minimizing either the cost functional $J_1(\lambda)$ defined in (121) (or, equivalently $J(\lambda_1, \lambda_2)$ defined in (98)) and $\tilde{J}(\lambda_1, \lambda_2)$ defined in (124). We have solved both the optimality system $(OS1)$ and that associated to the minimization of $\tilde{J}$ by Bi-CGStab iterations and by following the steps summarized in *Algorithm* 1 (see Sect. 3.1.4). We will name Bi-CGStab-VC this approach.

In Table 8 we report for both the functionals (121) and (124):

- The L^∞-norm on Γ^{out} of the jump of the solution, i.e., $[u]_{\Gamma^{out}} = \Vert u_1 - u_2 \Vert_{L^\infty(\Gamma^{out})}$
- The infimum of the minimized cost functional
- The number of Bi-CGStab-VC iterations to converge up a tolerance $\varepsilon = 10^{-8}$ versus the viscosity ν

A non-uniform spectral element discretization has been considered to solve the boundary-value problems in both Ω_1 and Ω_2. The domain Ω_1 (Ω_2, resp.) has been split in 3×6 (4×6, resp.) quadrilaterals with the same polynomial degree $N = 16$ in each spatial direction and in each element.

First of all we note that not only the solution (u_1, u_2), obtained by minimizing the cost functional J_1, features a jump on Γ^{out} (in fact we know that it is discontinuous on Γ^{out}), but also the solution obtained by minimizing the cost functional $\tilde{J}$ is discontinuous on Γ^{out}. Moreover, as pointed out in Remark 2, we observe that the value $\inf \tilde{J}$ is not null for any considered viscosity, however $\inf \tilde{J} \to 0$ as $\nu \to 0$. We have observed that the reached value $\inf J_1$ is independent of the viscosity and it is very close to the machine accuracy.

About the number of Bi-CGStab iterations needed to solve the variational equation $J_1'(\lambda) = 0$, we observe that the convergence rate linearly depends on the reciprocal of the viscosity, that the minimimiziation of $\tilde{J}$ requires twice the iterations to minimize J_1 and that the computational cost of each Bi-CGStab-VC iteration is the same for both the minimization problems. Then we conclude that the minimization of the cost functional $\tilde{J}$ costs twice that of J_1.

In Table 9 we report the number of BiCGStab-VC iterations needed to solve the optimality system (OS1) up to a tolerance $\varepsilon = 10^{-6}$, versus the polynomial degree N, for two different values of the viscosity: $\nu = 0.01$ and 0.005. It emerges that the convergence rate of Bi-CGStab-VC is independent of the polynomial degree.

Table 9 Number of Bi-CGStab-VC iterations for the minimization of $J_1(\lambda)$ on the test case #1

N	$\nu = 0.01$		N	$\nu = 0.005$	
	LB (1)	LB (SP^{-1})		LB (1)	LB (SP^{-1})
4	19	18	4	23	24
6	16	17	6	26	23
8	15	16	8	27	26
10	17	18	10	33	27
12	18	18	12	27	27
14	19	18	14	26	28
16	21	20	16	26	28

The acronym LB(1) stands for the implementation based on the computation of the square root of the discrete Laplace–Beltrami operator, while LB(SP^{-1}) stands for the implementation based on the inversion of the Steklov–Poincaré (or Dirichlet to Neumann) operator (see Remark 3.1.2).

3.2 *Domain Decomposition with Overlap*

Let us consider now a decomposition of Ω with overlap. Precisely, we introduce two subdomains Ω_1 and Ω_2, such that

$$\overline{\Omega} = \overline{\Omega}_1 \cup \overline{\Omega}_2, \quad \Omega_{12} = \Omega_1 \cap \Omega_2 \neq \emptyset, \quad \Gamma_i = \partial\Omega_i \setminus (\partial\Omega_i \cap \partial\Omega), \quad i = 1,2 \tag{127}$$

and we denote by $\mathbf{n}_{\Gamma_i}$ (for $i = 1,2$) the outward normal vector on Γ_i with respect to Ω_i.

In view of the considerations given at the beginning of this section, our aim is to investigate domain decomposition approaches alternative to the sharp-interface one which do not require a priori knowledge of interface conditions.

3.2.1 An Engineering Practice on Overlapping Subdomains

The simpler and, very likely, most largely used approach consists in extending the classical Schwarz method [40, 54] to the heterogeneous coupling, then iterating on the Dirichlet data on the interfaces Γ_1 and Γ_2.

For example, if A_1 and A_2 are the differential operators defined in (32) and (33), respectively, the *additive* (or sequential) version of the Schwarz method reads: given $u_1^{(0)}$ and $u_2^{(0)}$, for $k \geq 0$ do

$$\begin{cases} A_1 u_1^{(k+1)} = f & \text{in } \Omega_1 \\ u_1^{(k+1)} = g_1 & \text{on } (\partial\Omega_1 \setminus \Gamma_1)^{\text{in}} \\ u_1^{(k+1)} = u_2^{(k)} & \text{on } \Gamma_1^{\text{in}}, \end{cases} \qquad \begin{cases} A_2 u_2^{(k+1)} = f & \text{in } \Omega_2 \\ u_2^{(k+1)} = g_2 & \text{on } \partial\Omega_2 \setminus \Gamma_2 \\ u_2^{(k+1)} = u_1^{(k+1)} & \text{on } \Gamma_2. \end{cases}$$

If we replace the interface condition $u_2^{(k+1)} = u_1^{(k+1)}$ with $u_2^{(k+1)} = u_1^{(k)}$ on Γ_2, we obtain the so-called *multiplicative* (or parallel) version of the Schwarz method.

The convergence of the Schwarz method applied to the global advection–diffusion equation has been largely studied, see, e.g. [6, 7, 33, 47].

In [47], the analysis of the Schwarz alternating method is made for homogeneous singular perturbation problems in which the advection dominates. Precisely, the author proves that if the subdomains can be chosen to *follow the flow*, i.e., if the boundary interface of one of the subdomains corresponds to an outflow boundary for the streamlines of the flow, then the Schwarz iterates converge in the maximum norm with an error reduction factor per iteration that exponentially decays with increasing overlap or decreasing diffusion. On the contrary, if the flow is recirculating and the subdomains are not suitably chosen, numerical evidence shows that there can be some deterioration in the convergence factor of the Schwarz method. No theoretical results however are available in literature about the convergence of Schwarz method for heterogeneous decompositions.

3.2.2 Schwarz Method with Dirichlet/Robin Interface Conditions

In [34] a variant of the classical Schwarz method is proposed, always for homogeneous advection–diffusion problems, and it consists in replacing Dirichlet with Robin conditions only on one interface of the decomposition with the aim of accelerating the convergence.

Let us consider again the overlapping decomposition shown in Fig. 14. In [34], Houzeaux and Codina consider the homogenous problem (1) and propose to solve it by a two-domain approach as follows: find the pair (u_1, u_2) such that

$$\begin{cases} A_2 u_1 = f & \text{in } \Omega_1 \\ u_1 = g & \text{on } \partial\Omega_1 \setminus \Gamma_1 \\ u_1 = u_2 & \text{on } \Gamma_1 \\ \\ A_2 u_2 = f & \text{in } \Omega_2 \\ u_2 = g & \text{on } \partial\Omega_2 \setminus \Gamma_2 \\ \nu \dfrac{\partial u_1}{\partial n_{\Gamma_2}} - \dfrac{1}{2}(\mathbf{b}\cdot\mathbf{n}_{\Gamma_2})u_1 = \nu \dfrac{\partial u_2}{\partial n_{\Gamma_2}} - \dfrac{1}{2}(\mathbf{b}\cdot\mathbf{n}_{\Gamma_2})u_2 & \text{on } \Gamma_2. \end{cases} \tag{128}$$

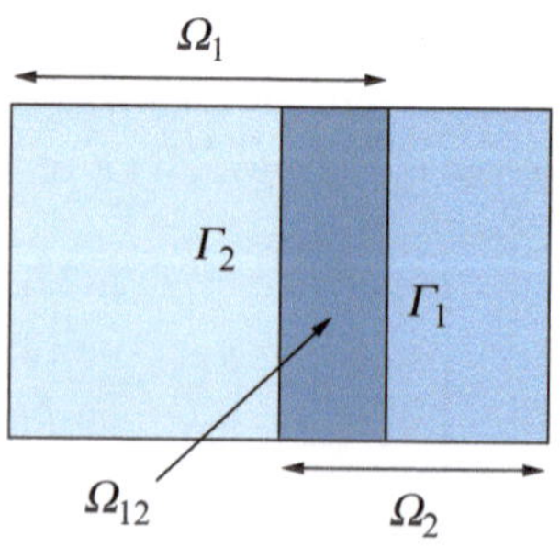

Fig. 14 The computational domain split in two overlapping subdomains

By introducing Steklov–Poincaré operators on the interfaces, they prove that problem (128) admits a unique solution (u_1, u_2) such that $u_1 = u_2$ on Ω_{12}. Moreover, the function

$$u = \begin{cases} u_1 & \text{in } \overline{(\Omega_1 \setminus \Omega_{12})} \\ u_2 & \text{in } \Omega_2 \end{cases}$$

coincides with the solution of (1).

However, in [34] an overlapping Dirichlet/Robin method is proposed for the solution of the two advection–diffusion problems, with the purpose of inheriting the robustness properties of the classical Schwarz method, yet allowing the limit case of zero (or extremely small) overlapping, for which Dirichlet/Dirichlet method fails. Note that the interface condition $(128)_6$ arises from writing the convective term in skew-symmetric form.

Problem (128) can be solved iterating by subdomains. The resulting method is called *Dirichlet–Robin method* and it reads: given $u_1^{(0)}$ and $u_2^{(0)}$, for $k \geq 0$ do

$$\begin{cases} A_2 u_1^{(k+1)} = f & \text{in } \Omega_1 \\ u_1^{(k+1)} = g & \text{on } \partial\Omega_1 \setminus \Gamma_1 \\ u_1^{(k+1)} = \vartheta u_2^{(k)} + (1-\vartheta) u_1^{(k)} & \text{on } \Gamma_1 \\ \\ A_2 u_2^{(k+1)} = f & \text{in } \Omega_2 \\ u_2^{(k+1)} = g & \text{on } \partial\Omega_2 \setminus \Gamma_2 \\ \nu \dfrac{\partial u_2^{(k+1)}}{\partial n_{\Gamma_2}} - \dfrac{1}{2}(\mathbf{b} \cdot \mathbf{n}_{\Gamma_2}) u_2^{(k+1)} = \nu \dfrac{\partial u_1^{(k+1)}}{\partial n_{\Gamma_2}} - \dfrac{1}{2}(\mathbf{b} \cdot \mathbf{n}_{\Gamma_2}) u_1^{(k+1)} & \text{on } \Gamma_2, \end{cases} \tag{129}$$

where $\vartheta > 0$ is a suitable relaxation parameter. As alternative to the relaxation of Dirichlet data $(129)_3$, the authors propose to relax the Robin data $(129)_6$. Under suitable choices of the relaxation parameter the Dirichlet–Robin algorithm (129) converges to the solution of the advection–diffusion problem (128).

When the heterogeneous coupling is considered, the Robin interface condition $(128)_6$ could be replaced by the following one:

$$-\frac{1}{2}(\mathbf{b} \cdot \mathbf{n}_{\Gamma_2}) u_1 = \nu \frac{\partial u_2}{\partial n_{\Gamma_2}} - \frac{1}{2}(\mathbf{b} \cdot \mathbf{n}_{\Gamma_2}) u_2 \qquad \text{on } \Gamma_2, \tag{130}$$

so that the iterative Dirichlet–Robin algorithm for the coupled advection/advection–diffusion problem should read:
given $u_1^{(0)}$ and $u_2^{(0)}$, for $k \geq 0$ do

$$
\begin{cases}
A_1 u_1^{(k+1)} = f & \text{in } \Omega_1 \\
u_1^{(k+1)} = g & \text{on } (\partial\Omega_1 \setminus \Gamma_1)^{\text{in}} \\
u_1^{(k+1)} = \vartheta u_2^{(k)} + (1-\vartheta) u_1^{(k)} & \text{on } \Gamma_1^{\text{in}} \\
\\
A_2 u_2^{(k+1)} = f & \text{in } \Omega_2 \\
u_2^{(k+1)} = g & \text{on } \partial\Omega_2 \setminus \Gamma_2 \\
\nu \dfrac{\partial u_2^{(k+1)}}{\partial n_{\Gamma_2}} - \dfrac{1}{2}(\mathbf{b}\cdot\mathbf{n}_{\Gamma_2}) u_2^{(k+1)} = -\dfrac{1}{2}(\mathbf{b}\cdot\mathbf{n}_{\Gamma_2}) u_1^{(k+1)} & \text{on } \Gamma_2.
\end{cases}
\tag{131}
$$

We note that algorithm (131) coincides with the Dirichlet–Neumann algorithm (47) when the overlap reduces to the empty set. We refer to Remark 1 in Sect. 3.3.2 about the classification of Neumann and Robin interface conditions.

3.3 Virtual Control Approach with Overlap for the Advection–Diffusion Equation

Let us consider an overlapping decomposition of Ω as in (127). As done for the non-overlapping situation presented in Sect. 3.1, we introduce the Dirichlet *virtual controls* $\lambda_1 \in L^2_{\mathbf{b}}(\Gamma_1^{\text{in}})$ and $\lambda_2 \in H^{1/2}(\Gamma_2)$ (Fig. 15) and we look for the solution of the following minimization problem:

$$
\inf_{\lambda_1,\lambda_2} \hat{J}(\lambda_1,\lambda_2), \tag{132}
$$

with

$$
\hat{J}(\lambda_1,\lambda_2) = \int_{\Omega_{12}} (u_1(\lambda_1) - u_2(\lambda_2))^2 \mathrm{d}\Omega, \tag{133}
$$

and $u_1 = u_1(\lambda_1)$, $u_2 = u_2(\lambda_2)$ solutions of

$$
\begin{cases}
A_1 u_1 = f & \text{in } \Omega_1 \\
u_1 = g_1 & \text{on } (\partial\Omega_1 \setminus \Gamma_1)^{\text{in}} \\
u_1 = \lambda_1 & \text{on } \Gamma_1^{\text{in}},
\end{cases}
\qquad
\begin{cases}
A_2 u_2 = f & \text{in } \Omega_2 \\
u_2 = g_2 & \text{on } \partial\Omega_2 \setminus \Gamma_2 \\
u_2 = \lambda_2 & \text{on } \Gamma_2.
\end{cases}
\tag{134}
$$

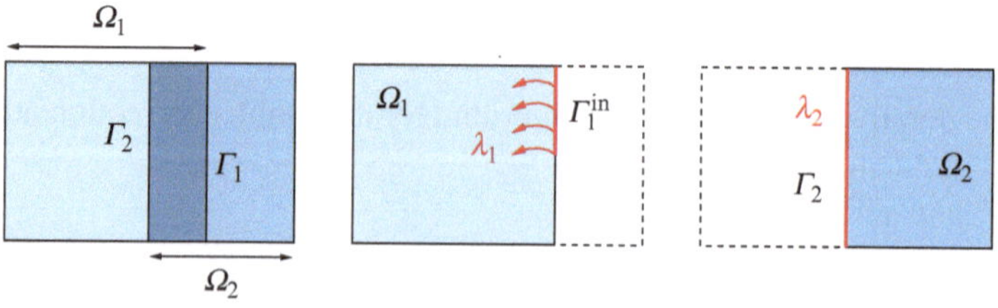

Fig. 15 Virtual control with overlap

The minimization problem (132) has been studied in [1, 26].
Along this section we set

$$\Lambda_1 = L^2_{\mathbf{b}}(\Gamma_1^{\mathrm{in}}),$$
$$\Lambda_2 = \left\{\lambda_2 \in H^{1/2}(\Gamma_2) : \exists \mu \in H^{1/2}(\partial\Omega_2) \text{ s.t. } \lambda_2 = \mu|_{\Gamma_2} \text{ and } g_2 = \mu|_{\partial\Omega_2\setminus\Gamma_2}\right\}, \tag{135}$$

The following result is stated in [26].

Proposition 1. *If the cost functional* $\hat{J}$ *can be written as the sum of a quadratic functional* $\hat{J}^0(\lambda_1, \lambda_2)$ *and an affine functional* $\hat{\mathscr{A}}(\lambda_1, \lambda_2)$ *(as done in Sect. 3.1), and if the seminorm*

$$\|(\lambda_1, \lambda_2)\| = \left(\hat{J}^0(\lambda_1, \lambda_2)\right)^{1/2} \tag{136}$$

is indeed a norm, then problem (132) admits a unique solution in the space obtained by completion of $\Lambda_1 \times \Lambda_2$ *with respect to the norm (136).*

The property of (136) being a norm, depends on problem data, i.e. on the convection field $\mathbf{b}$ and on the domain.

In [1], sufficient conditions which guarantee uniqueness of solution of the minimization problem (132) are furnished.
For simplicity, let us consider the decomposition of Ω in two subdomains, as described in (15) and we refer to [1] for more general situations where either the overlapping set $\Omega_{12} = \Omega_1 \cap \Omega_2$ is not connected or $\partial\Omega_{12} \cap \partial\Omega = \emptyset$. We denote by $\mathbf{n}_{12}$ the outward unit normal to Ω_{12}. The sufficient conditions (alternative one to each other) which guarantee uniqueness of solution for (132) are:

I. $\mathbf{b} \cdot \mathbf{n}_{12} \neq 0$ on $\partial\Omega_{12} \cap \partial\Omega$,

II. $\begin{cases} \mu = b_0 + \mathrm{div}\mathbf{b} \geq 0 \text{ on } \partial\Omega_{12}, \ \mu \not\equiv 0 \text{ on } \partial\Omega_{12}, \\ \text{the direction } \mathbf{b} \text{ at any point of } \partial\Omega_{12} \text{ forms with the outward normal} \\ \text{to } \partial\Omega_{12} \text{ an acute angle,} \end{cases}$

III. $\begin{cases} \mathbf{b} \cdot \mathbf{n}_{12} \neq 0 \text{ on } \partial\Omega_{12}, \ \dfrac{\mu}{b_n} - \dfrac{1}{2}\dfrac{\partial}{\partial\tau}\left(\dfrac{b_\tau}{b_n}\right) > 0 \text{ on } \partial\Omega_{12}, \\ \text{where } \dfrac{\partial}{\partial\tau} \text{ is the derivative along } \partial\Omega_{12}, \text{ while } b_n \text{ and } b_\tau \text{ are the normal} \\ \text{and tangential components, respectively, of } \mathbf{b} \text{ on } \partial\Omega_{12}. \end{cases}$

The previous proposition guarantees, under suitable assumptions, the uniqueness of the virtual controls and then that of the solution u_1 in Ω_1 and u_2 in Ω_2. However in general, $u_1 \neq u_2$ on the overlap Ω_{12}. A natural question is: *how do we recover in* Ω_{12} *a solution of the heterogeneous problem.*

The following result ensures that the difference between u_1 and u_2 in the $L^2(\Omega_{12})$ norm annihilates when the viscosity vanishes (see [26]).

Theorem 5. *If we set*

$$\varphi(\nu) = \inf_{\lambda_1,\lambda_2} \hat{J}(\lambda_1, \lambda_2) \tag{137}$$

and if we let $\nu \to 0$*, all other data being fixed, then*

$$\varphi(\nu) \to 0 \quad \text{as } \nu \to 0. \tag{138}$$

The optimality system associated to the minimization problem (132) can be derived by proceeding as in Sect. 3.1.

For any $\mu_1 \in \Lambda_1$, $\mu_2 \in \Lambda_2$, we introduce the auxiliary problems as follows:

$$\begin{cases} A_1 u_1^{\mu_1} = 0 & \text{in } \Omega_1 \\ u_1^{\mu_1} = 0 & \text{on } (\partial\Omega_1 \setminus \Gamma_1)^{\text{in}} \\ u_1^{\mu_1} = \mu_1 & \text{on } \Gamma_1^{\text{in}}, \end{cases} \qquad \begin{cases} A_2 u_2^{\mu_2} = 0 & \text{in } \Omega_2 \\ u_2^{\mu_2} = 0 & \text{on } \partial\Omega_2 \setminus \Gamma_2 \\ u_2^{\mu_2} = \mu_2 & \text{on } \Gamma_2, \end{cases} \tag{139}$$

and we differentiate the functional $\hat{J}$:

$$\begin{aligned} \left\langle \frac{\partial \hat{J}}{\partial \lambda_1}, \mu_1 \right\rangle &= \left(u_1(\lambda_1) - u_2(\lambda_2), u_1^{\mu_1}\right)_{L^2(\Omega_{12})} \qquad \forall \mu_1 \in \Lambda_1, \\ \left\langle \frac{\partial \hat{J}}{\partial \lambda_2}, \mu_2 \right\rangle &= -\left(u_1(\lambda_1) - u_2(\lambda_2), u_2^{\mu_2}\right)_{L^2(\Omega_{12})} \quad \forall \mu_2 \in \Lambda_2. \end{aligned} \tag{140}$$

We define the adjoint problems:

$$\begin{cases} A_1^* p_1 = \chi_{12}(u_1(\lambda_1) - u_2(\lambda_2)) & \text{in } \Omega_1 \\ p_1 = 0 & \text{on } \partial\Omega_1^{\text{out}} \end{cases} \tag{141}$$

and

$$\begin{cases} A_2^* p_2 = -\chi_{12}(u_1(\lambda_1) - u_2(\lambda_2)) & \text{in } \Omega_2 \\ p_2 = 0 & \text{on } \partial\Omega_2, \end{cases} \tag{142}$$

(where χ_{12} denotes the characteristic function of the overlapping set Ω_{12}) and, by Green's formulas and the boundary conditions set in (139), (141) and (142), the optimality system associated to the minimization problem (132) reads (in distributional sense):

$$(OS2) \qquad \begin{cases} \text{- State equations (134);} \\ \text{- Adjoint equations (141) and (142);} \\ \text{- Euler equations:} \\ (-\mathbf{b}\cdot\mathbf{n}_{\Gamma_1})p_1 = 0 & \text{on } \Gamma_1^{\text{in}} \\ \nu\dfrac{\partial p_2}{\partial \mathbf{n}_{\Gamma_2}} = 0 & \text{on } \Gamma_2. \end{cases}$$

The optimality system $(OS2)$ can be solved as described in Sect. 3.1.4 by a Bi-CGStab method.

3.3.1 Using Three Virtual Controls

In order to force the solutions u_1 and u_2 to coincide in Ω_{12}, a virtual control problem with three controls has been proposed and studied in [1]. Precisely, in addition to the Dirichlet controls λ_1 and λ_2, a distributed control $\lambda_3 \in L^2(\Omega_{12})$ is used as forcing term in the hyperbolic equation in Ω_1.

Let Λ_1 and Λ_2 the spaces defined in (135), then we set

$$\Lambda_3 = L^2(\Omega_{12}). \tag{143}$$

The *three virtual controls problem* is defined as follows. We seek $\boldsymbol{\lambda} = (\lambda_1, \lambda_2, \lambda_3) \in \Lambda_1 \times \Lambda_2 \times \Lambda_3$ solution of the regularized minimization problem

$$\inf_{\lambda_1,\lambda_2,\lambda_3} \hat{\hat{J}}_\alpha(\lambda_1, \lambda_2, \lambda_3), \tag{144}$$

where

$$\begin{aligned}\hat{\hat{J}}_\alpha(\lambda_1, \lambda_2, \lambda_3) = &\frac{1}{2}\int_{\Omega_{12}} (u_1(\lambda_1, \lambda_3) - u_2(\lambda_2))^2 d\Omega \\ &+\frac{\alpha}{2}\left(\|\lambda_1\|^2_{\Lambda_1} + \|\lambda_2\|^2_{\Lambda_2} + \|\omega\lambda_3\|^2_{\Lambda_3}\right),\end{aligned} \tag{145}$$

$u_1 = u_1(\lambda_1, \lambda_3)$ and $u_2 = u_2(\lambda_2)$ are the solutions of the state equations

$$\begin{cases} A_1 u_1 = f + \omega\lambda_3 & \text{in } \Omega_1 \\ u_1 = g & \text{on } (\partial\Omega_1 \setminus \Gamma_1)^{\text{in}} \\ u_1 = \lambda_1 & \text{on } \Gamma_1^{\text{in}} \end{cases} \qquad \begin{cases} A_2 u_2 = f & \text{in } \Omega_2 \\ u_2 = g & \text{on } \partial\Omega_2 \setminus \Gamma_2 \\ u_2 = \lambda_2 & \text{on } \Gamma_2, \end{cases} \tag{146}$$

$\alpha \geq 0$ is a penalization coefficient and, finally, ω is a smooth function in Ω such that

$$0 \leq \omega(\mathbf{x}) \leq 1 \text{ in } \Omega, \quad \omega = 0 \text{ in } \Omega\backslash\Omega_{12}, \quad \omega > 0 \text{ in } \Omega_{12}.$$

The optimality system associated to (145) reads (in variational form)

$$
(OS3) \quad \begin{cases} \text{- State equations (146);} \\ \text{- Adjoint equations (141) and (142);} \\ \text{- Euler–Lagrange equations:} \\ (-\mathbf{b}\cdot\mathbf{n}_{\Gamma_1})(p_1+\alpha\lambda_1)=0 & \text{on } \Gamma_1^{\text{in}} \\ \nu\dfrac{\partial p_2}{\partial \mathbf{n}_{\Gamma_2}}+\alpha\lambda_2=0 & \text{on } \Gamma_2 \\ \alpha\omega\lambda_3+\omega p_1=0 & \text{in } \Omega_{12}. \end{cases}
$$

The following Theorem is proved in [1]:

Theorem 6. *For any $\alpha > 0$, the minimization problem (144) has a unique solution depending on α, say $(\lambda_1,\lambda_2,\lambda_3)=(\lambda_1(\alpha),\lambda_2(\alpha),\lambda_3(\alpha))$, such that*

$$
\|u_1(\lambda_1(\alpha),\lambda_3(\alpha))-u_2(\lambda_2(\alpha))\|_{L^2(\Omega_{12})}\to 0 \quad \text{as } \alpha\to 0. \tag{147}
$$

Moreover, if there exists a solution $(\lambda_1^0,\lambda_2^0,\lambda_3^0)$ of the problem (146) such that the corresponding state functions coincide in Ω_{12}, i.e. $u_1^0(\lambda_1^0,\lambda_3^0)=u_2^0(\lambda_2^0)$ in Ω_{12}, then the solution $(\lambda_1^0,\lambda_2^0,\lambda_3^0)$ is unique and $\lambda_k(\alpha)\to\lambda_k^0$ as $\alpha\to 0$, for $k=1,2,3$.

Remark 4. The third control has been introduced to dump the difference between the hyperbolic and elliptic solutions on the overlap. It is important to highlight that it is added to the right hand side of the hyperbolic equation and not to the right hand side of the elliptic problem. This choice guarantees the uniqueness of solution of the minimization problem (144) when $\alpha = 0$, through the application of the uniqueness continuation theorem.

3.3.2 Numerical Results on Virtual Control Approaches

In this section we present some numerical results obtained by solving the coupled advection/advection–diffusion problem by two- and three-virtual controls approaches. First of all, we consider the one-dimensional problem

$$
\begin{cases} -\nu u''(x)+u'(x)=1 & 0<x<1 \\ u(0)=u(1)=0, \end{cases} \tag{148}
$$

and we set $\Omega_1=(0,0.6)$, $\Omega_2=(0.3,1)$. In Fig. 16 we show the numerical solution obtained with both two-controls (dashed line) and three-controls (solid line), for $\nu=1$, at left and $\nu=10^{-2}$ at right. The regularization parameter in $\hat{J}_\alpha$ is $\alpha=0$. The discretization is performed by spectral elements, precisely, we have decomposed both Ω_1 and Ω_2 in two spectral elements and the common element

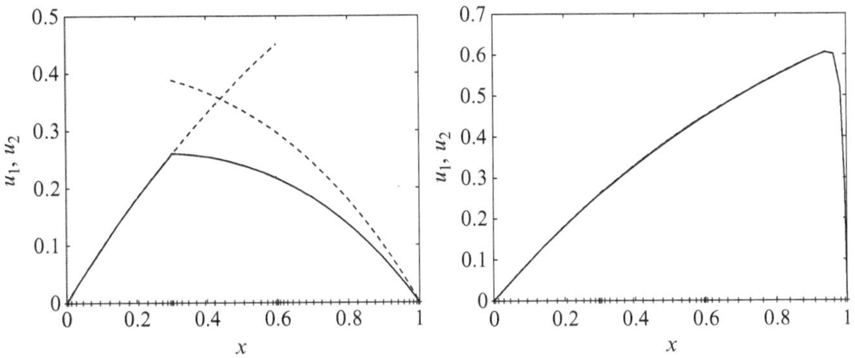

Fig. 16 Numerical solutions of (148) obtained with two controls (*dashed line*) and three controls (*solid line*) for $\nu = 1$ at *left* and for $\nu = 10^{-2}$ at *right*. $\Omega_1 = (0, 0.6)$, $\Omega_2 = (0.3, 1)$

Table 10 Test case #1. The number of Bi-CGStab iterations to solve the optimality systems (OS2) and (OS3) and the infimum of the cost functionals $\hat{J}$ and $\hat{\hat{J}}_0$ versus the viscosity ν

ν	Two-controls		Three-controls	
	#it	$\inf \hat{J}(\lambda_1, \lambda_2)$	#it	$\inf \hat{\hat{J}}_0(\lambda_1, \lambda_2, \lambda_3)$
0.1	18	8.71×10^{-4}	319	2.83×10^{-11}
0.01	15	5.85×10^{-5}	276	1.97×10^{-11}
0.001	18	4.92×10^{-7}	220	5.81×10^{-11}
0.0001	18	9.79×10^{-9}	190	2.45×10^{-11}

discretizes the overlap Ω_{12}. The polynomial degree used is $N = 16$ in each element of both Ω_1 and Ω_2 when $\nu = 1$, while it is $N = 16$ in each element of Ω_1 and $N = 24$ in $\Omega_2 \setminus \Omega_{12}$ when $\nu = 10^{-2}$. As we can see the solution obtained with three-controls matches on the overlap Ω_{12} also with large viscosity $\nu = 1$.

Note that the interface Γ_1 is an outflow boundary for the hyperbolic problem, so that the control λ_1 is not needed. The number of degrees of freedom (i.e. the dimension of the system solved by Bi-CGStab) is one for the two controls approach, while it is of the same order of the number of discretization nodes on the overlap (about N) for the three controls approach.

Let us consider now the 2D problem described in the Test case # 1 and let $\hat{\hat{J}}_0$ denotes the cost functional $\hat{\hat{J}}_\alpha$ with $\alpha = 0$ (i.e. without regularization). In the following table the infimum reached by both the cost functionals $\hat{J}$ and $\hat{\hat{J}}_0$ is shown for different values of the viscosity ν. It is evident that the minimization of the cost functional with three controls provides a better solution with respect to the two virtual controls approach. Nevertheless, the cost of the three virtual controls approach (in terms of BiCG-Stab iterations needed to solve the optimality system) is very large, as shown in Table 10.

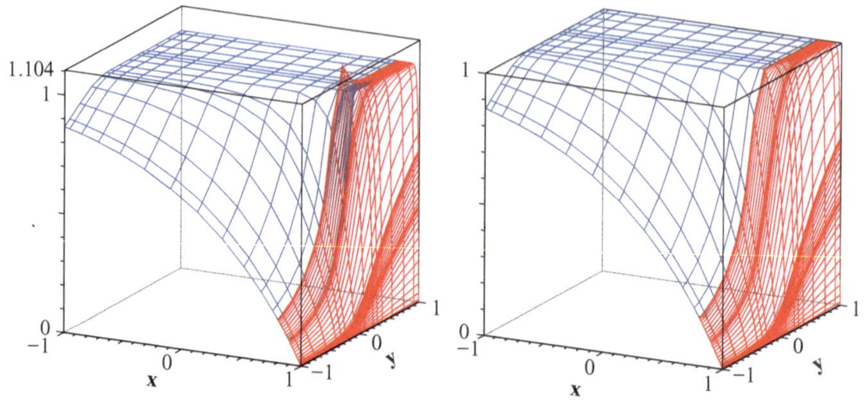

Fig. 17 Test case #1. $\nu = 0.01$. *Left*: the solution obtained by minimizing $\hat{J}(\lambda_1, \lambda_2)$. *Right*: the solution obtained by minimizing $\hat{\hat{J}}_\alpha(\lambda_1, \lambda_2, \lambda_3)$ with $\alpha = 0$

The stopping test for Bi-CGStab iterations is performed on the norm of the relative residual with tolerance $\varepsilon = 10^{-6}$. We observe that the number of iterations is small and is independent of the viscosity in the case of two virtual controls, while it is very large for the three virtual controls approach, even if it decreases when $\nu \to 0$.

In Fig. 17 we can appreciate the difference between the hyperbolic solution u_1 and the elliptic one u_2 inside the overlapping region Ω_{12} for the two-virtual controls approach (left), and the goodness of the solution of the three virtual controls approach (right) when the viscosity is $\nu = 0.01$.

Remark 5. We conclude this Section by highlighting some features of the virtual control approach with overlap.

The analysis carried out on the virtual control approach with overlap represents a formal mathematical justification to engineering practice, that is to the Schwarz method applied to heterogeneous problems.

The virtual control approach with overlap is more "indifferent" with respect to interface conditions (no a priori information are required, contrary to the virtual control approach without overlap (see Remark 3).)

However, some open questions remain about the setting of the cost functional. In particular it is interesting to know if a "best" functional exists, if it is problem dependent or, again, if it depends on the characteristic parameters of the problem itself.

3.4 Virtual Control with Overlap for the Stokes–Darcy Coupling

In this section we apply the virtual control approach with overlap introduced in Sect. 3.3 to the coupled Stokes–Darcy problem that we have considered in Sect. 2.6.

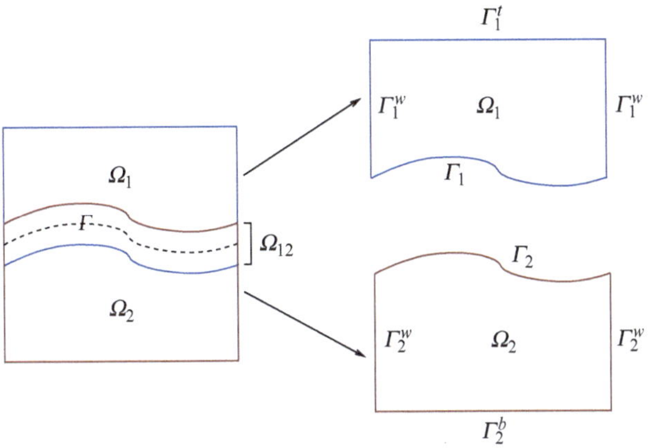

Fig. 18 Schematic representation of the computational domain

Figure 18 shows our computational domain. In the subdomain Ω_1 we consider the following Stokes problem: find $(\mathbf{u}, p) \in [H^1(\Omega_1)]^2 \times L^2(\Omega_1)$ such that

$$\begin{cases} -\nu\Delta\mathbf{u} + \nabla p = \mathbf{f} & \text{in } \Omega_1 \\ \text{div}\,\mathbf{u} = 0 & \text{in } \Omega_1 \\ \nu\nabla\mathbf{u}\cdot\mathbf{n}_1 - p\mathbf{n}_1 = \mathbf{g} & \text{on } \Gamma_1^t \\ \mathbf{u} = \mathbf{u}^* & \text{on } \Gamma_1^w \\ \mathbf{u} = \boldsymbol{\lambda}_1 & \text{on } \Gamma_1, \end{cases} \tag{149}$$

where $\mathbf{f}$, $\mathbf{g}$ and $\mathbf{u}^*$ are suitably chosen enough regular data.

On the other hand, in the subdomain Ω_2, we consider the Darcy problem: find the piezometric head $\varphi \in H^1(\Omega_2)$ such that

$$\begin{cases} -\text{div}\,(K\nabla\varphi) = 0 & \text{in } \Omega_2 \\ K\nabla\varphi\cdot\mathbf{n}_2 = \boldsymbol{\psi}_N & \text{on } \Gamma_2^w \\ \varphi = \psi_D & \text{on } \Gamma_2^b \\ \varphi = \lambda_2 & \text{on } \Gamma_2, \end{cases} \tag{150}$$

where $\boldsymbol{\psi}_N$ and ψ_D are suitable boundary data.

We refer to Fig. 18 for the notation of the boundaries.

$\boldsymbol{\lambda}_1$ and λ_2 are the controls variables which have to be seeked in the following spaces, respectively:

$$\Lambda_1 = \left\{\boldsymbol{\mu} \in [H^{1/2}(\Gamma_1)]^2 : \exists\mathbf{v} \in [H^1(\Omega_1)]^2,\ \mathbf{v} = \boldsymbol{\mu} \text{ on } \Gamma_1,\ \mathbf{v} = \mathbf{0} \text{ on } \Gamma_1^w\right\}, \tag{151}$$

$$\Lambda_2 = \left\{\mu \in H^{1/2}(\Gamma_2) : \exists\psi \in H^1(\Omega_2),\ \psi = \mu \text{ on } \Gamma_2,\ \nabla\varphi\cdot\mathbf{n}_2 = 0 \text{ on } \Gamma_2^w,\ \psi = 0 \text{ on } \Gamma_2^b\right\}. \tag{152}$$

$\boldsymbol{\lambda}_1$ and λ_2 are the solutions of the following minimization problem:

$$\inf_{\boldsymbol{\lambda}_1,\lambda_2} J(\boldsymbol{\lambda}_1,\lambda_2) \quad \text{with} \quad J(\boldsymbol{\lambda}_1,\lambda_2) = \frac{1}{2}\int_{\Omega_{12}} (\mathbf{u} + K\nabla\varphi)^2 . \tag{153}$$

Remark 6. Other functionals may be considered for the minimization problem (153) instead of J. For example, we may minimize the jump of pressures in the overlapping region, thus considering

$$\inf_{\boldsymbol{\lambda}_1,\lambda_2} \overline{J}(\boldsymbol{\lambda}_1,\lambda_2) \quad \text{with} \quad \overline{J}(\boldsymbol{\lambda}_1,\lambda_2) = \frac{1}{2}\int_{\Omega_{12}} (p - g\varphi)^2 . \tag{154}$$

Moreover, we could take into account some continuity condition (i.e., the continuity of the normal velocities) on the physical interface $\Gamma \subset \Omega_{12}$ between the fluid and the porous-media regions. In this case we consider the functional

$$\tilde{J}(\boldsymbol{\lambda}_1,\lambda_2) = \frac{1}{2}\int_{\Gamma} (\mathbf{u}\cdot\mathbf{n} + K\nabla\varphi\cdot\mathbf{n})^2 + \frac{1}{2}\int_{\Omega_{12}} (p - g\varphi)^2 , \tag{155}$$

where $\mathbf{n}$ is the normal unit vector on Γ directed outwards of the fluid domain.

We introduce now the following auxiliary problems:
find $(\mathbf{u}^f, p^f) \in [H^1(\Omega_1)]^2 \times L^2(\Omega_1)$ such that

$$\begin{cases} -\nu\Delta\mathbf{u}^f + \nabla p^f = \mathbf{f} & \text{in } \Omega_1 \\ \operatorname{div}\mathbf{u}^f = 0 & \text{in } \Omega_1 \\ \nu\nabla\mathbf{u}^f\cdot\mathbf{n}_1 - p^f\mathbf{n}_1 = \mathbf{g} & \text{on } \Gamma_1^t \\ \mathbf{u}^f = \mathbf{u}^* & \text{on } \Gamma_1^w \\ \mathbf{u}^f = \mathbf{0} & \text{on } \Gamma_1, \end{cases} \tag{156}$$

and find $\varphi^* \in H^1(\Omega_2)$ such that

$$\begin{cases} -\operatorname{div}(K\nabla\varphi^*) = 0 & \text{in } \Omega_2 \\ K\nabla\varphi^*\cdot\mathbf{n}_2 = \boldsymbol{\psi}_N & \text{on } \Gamma_2^w \\ \varphi^* = \psi_D & \text{on } \Gamma_2^b \\ \varphi^* = 0 & \text{on } \Gamma_2. \end{cases} \tag{157}$$

Moreover, we consider the following problems depending only on the control variables:

find $(\mathbf{u}^{\lambda_1}, p^{\lambda_1}) \in [H^1(\Omega_1)]^2 \times L^2(\Omega_1)$ such that

$$\begin{cases} -\nu\Delta\mathbf{u}^{\lambda_1} + \nabla p^{\lambda_1} = \mathbf{0} & \text{in } \Omega_1 \\ \operatorname{div}\mathbf{u}^{\lambda_1} = 0 & \text{in } \Omega_1 \\ \nu\nabla\mathbf{u}^{\lambda_1}\cdot\mathbf{n}_1 - p^{\lambda_1}\mathbf{n}_1 = \mathbf{0} & \text{on } \Gamma_1^t \\ \mathbf{u}^{\lambda_1} = \mathbf{0} & \text{on } \Gamma_1^w \\ \mathbf{u}^{\lambda_1} = \boldsymbol{\lambda}_1 & \text{on } \Gamma_1, \end{cases} \tag{158}$$

and find $\varphi^{\lambda_2} \in H^1(\Omega_2)$ such that

$$\begin{cases} -\text{div}\,(K\nabla\varphi^{\lambda_2}) = 0 & \text{in } \Omega_2 \\ K\nabla\varphi^{\lambda_2}\cdot\mathbf{n}_2 = \mathbf{0} & \text{on } \Gamma_2^w \\ \varphi^{\lambda_2} = 0 & \text{on } \Gamma_2^b \\ \varphi^{\lambda_2} = \lambda_2 & \text{on } \Gamma_2. \end{cases} \tag{159}$$

Then, we can split

$$\mathbf{u} = \mathbf{u}^f + \mathbf{u}^{\lambda_1}, \quad p = p^f + p^{\lambda_1}, \quad \varphi = \varphi^* + \varphi^{\lambda_2}. \tag{160}$$

In this way we can rewrite the functional $J(\boldsymbol{\lambda}_1, \lambda_2)$ in (153) as

$$J(\boldsymbol{\lambda}_1, \lambda_2) = J^0(\boldsymbol{\lambda}_1, \lambda_2) + \mathscr{A}(\boldsymbol{\lambda}_1, \lambda_2), \tag{161}$$

where $J^0(\boldsymbol{\lambda}_1, \lambda_2)$ is the quadratic functional

$$J^0(\boldsymbol{\lambda}_1, \lambda_2) = \frac{1}{2}\int_{\Omega_{12}} (\mathbf{u}^{\lambda_1} + K\nabla\varphi^{\lambda_2})^2 \tag{162}$$

while $\mathscr{A}(\boldsymbol{\lambda}_1, \lambda_2)$ is the affine functional

$$\mathscr{A}(\boldsymbol{\lambda}_1, \lambda_2) = \frac{1}{2}\int_{\Omega_{12}} (\mathbf{u}^f + K\nabla\varphi^*)^2 + \int_{\Omega_{12}} (\mathbf{u}^{\lambda_1} + K\nabla\varphi^{\lambda_2})\cdot(\mathbf{u}^f + K\nabla\varphi^*). \tag{163}$$

We compute now $\nabla J = \nabla J^0 + \nabla\mathscr{A}$.
We have

$$\langle\frac{\partial J^0}{\partial\boldsymbol{\lambda}_1}, \boldsymbol{\mu}_1\rangle = \int_{\Omega_{12}} \mathbf{u}^{\mu_1}\cdot(\mathbf{u}^{\lambda_1} + K\nabla\varphi^{\lambda_2}). \tag{164}$$

Considering the dual problem

$$\begin{cases} -\nu\Delta\mathbf{v} + \nabla q = (\mathbf{u}^{\lambda_1} + K\nabla\varphi^{\lambda_2})\chi_{\Omega_{12}} & \text{in } \Omega_1 \\ \text{div}\,\mathbf{v} = 0 & \text{in } \Omega_1 \\ \nu\nabla\mathbf{v}\cdot\mathbf{n}_1 - q\mathbf{n}_1 = \mathbf{0} & \text{on } \Gamma_1^t \\ \mathbf{v} = \mathbf{0} & \text{on } \Gamma_1^w \\ \mathbf{v} = \mathbf{0} & \text{on } \Gamma_1, \end{cases} \tag{165}$$

we can characterize the operator (164) as

$$\left\langle\frac{\partial J^0}{\partial\boldsymbol{\lambda}_1}, \boldsymbol{\mu}_1\right\rangle = -\int_{\Gamma_1} (\nu\nabla\mathbf{v}\cdot\mathbf{n}_1 - q\mathbf{n}_1)\cdot\boldsymbol{\mu} \qquad \forall\boldsymbol{\mu} \in \Lambda_1. \tag{166}$$

On the other hand, we have

$$\left\langle\frac{\partial J^0}{\partial\lambda_2}, \mu_2\right\rangle = \int_{\Omega_{12}} -\text{div}(K(\mathbf{u}^{\lambda_1} + K\nabla\varphi^{\lambda_2})\chi_{\Omega_{12}})\varphi^{\mu_2}, \tag{167}$$

and, using the dual problem:

$$\begin{cases} -\text{div}\,(K\nabla\psi) = -\text{div}(K(\mathbf{u}^{\lambda_1} + \nabla\varphi^{\lambda_2})\chi_{\Omega_{12}}) & \text{in } \Omega_2 \\ K\nabla\psi \cdot \mathbf{n}_2 = \mathbf{0} & \text{on } \Gamma_2^w \\ \psi = 0 & \text{on } \Gamma_2^b \\ \psi = 0 & \text{on } \Gamma_2, \end{cases} \tag{168}$$

we obtain

$$\left\langle \frac{\partial J^0}{\partial \lambda_2}, \mu_2 \right\rangle = -\int_{\Gamma_2} K\nabla\psi \cdot \mathbf{n}_2\, \mu_2 \qquad \forall \mu_2 \in \Lambda_2. \tag{169}$$

We proceed in a similar way to characterize the affine functional $\mathscr{A}$. In this case, we have

$$\left\langle \frac{\partial \mathscr{A}}{\partial \boldsymbol{\lambda}_1}, \boldsymbol{\mu}_1 \right\rangle = -\int_{\Gamma_1} (\nu\nabla\tilde{\mathbf{v}} \cdot \mathbf{n}_1 - \tilde{q}\mathbf{n}_1) \cdot \boldsymbol{\mu} \qquad \forall \boldsymbol{\mu} \in \Lambda_1, \tag{170}$$

$$\left\langle \frac{\partial \mathscr{A}}{\partial \lambda_2}, \mu_2 \right\rangle = -\int_{\Gamma_2} K\nabla\tilde{\psi} \cdot \mathbf{n}_2\, \mu_2 \qquad \forall \mu_2 \in \Lambda_2. \tag{171}$$

$(\tilde{\mathbf{v}}, \tilde{q}) \in [H^1(\Omega_1)]^2 \times L^2(\Omega_1)$ is the solution of the dual problem (165) with forcing term $(\mathbf{u}^f + K\nabla\varphi^*)\chi_{\Omega_{12}}$, while $\tilde{\psi} \in H^1(\Omega_2)$ is the solution of the dual problem (168) with forcing term $-\text{div}(K(\mathbf{u}^f + K\nabla\varphi^*)\chi_{\Omega_{12}})$.

To solve the minimization problem (153) we use the following algorithm:

1. Solve (156) and (157) to get $\mathbf{u}^f$, p^f and φ^*.
2. Compute $\nabla\mathscr{A}$:
   - solve (165) with forcing term $(\mathbf{u}^f + K\nabla\varphi^*)\chi_{\Omega_{12}}$ and compute (170);
   - solve (168) with forcing term $-\text{div}(K(\mathbf{u}^f + K\nabla\varphi^*)\chi_{\Omega_{12}})$ and compute (171).
3. Find $(\boldsymbol{\lambda}_1, \lambda_2) \in \Lambda_1 \times \Lambda_2$ such that $\nabla J^0 = -\nabla\mathscr{A}$. To this aim we use an iterative method like Bi-CGStab. At each iteration, to compute $\nabla J^0(\boldsymbol{\lambda}_1, \lambda_2)$ we do
   - solve (158) and (159);
   - compute $\mathbf{u}^{\lambda_1} + K\nabla\varphi^{\lambda_2}$ in Ω_{12};
   - solve (165) to get (166);
   - solve (168) to get (169).
4. Finally, solve (158) and (159) using the functions $\boldsymbol{\lambda}_1$ and λ_2 computed at step 3 and use (160) to obtain the desired solutions.

3.4.1 Stokes/Darcy Coupling with Three Virtual Controls

A three virtual controls approach for the Stokes/Darcy coupling with overlap can be formulated as follows:

$$\begin{cases} \alpha\mathbf{u} - \nu\Delta\mathbf{u} + (\mathbf{u}\cdot\nabla)\mathbf{u} + \nabla p = \mathbf{0} & \text{in } \Omega_1 \\ \mathrm{div}\mathbf{u} = 0 & \text{in } \Omega_1 \\ \nu\nabla\mathbf{u}\cdot\mathbf{n}_1 - p\mathbf{n}_1 = \mathbf{g} & \text{on } \Gamma_1^t \\ \mathbf{u} = \mathbf{u}^* & \text{on } \Gamma_1^w \\ \mathbf{u} = \boldsymbol{\lambda}_1 & \text{on } \Gamma_1 \\ -\mathrm{div}(K\nabla\varphi) = \chi_{\Omega_{12}}\lambda_3 & \text{in } \Omega_2 \\ K\nabla\varphi\cdot\mathbf{n}_2 = \boldsymbol{\psi}_N & \text{on } \Gamma_2^w \\ \varphi = \psi_D & \text{on } \Gamma_2^b \\ \varphi = \lambda_2 & \text{on } \Gamma_2, \end{cases}$$

where λ_3 is the third control, while other notations are those introduced in the previous section. It turns out that the virtual controls $\boldsymbol{\lambda}_1$, λ_2, and λ_3 are solutions of the minimization problem

$$\inf_{\boldsymbol{\lambda}_1,\lambda_2,\lambda_3} J(\boldsymbol{\lambda}_1,\lambda_2,\lambda_3).$$

Several possible choices can be made for the cost functional J, e.g.,

$$J(\boldsymbol{\lambda}_1,\lambda_2,\lambda_3) = \int_{\Omega_{12}} (K\nabla\varphi + \mathbf{u})^2 \mathrm{d}\Omega.$$

A discussion about this approach (and related ones) is given in [14].

3.5 *Coupling for Incompressible Flows*

The Navier–Stokes/potential coupling introduced in Sect. 2.5 has been considered by Glowinski et al. [12, 13] in the framework of virtual controls with overlapping decomposition.

We denote by Ω_1 the extended subdomains where we consider the potential model, while let Ω_2 be the extended subregion where we consider the full Navier–Stokes equations. Finally, $\Omega_{12} = \Omega_1 \cap \Omega_2$ is the overlapping region, and $\Gamma_i = \partial\Omega_i \setminus (\partial\Omega_i \cap \partial\Omega)$, for $i = 1, 2$. See Fig. 19.

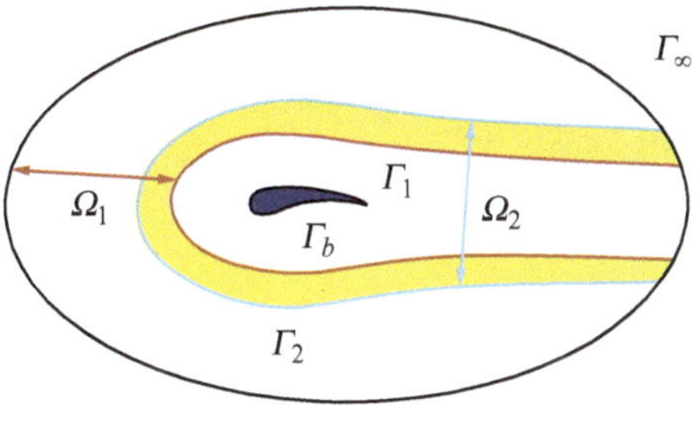

Fig. 19 Splitting of the computational domain in two overlapping regions for the Navier–Stokes/potential coupling

We consider two control variables λ_1 and $\boldsymbol{\lambda}_2$ in the following spaces, respectively:

$$
\begin{aligned}
\Lambda_1 &= \left\{\mu \in H^{1/2}(\Gamma_1) \,:\, \exists \psi \in H^1(\Omega_1),\ \psi = \mu \text{ on } \Gamma_1,\ \tfrac{\partial \psi}{\partial \mathbf{n}_\infty} = 0 \text{ on } \Gamma_\infty \right\}, \\
\Lambda_2 &= \{\boldsymbol{\mu} \in [H^{1/2}(\Gamma_2)]^d \,:\, \exists \mathbf{v} \in [H^1(\Omega_2)]^d,\ \mathbf{v} = \boldsymbol{\mu} \text{ on } \Gamma_2, \\
&\qquad\qquad \mathbf{v} = \mathbf{0} \text{ on } \Gamma_b \cup (\Gamma_\infty \cap \partial\Omega_2),\ d = 2,3\}.
\end{aligned}
$$

λ_1 and $\boldsymbol{\lambda}_2$ represent Dirichlet interface conditions for the two subproblems. Indeed, we consider:

$$
\begin{cases}
\Delta\varphi = 0 & \text{in } \Omega_1 \\
\dfrac{\partial\varphi}{\partial \mathbf{n}_\infty} = \mathbf{u}_\infty \cdot \mathbf{n}_\infty & \text{on } \Gamma_\infty \cap \partial\Omega_1 \\
\varphi = \lambda_1 & \text{on } \Gamma_1
\end{cases}
\tag{172}
$$

and

$$
\begin{cases}
\alpha\mathbf{u} - \nu\Delta\mathbf{u} + (\mathbf{u}\cdot\nabla)\mathbf{u} + \nabla p = \mathbf{f} & \text{in } \Omega_2 \\
\operatorname{div} \mathbf{u} = 0 & \text{in } \Omega_2 \\
\mathbf{u} = \mathbf{0} & \text{on } \Gamma_b \cup (\Gamma_\infty \cap \partial\Omega_2) \\
\mathbf{u} = \boldsymbol{\lambda}_2 & \text{on } \Gamma_2.
\end{cases}
\tag{173}
$$

The unknown Dirichlet data λ_1 and $\boldsymbol{\lambda}_2$ are the solutions of the minimization problem:

$$
\inf_{\lambda_1, \boldsymbol{\lambda}_2} J(\lambda_1, \boldsymbol{\lambda}_2) \quad \text{with} \quad J(\lambda_1, \boldsymbol{\lambda}_2) = \frac{1}{2}\int_{\Omega_{12}} (\nabla\varphi - \mathbf{u})^2 d\Omega \tag{174}
$$

and satisfying the condition

$$
\int_{\Gamma_2} \boldsymbol{\lambda}_2 \cdot \mathbf{n}_{\Gamma_2} d\Gamma + \int_{\Gamma_\infty \cap \partial\Omega_2} \mathbf{u}_\infty \cdot \mathbf{n}_\infty d\Gamma = 0.
$$

We refer the interested reader to [12, 13]. A similar approach for the case of compressible flows is presented in [28].

References

1. V.I. Agoshkov, P. Gervasio, A. Quarteroni, Optimal control in heterogeneous domain decomposition methods for advection-diffusion equations. Mediterr. J. Math. **3**(2), 147–176 (2006)
2. G. Aguilar, F. Lisbona, Interface conditions for a kind of non linear elliptic hyperbolic problems, in *Domain Decomposition Methods for Partial Differential Equations*, ed. by A. Quarteroni, J. Périeaux, Y.A. Kuznetsov, O.B. Widlund (American Mathematical Society, Providence, 1994), pp. 89–95

3. F. Brezzi, C. Canuto, A. Russo, A self-adaptive formulation for the Euler/Navier–Stokes coupling. Comput. Meth. Appl. Mech. Eng. **73**, 317–330 (1989)
4. L. Badea, M. Discacciati, A. Quarteroni, Numerical analysis of the Navier–Stokes/Darcy coupling. Numer. Math. **115**(2), 195–227 (2010)
5. G.S. Beavers, D.D. Joseph, Boundary conditions at a naturally permeable wall. J. Fluid Mech. **30**, 197–207 (1967)
6. X.C. Cai, Some Domain Decompostion Algorithms for Nonselfadjoint Elliptic and Parabolic Partial Differential Equations. Technical Report TR-461, Department of Computer Science, Courant Institute, 1989
7. X.-C. Cai, Additive Schwarz algorithms for parabolic convection-diffusion equations. Numer. Math. **60**(1), 41–61 (1991)
8. C. Canuto, M.Y. Hussaini, A. Quarteroni, T.A. Zang, *Spectral Methods. Evolution to Complex Geometries and Applications to Fluid Dynamics* (Springer, Heidelberg, 2007)
9. T.F. Chan, T.P. Mathew, Domain decomposition algorithms, in *Acta Numer.* (Cambridge University Press, Cambridge, 1994), pp. 61–143
10. C. Carlenzoli, A. Quarteroni, Adaptive domain decomposition methods for advection-diffusion problems, in *Modeling, Mesh Generation, and Adaptive Numerical Methods for Partial Differential Equations (Minneapolis, MN, 1993), IMA Volumes in Mathematics and its Applications*, vol. 75 (Springer, New York, 1995), pp. 165–186
11. K. Deckelnick, G. Dziuk, C.M. Elliott, Computation of geometric partial differential equations and mean curvature flow. Acta Numer. **14**, 139–232 (2005)
12. Q.V. Dinh, R. Glowinski, J. Périaux, Applications of domain decomposition techniques to the numerical solution of the Navier–Stokes equations, in *Numerical Methods for Engineering*, vol. 1 (Dunod, Paris, 1980), pp. 383–404
13. Q.V. Dinh, R. Glowinski, J. Périaux, G. Terrasson, On the coupling of viscous and inviscid models for incompressible fluid flows via domain decomposition, in *First Conf. on Domain Decomposition Methods for Partial Differential Equations*, ed. by G.A. Meurant, R. Glowinski, G.H. Golub, J. Périaux (SIAM, Philadelphia, 1988), pp. 350–368
14. M. Discacciati, P. Gervasio, A. Quarteroni, Virtual controls for heterogeneous problems. Technical Report, in progress.
15. M. Discacciati, *Domain Decomposition Methods for the Coupling of Surface and Groundwater Flows*. Ph.D. thesis, Ecole Polytechnique Fédérale de Lausanne, Switzerland, 2004
16. M. Discacciati, Iterative methods for Stokes/Darcy coupling, in *Domain Decomposition Methods in Science and Engineering. Lecture Notes in Computational Science and Engineering (40)*, ed. by R. Kornhuber et al. (Springer, Berlin and Heidelberg, 2004) pp. 563–570
17. M. Discacciati, E. Miglio, A. Quarteroni, Mathematical and numerical models for coupling surface and groundwater flows. Appl. Numer. Math. **43**, 57–74 (2002)
18. M. Discacciati, A. Quarteroni, Analysis of a domain decomposition method for the coupling of Stokes and Darcy equations, in *Numerical Mathematics and Advanced Applications, ENUMATH 2001* ed. by F. Brezzi, A. Buffa, S. Corsaro, A. Murli (Springer, Milan, 2003), pp. 3–20
19. M. Discacciati, A. Quarteroni, Navier–Stokes/Darcy coupling: modeling, analysis, and numerical approximation. Rev. Mat. Complut. **22**(2), 315–426 (2009)
20. M. Discacciati, A. Quarteroni, A. Valli, Robin–Robin domain decomposition methods for the Stokes–Darcy coupling. SIAM J. Numer. Anal. **45**(3), 1246–1268 (electronic) (2007)
21. E. Dubach, Contribution à la résolution des equations fluides en domaine non borné. Ph.D. thesis, Universitè Paris 13, 1993
22. A. Frati, F. Pasquarelli, A. Quarteroni, Spectral approximation to advection-diffusion problems by the fictitious interface method. J. Comput. Phys. **107**(2), 201–211 (1993)
23. M.J. Gander, L. Halpern, C. Japhet, Optimized Schwarz algorithms for coupling convection and convection-diffusion problems, in *Domain Decomposition Methods in Science and Engineering (Lyon, 2000)*, ed. by N. Debit, M. Garbey, R. Hoppe, J. Périaux, D. Keyes, Y. Kuznetsov, Theory Eng. Appl. Comput. Methods, Internat. Center Numer. Methods Eng. (CIMNE, Barcelona, 2002), pp. 255–262.

24. M.J. Gander, L. Halpern, C. Japhet, V. Martin, Advection diffusion problems with pure advection approximation in subregions, in *Domain Decomposition Methods in Science and Engineering XVI*, *Lecture Notes in Computer Science and Engineering*, vol. 55 ed. by O. Widlund and D. Keyes, (Springer, Berlin, 2007), pp. 239–246
25. M.J. Gander, L. Halpern, C. Japhet, V. Martin, Viscous problems with a vanishing viscosity approximation in subregions: a new approach based on operator factorization, in *Proceedings of ESAIM*, CANUM2008, Saint Jean de Monts, Vendée, France, vol. 27, 2009, pp. 272–288
26. P. Gervasio, J.-L. Lions, A. Quarteroni, Heterogeneous coupling by virtual control methods. Numer. Math. **90**(2), 241–264 (2001)
27. R. Glowinski, J. Périaux, Q.V. Dinh, Domain decomposition methods for non linear problems in fluid dynamics. Technical Report RR-0147, INRIA, 07 1982
28. R. Glowinski, J. Périaux, G. Terrasson, On the coupling of viscous and inviscid models for compressible fluid flows via domain decomposition, in *Third International Symposium on Domain Decomposition Methods for Partial Differential Equations*, Houston, TX, 1989, ed. by T.F. Chan, R. Glowinski, J. Périaux, O.B. Widlund, (SIAM, Philadelphia, PA, 1990), pp. 64–97
29. F. Gastaldi, A. Quarteroni, On the coupling of hyperbolic and parabolic systems: analytical and numerical approach. Appl. Numer. Math. **6**(1), 3–31 (1989)
30. F. Gastaldi, A. Quarteroni, G. Sacchi Landriani, On the coupling of two dimensional hyperbolic and elliptic equations: analytical and numerical approach, in *Third International Symposium on Domain Decomposition Methods for Partial Differential Equations*, Houston, TX, 1989, ed. by J. Périeaux, T.F.Chan, R.Glowinski, O.B.Widlund, (SIAM, Philadelphia, 1990), pp. 22–63
31. V. Girault, P.A. Raviart, *Finite Element Methods for Navier–Stokes Equations. Theory and Algorithms* (Springer, Berlin, 1986)
32. V. Girault, B. Rivière, DG approximation of coupled Navier–Stokes and Darcy equations by Beaver–Joseph–Saffman interface condition. Technical Report TR-MATH 07-09, University of Pittsburgh, Department of Mathematics, 2007
33. G. Houzeaux, R. Codina, *A Geometrical Domain Decomposition Method in Computational Fluid Dynamics*, *Monograph CIMNE*, vol. 70 (International Center for Numerical Methods in Engineering (CIMNE), Barcelona, 2002)
34. G. Houzeaux, R. Codina, An iteration-by-subdomain overlapping Dirichlet/Robin domain decomposition method for advection-diffusion problems. J. Comput. Appl. Math. **158**(2), 243–276 (2003)
35. A. Iafrati, E.F. Campana, A domain decomposition approach to compute wave breaking (wave-breaking flows). Int. J. Numer. Meth. Fluids **41**, 419–445 (2003)
36. W. Jäger and A. Mikelić, On the boundary conditions at the contact interface between a porous medium and a free fluid. Ann. Scuola Norm. Sup. Pisa Cl. Sci. **23**, 403–465 (1996)
37. W. Jäger, A. Mikelić, On the interface boundary condition of Beavers, Joseph and Saffman. SIAM J. Appl. Math., 60(4):1111–1127 (2000)
38. W. Jäger, A. Mikelić, N. Neuss, Asymptotic analysis of the laminar viscous flow over a porous bed. SIAM J. Sci. Comput., 22(6):2006–2028 (2001)
39. J.-L. Lions, *Optimal Control of Systems Governed by Partial Differential Equations.* (Springer, New York, 1971)
40. P.-L. Lions, On the Schwarz alternating method. I, in *First International Symposium on Domain Decomposition Methods for Partial Differential Equations*, Paris, 1987 (SIAM, Philadelphia, PA, 1988), pp. 1–42
41. P.L. Lions, On the Schwarz alternating method III: a variant for non-overlapping subdomains, in *Third International Symposium on Domain Decomposition Methods for Partial Differential Equations* ed. by T.F. Chan et al., (SIAM, Philadelphia, 1990), pp. 202–231
42. J.L. Lions, E. Magenes, *Problèmes aux Limites Non Homogènes et Applications*, vol. 1 (Dunod, Paris, 1968)
43. J.-L. Lions, O. Pironneau, Algorithmes parallèles pour la solution de problèmes aux limites. C. R. Acad. Sci. Paris Sér. I Math. **327**, 947–952 (1998)
44. J.-L. Lions, O. Pironneau, Sur le contrôle parallèle des systèmes distribués. C. R. Acad. Sci. Paris Sér. I Math. **327**, 993–998 (1998)

45. J.-L. Lions, O. Pironneau, Domain decomposition methods for CAD. C. R. Acad. Sci. Paris Sér. I Math. **328**, 73–80 (1999)
46. W.L. Layton, F. Schieweck, I. Yotov, Coupling fluid flow with porous media flow. SIAM J. Num. Anal. **40**, 2195–2218 (2003)
47. T.P. Mathew, Uniform convergence of the Schwarz alternating method for solving singularly perturbed advection-diffusion equations. SIAM J. Numer. Anal. **35**(4), 1663–1683 (1998)
48. D. Peaceman, H. Rachford, The numerical solution of parabolic and elliptic differential equations. J. SIAM **3**, 28–41 (1955)
49. L.E. Payne, B. Straughan, Analysis of the boundary condition at the interface between a viscous fluid and a porous medium and related modelling questions. J. Math. Pures Appl. **77**, 317–354 (1998)
50. A. Quarteroni, *Numerical Models for Differential Problems. Series MS&A*. vol. 2 (Springer, Milan, 2009)
51. A. Quarteroni, A. Valli, *Domain Decomposition Methods for Partial Differential Equations. Numerical Mathematics and Scientific Computation*. (The Clarendon Press, Oxford University Press, New York, 1999)
52. B. Rivière, I. Yotov, Locally conservative coupling of Stokes and Darcy flows. SIAM J. Numer. Anal. **42**(5), 1959–1977 (2005)
53. P.G. Saffman, On the boundary condition at the interface of a porous medium. *Stud. Appl. Math.* **1**, 93–101 (1971)
54. H.A. Schwarz, Über einige abbildungsaufgaben. *Ges. Math. Abh.* **11**, 65–83 (1869)
55. J.S. Scroggs, A parallel algorithm for nonlinear convection-diffusion equations, in *Third International Symposium on Domain Decomposition Methods for Partial Differential Equations*, Houston, TX, 1989 (SIAM, Philadelphia, PA, 1990), pp. 373–384
56. K. Schenk and F.K. Hebeker, Coupling of two dimensional viscous and inviscid incompressible Stokes equtions. Technical Report Preprint 93-68 (SFB 359), Heidelberg University, 1993
57. H.A. van der Vorst, Bi-CGSTAB: a fast and smoothly converging variant of Bi-CG for the solution of nonsymmetric linear systems. SIAM J. Sci. Statist. Comput. **13**(2), 631–644 (1992)

Primer of Adaptive Finite Element Methods

Ricardo H. Nochetto and Andreas Veeser

Abstract Adaptive finite element methods (AFEM) are a fundamental numerical instrument in science and engineering to approximate partial differential equations. In the 1980s and 1990s a great deal of effort was devoted to the design of a posteriori error estimators, following the pioneering work of Babuška. These are computable quantities, depending on the discrete solution(s) and data, that can be used to assess the approximation quality and improve it adaptively. Despite their practical success, adaptive processes have been shown to converge, and to exhibit optimal cardinality, only recently for dimension $d > 1$ and for linear elliptic PDE. These series of lectures presents an up-to-date discussion of AFEM encompassing the derivation of upper and lower a posteriori error bounds for residual-type estimators, including a critical look at the role of oscillation, the design of AFEM and its basic properties, as well as a complete discussion of convergence, contraction property and quasi-optimal cardinality of AFEM.

1 Piecewise Polynomial Approximation

We start with a discussion of piecewise polynomial approximation in W_p^k Sobolev spaces and graded meshes in any dimension d. We first compare pointwise approximation over uniform and graded meshes for $d = 1$ in Sect. 1.1, which reveals the

---

R.H. Nochetto (✉)
Department of Mathematics and Institute of Physical Science and Technology,
University of Maryland, College Park, MD 20742, USA
e-mail: rhn@math.umd.edu

A. Veeser
Dipartimento di Matematica, Università degli Studi di Milano, Via C. Saldini 50,
I-20133 Milano, Italy
e-mail: andreas.veeser@unimi.it

S. Bertoluzza et al., *Multiscale and Adaptivity: Modeling, Numerics and Applications*,
Lecture Notes in Mathematics 2040, DOI 10.1007/978-3-642-24079-9_3,

advantages of the latter over the former and sets the tone for the rest of the paper. We continue with the concept of Sobolev number in Sect. 1.2.

We explore the geometric aspects of mesh refinement for conforming meshes in Sect. 1.3 and nonconforming meshes in Sect. 1.7, but postpone a full discussion until Sect. 6. We include a statement about complexity of the refinement procedure, which turns out to be instrumental later.

We briefly discuss the construction of finite element spaces in Sect. 1.4, along with polynomial interpolation of functions in Sobolev spaces in Sect. 1.5. This provides local estimates adequate for comparison of quasi-uniform and graded meshes for $d > 1$. We exploit them in developing the so-called error equidistribution principle and the construction of suitably graded meshes via thresholding in Sect. 1.6. We conclude that graded meshes can deliver optimal interpolation rates for certain classes of singular functions, and thus supersede quasi-uniform refinement.

1.1 Classical vs Adaptive Pointwise Approximation

We start with a simple motivation in 1d for the use of adaptive procedures, due to DeVore [22]. Given $\Omega = (0,1)$, a partition $\mathscr{T}_N = \{x_n\}_{n=0}^N$ of Ω

$$0 = x_0 < x_1 < \cdots < x_n < \cdots < x_N = 1$$

and a continuous function $u : \Omega \to \mathbb{R}$, we consider the problem of *interpolating* u by a *piecewise constant* function U_N over $\mathscr{T}_N$. To quantify the difference between u and U_N we resort to the *maximum norm* and study two cases depending on the regularity of u.

Case 1: W^1_∞-regularity. Suppose that u is Lipschitz in $[0,1]$. We consider the approximation

$$U_N(x) := u(x_{n-1}) \quad \text{for all } x_{n-1} \le x < x_n.$$

Since

$$|u(x) - U_N(x)| = |u(x) - u(x_{n-1})| = \left| \int_{x_{n-1}}^{x} u'(t)dt \right| \le h_n \|u'\|_{L^\infty(x_{n-1},x_n)},$$

we conclude that

$$\|u - U_N\|_{L^\infty(\Omega)} \le \frac{1}{N} \|u'\|_{L^\infty(\Omega)}, \tag{1}$$

provided the local mesh-size h_n is about constant (*quasi-uniform* mesh), and so proportional to N^{-1} (the reciprocal of the number of degrees of freedom). Note that the same integrability is used on both sides of (1). A natural question arises: *Is it possible to achieve the same asymptotic decay rate N^{-1} with weaker regularity demands?*

Case 2: W_1^1-regularity. To answer this question, we suppose $\|u'\|_{L^1(\Omega)} = 1$ and consider the non-decreasing function

$$\varphi(x) := \int_0^x |u'(t)|dt,$$

which satisfies $\varphi(0) = 0$ and $\varphi(1) = 1$. Let $\mathscr{T}_N = \{x_n\}_{n=0}^N$ be the partition given by

$$\int_{x_{n-1}}^{x_n} |u'(t)|dt = \varphi(x_n) - \varphi(x_{n-1}) = \frac{1}{N}.$$

Then, for $x \in [x_{n-1}, x_n]$,

$$|u(x) - u(x_{n-1})| = \left| \int_{x_{n-1}}^{x} u'(t)dt \right| \leq \int_{x_{n-1}}^{x} |u'(t)|dt \leq \int_{x_{n-1}}^{x_n} |u'(t)|dt = \frac{1}{N},$$

whence

$$\|u - U_N\|_{L^\infty(\Omega)} \leq \frac{1}{N} \|u'\|_{L^1(\Omega)}. \tag{2}$$

We thus conclude that we could achieve the same rate of convergence N^{-1} for rougher functions with just $\|u'\|_{L^1(\Omega)} < \infty$. The following comments are in order for Case 2.

Remark 1 (Equidistribution). The optimal mesh $\mathscr{T}_N$ *equidistributes* the max-error. This mesh is graded instead of uniform but, in contrast to a uniform mesh, such a partition may not be adequate for another function with the same basic regularity as u. It is instructive to consider the singular function $u(x) = x^\gamma$ with $\gamma = 0.1$ and error tolerance 10^{-2} to quantify the above computations: if N_1 and N_2 are the number of degrees of freedom with uniform and graded partitions, we obtain $N_1/N_2 = 10^{18}$.

Remark 2 (Nonlinear approximation). The regularity of u in (2) is measured in $W_1^1(\Omega)$ instead of $W_\infty^1(\Omega)$ and, consequently, the fractional γ regularity measured in $L^\infty(\Omega)$ increases to one full derivative when expressed in $L^1(\Omega)$. This exchange of integrability between left and right-hand side of (2), and gain of differentiability, is at the heart of the matter and the very reason why suitably graded meshes achieve optimal asymptotic error decay for singular functions. By those we mean functions which are not in the usual linear Sobolev scale, say $W_\infty^1(\Omega)$ in this example, but rather in a nonlinear scale [22]. We will get back to this issue in Sect. 7.

1.2 The Sobolev Number: Scaling and Embedding

In order to make Remark 2 more precise, we introduce the Sobolev number. Let $\Omega \subset \mathbb{R}^d$ with $d > 1$ be a Lipschitz and bounded domain, and let $k \in \mathbb{N}$,

$1 \le p \le \infty$. The Sobolev space $W_p^k(\Omega)$ is defined by

$$W_p^k(\Omega) := \{v : \Omega \to \mathbb{R} \mid D^\alpha v \in L^p(\Omega) \ \ \forall |\alpha| \le k\}.$$

If $p = 2$ we set $H^k(\Omega) = W_2^k(\Omega)$ and note that this is a Hilbert space. The *Sobolev number* of $W_p^k(\Omega)$ is given by

$$\mathrm{sob}(W_p^k) := k - \frac{d}{p}. \tag{3}$$

This number governs the scaling properties of the semi-norm

$$|v|_{W_p^k(\Omega)} := \left(\sum_{|\alpha|=k} \|D^\alpha v\|_{L^p(\Omega)}^p \right)^{1/p},$$

because rescaling variables $\hat{x} = \frac{1}{h}x$ for all $x \in \Omega$, transforms Ω into $\hat{\Omega}$ and v into $\hat{v}$, while the corresponding norms scale as

$$|\hat{v}|_{W_p^k(\hat{\Omega})} = h^{\mathrm{sob}(W_p^k)} |v|_{W_p^k(\Omega)}.$$

In addition, we have the following *compact embedding*: if $m > k$ and $\mathrm{sob}(W_q^m) > \mathrm{sob}(W_p^k)$, then

$$W_q^m(\Omega) \subset W_p^k(\Omega).$$

We say that two Sobolev spaces are in the same nonlinear Sobolev scale if they have the same Sobolev number. We note that for compactness the space $W_q^m(\Omega)$ must be above the Sobolev scale of $W_p^k(\Omega)$. A relevant example for $d = 2$ are the pair $H^1(\Omega)$ and $L^\infty(\Omega)$ which have the same Sobolev number, in fact $\mathrm{sob}(H^1) = \mathrm{sob}(L^\infty) = 0$, but the former is not even contained in the latter: in fact

$$v(x) = \log\log\frac{|x|}{2} \in H^1(\Omega)\backslash L^\infty(\Omega)$$

in the unit ball. This is a source of difficulties for polynomial interpolation theory and the need for quasi-interpolation operators. This is discussed in Sect. 1.5.

We conclude with a comment about Remark 2. We see that $d = 1$ and $\mathrm{sob}(W_1^1) = \mathrm{sob}(L^\infty) = 0$ but $W_1^1(\Omega)$ is compactly embedded in $L^\infty(\Omega)$ in this case. This shows that these two spaces are in the same nonlinear Sobolev scale and that the above inequality between Sobolev numbers for a compact embedding is only sufficient.

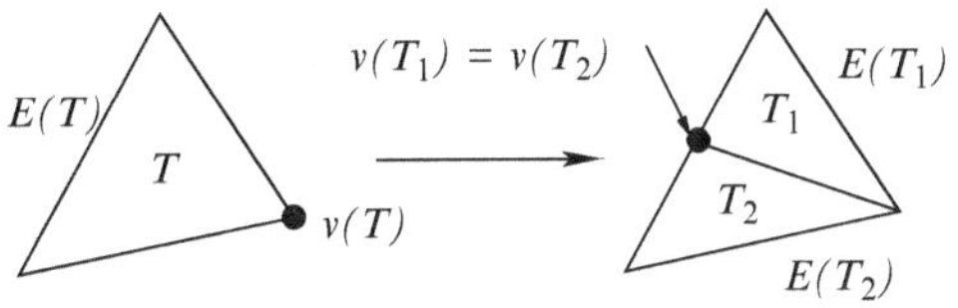

Fig. 1 Triangle $T \in \mathscr{T}$ with vertex $v(T)$ and opposite refinement edge $E(T)$. The bisection rule for $d = 2$ consists of connecting $v(T)$ with the midpoint of $E(T)$, thereby giving rise to children T_1, T_2 with common vertex $v(T_1) = v(T_2)$, the newly created vertex, and opposite refinement edges $E(T_1), E(T_2)$

1.3 Conforming Meshes: The Bisection Method

In order to approximate functions in $W_p^k(\Omega)$ by piecewise polynomials, we decompose Ω into simplices. We briefly discuss the *bisection* method, the most elegant and successful technique for subdividing Ω in any dimension into a conforming mesh. We also discuss briefly nonconforming meshes in Sect. 1.7. We present complete proofs, especially of the complexity of bisection, later in Sect. 6.

We focus on $d = 2$ and follow Binev et al. [7], but the results carry over to any dimension $d > 1$ (see Stevenson [53]). We refer to Nochetto et al. [45] for a rather complete discussion for $d > 1$.

Let $\mathscr{T}$ denote a *mesh* (triangulation or grid) made of simplices T, and let $\mathscr{T}$ be *conforming* (edge-to-edge). Each element is labeled, namely it has an edge $E(T)$ assigned for refinement (and an opposite vertex $v(T)$ for $d = 2$); see Fig. 1.

The bisection method consists of a suitable *labeling* of the initial mesh $\mathscr{T}_0$ and a rule to assign the refinement edge to the two children. For $d = 2$ we consider the *newest vertex bisection* as depicted in Fig. 1. For $d > 2$ the situation is more complicated and one needs the concepts of type and vertex order [45, 53].

Bisection creates a *unique* master forest $\mathbb{F}$ of binary trees with infinite depth, where each node is a simplex (triangle in 2d), its two successors are the two children created by bisection, and the roots of the binary trees are the elements of the initial conforming partition $\mathscr{T}_0$. It is important to realize that, no matter how an element arises in the subdivision process, its associated newest vertex is unique and only depends on the labeling of $\mathscr{T}_0$: so $v(T)$ and $E(T)$ are independent of the order of the subdivision process for all $T \in \mathbb{F}$; see Lemma 16 in Sect. 6. Therefore, $\mathbb{F}$ is unique.

A finite subset $\mathscr{F} \subset \mathbb{F}$ is called a *forest* if $\mathscr{T}_0 \subset \mathscr{F}$ and the nodes of $\mathscr{F}$ satisfy:

- All nodes of $\mathscr{F} \setminus \mathscr{T}_0$ have a predecessor.
- All nodes in $\mathscr{F}$ have either two successors or none.

Any node $T \in \mathscr{F}$ is thus uniquely connected with a node T_0 of the initial triangulation $\mathscr{T}_0$, i.e. T belongs to the infinite tree $\mathbb{F}(T_0)$ emanating from T_0. Furthermore, any forest may have *interior nodes*, i.e. nodes with successors, as well as *leaf nodes*, i.e. nodes without successors. The set of leaves corresponds to a mesh (or triangulation, grid, partition) $\mathscr{T} = \mathscr{T}(\mathscr{F})$ of $\mathscr{T}_0$ which may not be conforming or edge-to-edge.

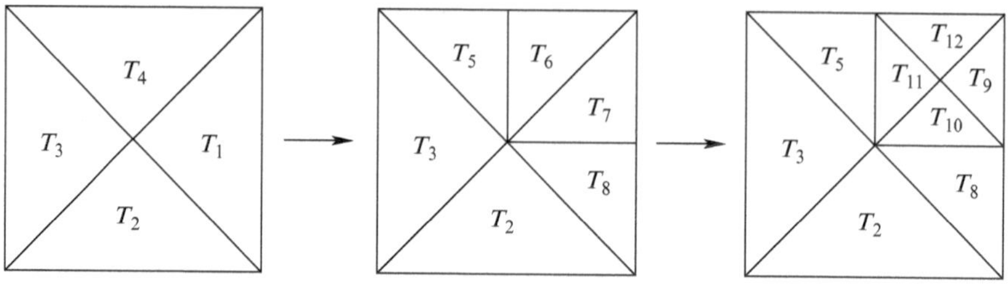

Fig. 2 Sequence of bisection meshes $\{\mathscr{T}_k\}_{k=0}^2$ starting from the initial mesh $\mathscr{T}_0 = \{T_i\}_{i=1}^4$ with longest edges labeled for bisection. Mesh $\mathscr{T}_1$ is created from $\mathscr{T}_0$ upon bisecting T_1 and T_4, whereas mesh $\mathscr{T}_2$ arises from $\mathscr{T}_1$ upon refining T_6 and T_7. The bisection rule is described in Fig. 1

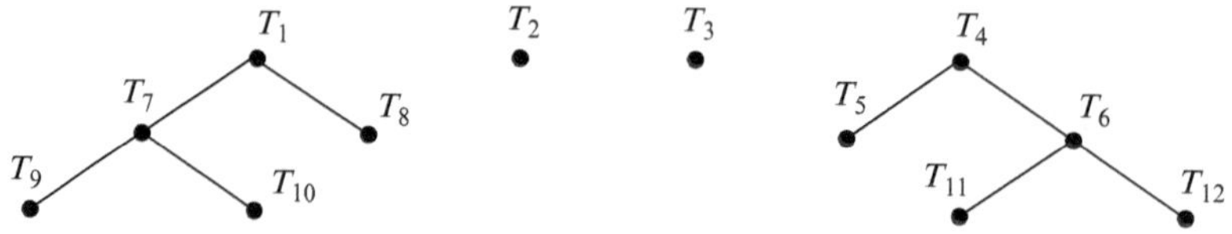

Fig. 3 Forest $\mathscr{F}_2$ corresponding to the grid sequence $\{\mathscr{T}_k\}_{k=0}^2$ of Fig. 2. The roots of $\mathscr{F}_2$ form the initial mesh $\mathscr{T}_0$ and the leaves of $\mathscr{F}_2$ constitute the conforming bisection mesh $\mathscr{T}_2$. Moreover, each level of $\mathscr{F}_2$ corresponds to all elements with generation equal to the level

We thus introduce the set $\mathbb{T}$ of all conforming refinements of $\mathscr{T}_0$:

$$\mathbb{T} := \{\mathscr{T} = \mathscr{T}(\mathscr{F}) \mid \mathscr{F} \subset \mathbb{F} \text{ is finite and } \mathscr{T}(\mathscr{F}) \text{ is conforming}\}.$$

If $\mathscr{T}_* = \mathscr{T}(\mathscr{F}_*) \in \mathbb{T}$ is a conforming refinement of $\mathscr{T} = \mathscr{T}(\mathscr{F}) \in \mathbb{T}$, we write $\mathscr{T}_* \geq \mathscr{T}$ and understand this inequality in the sense of trees, namely $\mathscr{F} \subset \mathscr{F}_*$.

Example. Consider $\mathscr{T}_0 = \{T_i\}_{i=1}^4$ and the longest edge to be the refinement edge. Figure 2 displays a sequence of conforming meshes $\mathscr{T}_k \in \mathbb{T}$ created by bisection.

Each element T_i of $\mathscr{T}_0$ is a root of a finite tree emanating from T_i, which together form the forest $\mathscr{F}_2$ corresponding to mesh $\mathscr{T}_2 = \mathscr{T}(\mathscr{F}_2)$. Figure 3 displays $\mathscr{F}_2$, whose leaf nodes are the elements of $\mathscr{T}_2$.

Properties of bisection: We now discuss several crucial geometric properties of bisection. We start with the concept of shape regularity. For any $T \in \mathscr{T}$, we define

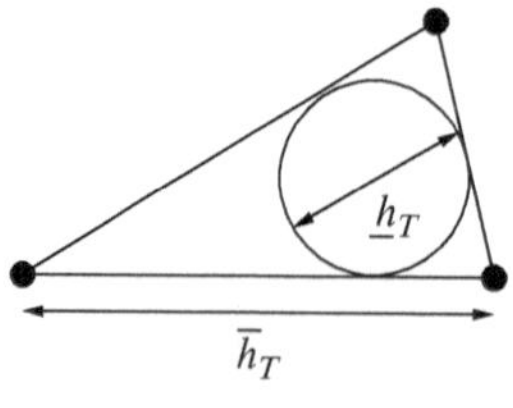

$$\overline{h}_T = \operatorname{diam}(T)$$
$$h_T = |T|^{1/d}$$
$$\underline{h}_T = 2\sup\{r > 0 \mid B(x,r) \subset T \text{ for } x \in T\}.$$

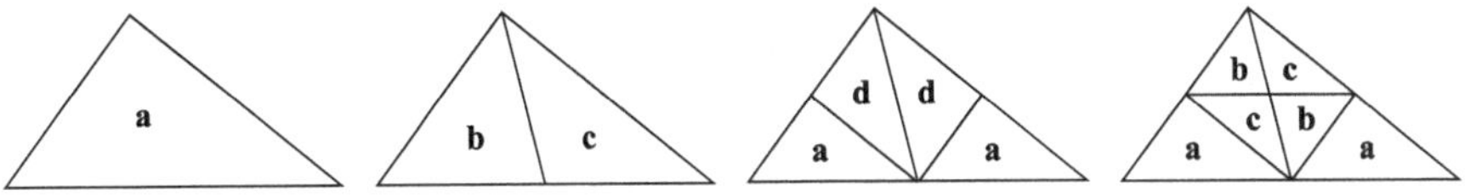

Fig. 4 Bisection produces at most four similarity classes for any triangle

Then

$$\underline{h}_T \leq h_T \leq \overline{h}_T \leq \sigma \underline{h}_T \qquad \forall T \in \mathscr{T},$$

where $\sigma > 1$ is the shape regularity constant. We say that a sequence of meshes is *shape regular* if σ is uniformly bounded, or in other words that the element shape does not degenerate with refinement. The next lemma guarantees that bisection keeps σ bounded.

Lemma 1 (Shape regularity). *The partitions $\mathscr{T}$ generated by newest vertex bisection satisfy a uniform minimal angle condition, or equivalently σ is uniformly bounded, only depending on the initial partition $\mathscr{T}_0$.*

Proof. Each $T \in \mathscr{T}_0$ gives rise to a fixed number of similarity classes, namely 4 for $d = 2$ according to Fig. 4. This, combined with the fact that $\#\mathscr{T}_0$ is finite, yields the assertion. □

We define the *generation (or level)* $g(T)$ of an element $T \in \mathscr{T}$ as the number of bisections needed to create T from its ancestor $T_0 \in \mathscr{T}_0$. Since bisection splits an element into two children with equal measure, we realize that

$$h_T = 2^{-g(T)/2} h_{T_0} \qquad \forall T \in \mathscr{T}. \tag{4}$$

Referring to Fig. 3 we observe that the leaf nodes $T_9, T_{10}, T_{11}, T_{12}$ have generation 2, whereas T_5, T_8 have generation 1 and T_2, T_3 have generation 0.
The following geometric property is a simple consequence of (4).

Lemma 2 (Element size vs generation). *There exist constants $0 < D_1 < D_2$, only depending on $\mathscr{T}_0$, such that*

$$D_1 2^{-g(T)/2} \leq h_T < \overline{h}_T \leq D_2 2^{-g(T)/2} \quad \forall T \in \mathscr{T}. \tag{5}$$

Labeling and bisection rule: Whether the recursive application of bisection does not lead to inconsistencies depends on a suitable initial labeling of edges and a bisection rule. For $d = 2$ they are simple to state [7], but for $d > 2$ we refer to Condition (b) of Sect. 4 of [53]. Given $T \in \mathscr{T}$ with generation $g(T) = i$, we assign the label $(i+1, i+1, i)$ to T with i corresponding to the refinement edge $E(T)$. The following rule dictates how the labeling changes with refinement: the side i is bisected and both new sides as well as the bisector are labeled $i+2$ whereas the remaining labels do not change. To guarantee that the label of an edge is independent of the elements sharing this edge, we need a special labeling for $\mathscr{T}_0$ [7]:

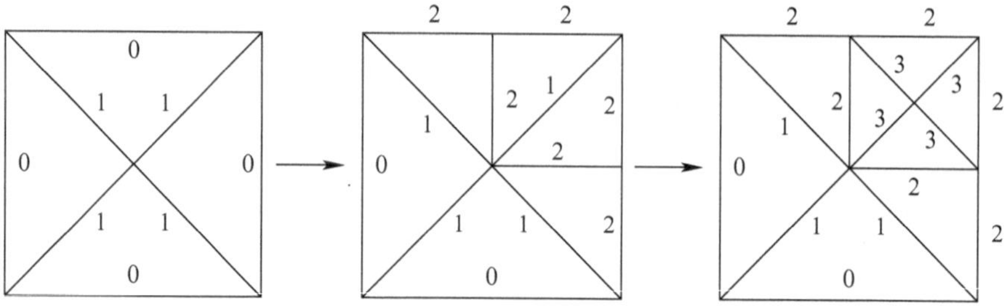

Fig. 5 Initial labeling and its evolution for the sequence of conforming refinements $\mathcal{T}_0 \leq \mathcal{T}_1 \leq \mathcal{T}_2$ of Fig. 2

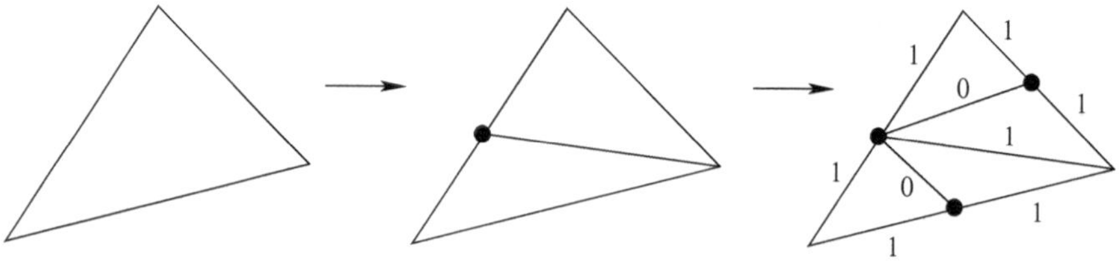

Fig. 6 Bisecting each triangle of $\mathcal{T}_0$ twice and labeling edges in such a way that all boundary edges have label 1 yields an initial mesh satisfying (6)

Edges of $\mathcal{T}_0$ *have labels* 0 *or* 1 *and all elements* $T \in \mathcal{T}_0$ *have exactly two edges with label* 1 *and one with label* 0. (6)

It is not obvious that such a labeling exists, but if it does then all elements of $\mathcal{T}_0$ can be split into pairs of compatibly divisible elements. We refer to Fig. 5 for an example of initial labeling of $\mathcal{T}_0$ satisfying (6) and the way it evolves for two successive refinements $\mathcal{T}_2 \geq \mathcal{T}_1 \geq \mathcal{T}_0$ corresponding to Fig. 2.

To guarantee (6) we can proceed as follows: given a coarse mesh of elements T we can bisect twice each T and label the four grandchildren, as indicated in Fig. 6 for the resulting mesh $\mathcal{T}_0$ to satisfy the initial labeling [7]. A similar, but much trickier, construction can be made in any dimension $d > 2$ (see Stevenson [53]). For $d = 3$ the number of elements increases by an order of magnitude, which indicates that (6) is a severe restriction in practice. Finding alternative, more practical, conditions is an open and important problem.

The procedure **REFINE**: Given $\mathcal{T} \in \mathbb{T}$ and a subset $\mathcal{M} \subset \mathcal{T}$ of marked elements, the procedure

$$\mathcal{T}_* = \mathsf{REFINE}(\mathcal{T}, \mathcal{M})$$

creates a new conforming refinement $\mathcal{T}_*$ of $\mathcal{T}$ by bisecting all elements of $\mathcal{M}$ at least once and perhaps additional elements to keep conformity.

Conformity is a constraint in the refinement procedure that prevents it from being completely local. The propagation of refinement beyond the set of marked elements $\mathcal{M}$ is a rather delicate matter, which we discuss later in Sect. 6. For instance, we show that a naive estimate of the form

$$\#\mathcal{T}_* - \#\mathcal{T} \leq \Lambda_0 \, \#\mathcal{M}$$

is *not* valid with an absolute constant Λ_0 independent of the refinement level. This can be repaired upon considering the cumulative effect for a sequence of conforming bisection meshes $\{\mathcal{T}_k\}_{k=0}^{\infty}$. This is expressed in the following crucial complexity result due to Binev et al. [7] for $d = 2$ and Stevenson [53] for $d > 2$. We present a complete proof later in Sect. 6.

Theorem 1 (Complexity of REFINE). *If $\mathcal{T}_0$ satisfies the initial labeling* (6) *for $d = 2$, or that in [53, Sect. 4] for $d > 2$, then there exists a constant $\Lambda_0 > 0$ only depending on $\mathcal{T}_0$ and d such that for all $k \geq 1$*

$$\#\mathcal{T}_k - \#\mathcal{T}_0 \leq \Lambda_0 \sum_{j=0}^{k-1} \#\mathcal{M}_j .$$

If elements $T \in \mathcal{M}$ are to be bisected $b \geq 1$ times, then the procedure REFINE can be applied recursively, and Theorem 1 remains valid with Λ_0 also depending on b.

1.4 Finite Element Spaces

Given a conforming mesh $\mathcal{T} \in \mathbb{T}$ we define the finite element space of continuous piecewise polynomials of degree $n \geq 1$

$$\mathbb{S}^{n,0}(\mathcal{T}) := \{v \in C^0(\overline{\Omega}) |\ v|_T \in \mathbb{P}_n(T) \ \ \forall T \in \mathcal{T}\};$$

note that $\mathbb{S}^{n,0}(\mathcal{T}) \subset H^1(\Omega)$. We refer to Braess [10], Brenner–Scott [11], Ciarlet [19] and Siebert [50] for a discussion on the local construction of this space along with its properties.

We focus on the piecewise linear case $n = 1$ (Courant elements). Global continuity can be simply enforced by imposing continuity at the vertices z of $\mathcal{T}$, the so-called *nodal values*. We denote by $\mathcal{N}$ the set of vertices z of $\mathcal{T}$.

However, the following local construction leads to global continuity. If T is a generic simplex of $\mathcal{T}$, namely the convex hull of $\{z_i\}_{i=0}^{d}$, then we associate to each vertex z_i a *barycentric coordinate* λ_i^T, which is the linear function in T with nodal value 1 at z_i and 0 at the other vertices of T. Upon pasting together the barycentric coordinates λ_z^T of all simplices T containing vertex $z \in \mathcal{N}$, we obtain a continuous piecewise linear function $\varphi_z \in \mathbb{S}^{1,0}(\mathcal{T})$ as depicted in Fig. 7 for $d = 2$: The set $\{\varphi_z\}_{z\in\mathcal{N}}$ of all such functions is the nodal basis of $\mathbb{S}^{1,0}(\mathcal{T})$, or Courant basis. We denote by $\omega_z := \text{supp}(\varphi_z)$ the support of φ_z, from now on called *star* associated to z, and by γ_z the scheleton of ω_z, namely all the sides containing z.

We denote functions in $\mathbb{S}^{n,0}(\mathcal{T})$ with capital letters. In view of the definition of φ_z, we have the following unique representation of any function $V \in \mathbb{S}^{n,0}(\mathcal{T})$:

$$V(x) = \sum_{z\in\mathcal{N}} V(z)\varphi_z(x).$$

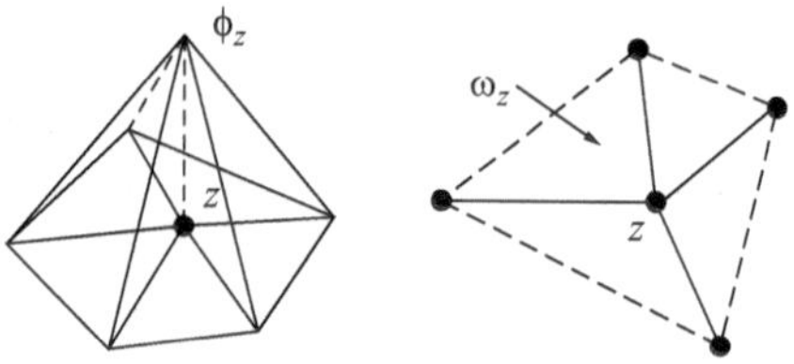

Fig. 7 Piecewise linear basis function φ_z corresponding to interior node z, support ω_z of φ_z and scheleton γ_z, the latter being composed of all sides within the interior of ω_z

If we further impose $V(z) = 0$ for all $z \in \partial\Omega \cap \mathcal{N}$, then $V \in H_0^1(\Omega)$. We denote by

$$\mathbb{V}(\mathcal{T}) := \mathbb{S}^{n,0}(\mathcal{T}) \cap H_0^1(\Omega),$$

the subspace of finite element functions which vanish on $\partial\Omega$. Note that we do not explicitly refer to the polynomial degree, which will be clear in each context.

For each simplex $T \in \mathcal{T}$, generated by vertices $\{z_i\}_{i=0}^d$, the *dual functions* $\{\lambda_i^*\}_{i=0}^d \subset \mathbb{P}_1(T)$ to the barycentric coordinates $\{\lambda_i\}_{i=0}^d$ satisfy the bi-orthogonality relation $\int_T \lambda_i^* \lambda_j = \delta_{ij}$, and are given by

$$\lambda_i^* = \frac{(1+d)^2}{|T|}\lambda_i - \frac{1+d}{|T|}\sum_{j \neq i} \lambda_j \qquad \forall\, 0 \leq i \leq d.$$

The *Courant dual basis* $\varphi_z^* \in \mathbb{S}^{n,-1}(\mathcal{T})$ are the discontinuous piecewise linear functions over $\mathcal{T}$ given by

$$\varphi_z^* = \frac{1}{\nu_z}\sum_{T \ni z} (\lambda_z^T)^* \chi_T \qquad \forall\, z \in \mathcal{N},$$

where $\nu_z \in \mathbb{N}$ is the valence of z (number of elements of $\mathcal{T}$ containing z) and χ_T is the characteristic function of T. These functions have the same support ω_z as the nodal basis φ_z and satisfy the global bi-orthogonality relation

$$\int_\Omega \varphi_z^* \varphi_y = \delta_{zy} \qquad \forall\, z, y \in \mathcal{N}.$$

1.5 Polynomial Interpolation in Sobolev Spaces

If $v \in C^0(\overline{\Omega})$ we define the *Lagrange interpolant* $I_{\mathcal{T}} v$ of v as follows:

$$I_{\mathcal{T}} v(x) = \sum_{z \in \mathcal{N}} v(z)\varphi_z(x).$$

For functions without point values, such as functions in $H^1(\Omega)$ for $d > 1$, we need to determine nodal values by averaging. For any conforming refinement $\mathscr{T} \geq \mathscr{T}_0$ of $\mathscr{T}_0$, the averaging process extends beyond nodes and so gives rise to the discrete neighborhood

$$N_{\mathscr{T}}(T) := \{T' \in \mathscr{T} \mid T' \cap T \neq \emptyset\}$$

for each element $T \in \mathscr{T}$ along with the *local quasi-uniformity* properties

$$\max_{T \in \mathscr{T}} \#N_{\mathscr{T}}(T) \leq C(\mathscr{T}_0), \qquad \max_{T' \in N_{\mathscr{T}}(T)} \frac{|T|}{|T'|} \leq C(\mathscr{T}_0),$$

where $C(\mathscr{T}_0)$ depends only on the shape coefficient of $\mathscr{T}_0$ given by

$$\sigma(\mathscr{T}_0) := \max_{T \in \mathscr{T}_0} \frac{\overline{h}_T}{\underline{h}_T}.$$

We introduce now one such operator $I_{\mathscr{T}}$ due to Scott–Zhang [11, 48], from now on called *quasi-interpolation operator*. We focus on polynomial degree $n = 1$, but the construction is valid for any n; see [11, 48] for details. We recall that $\{\varphi_z\}_{z \in \mathscr{N}}$ is the global Lagrange basis of $\mathbb{S}^{1,0}(\mathscr{T})$, $\{\varphi_z^*\}_{z \in \mathscr{N}}$ is the global dual basis, and $\operatorname{supp} \varphi_z^* = \operatorname{supp} \varphi_z$ for all $z \in \mathscr{N}$. We thus define $I_{\mathscr{T}} : L^1(\Omega) \to \mathbb{S}^{1,0}(\mathscr{T})$ to be

$$I_{\mathscr{T}} v = \sum_{z \in \mathscr{N}} \langle v, \varphi_z^* \rangle \varphi_z.$$

If $0 \leq s \leq 2$ is a regularity index and $1 \leq p \leq \infty$ is an integrability index, then we would like to prove the *quasi-local error estimate*

$$\|D^t(v - I_{\mathscr{T}} v)\|_{L^q(T)} \lesssim h_T^{\operatorname{sob}(W_p^s) - \operatorname{sob}(W_q^t)} \|D^s v\|_{L^p(N_{\mathscr{T}}(T))} \tag{7}$$

for all $T \in \mathscr{T}$, provided $0 \leq t \leq s$, $1 \leq q \leq \infty$ are such that $\operatorname{sob}(W_p^s) > \operatorname{sob}(W_q^t)$.

We first observe that by construction $I_{\mathscr{T}}$ is invariant in $\mathbb{S}^{1,0}(\mathscr{T})$, namely,

$$I_{\mathscr{T}} P = P \quad \text{for all} \quad P \in \mathbb{S}^{1,0}(\mathscr{T}).$$

Since the averaging process giving rise to the values of $I_{\mathscr{T}} v$ for each element $T \in \mathscr{T}$ takes place in the neighborhood $N_{\mathscr{T}}(T)$, we also deduce the local invariance

$$I_{\mathscr{T}} P|_T = P \quad \text{for all} \quad P \in \mathbb{P}_1(N_{\mathscr{T}}(T))$$

as well as the local stability estimate for any $1 \leq q \leq \infty$

$$\|I_{\mathscr{T}} v\|_{L^q(T)} \lesssim \|v\|_{L^q(N_{\mathscr{T}}(T))}.$$

We thus may write

$$v - I_{\mathscr{T}} v|_T = (v - P) - I_{\mathscr{T}}(v - P)|_T \quad \text{for all } T \in \mathscr{T},$$

where $P \in \mathbb{P}_{s-1}$ is arbitrary ($P = 0$ if $s = 0$). It suffices now to prove (7) in the reference element $\widehat{T}$ and scale back and forth to T; the definition (3) of Sobolev number accounts precisely for this scaling. We keep the notation T for $\widehat{T}$, apply the inverse estimate for linear polynomials $\|D^t(I_{\mathscr{T}} v)\|_{L^q(T)} \lesssim \|I_{\mathscr{T}} v\|_{L^q(T)}$ to $v - P$ instead of v, and use the above local stability estimate, to infer that

$$\|D^t(v - I_{\mathscr{T}} v)\|_{L^q(T)} \lesssim \|v - P\|_{W^t_q(N_{\mathscr{T}}(T))} \lesssim \|v - P\|_{W^s_p(N_{\mathscr{T}}(T))}.$$

The last inequality is a consequence $W^s_p(N_{\mathscr{T}}(T)) \subset W^t_q(N_{\mathscr{T}}(T))$ because sob $(W^s_p) > \operatorname{sob}(W^t_q)$ and $t \leq s$. Estimate (7) now follows from the Bramble–Hilbert lemma [11, Lemma 4.3.8], [19, Theorem 3.1.1]

$$\inf_{P \in \mathbb{P}_{s-1}(N_{\mathscr{T}}(T))} \|v - P\|_{W^s_p(N_{\mathscr{T}}(T))} \lesssim \|D^s v\|_{L^p(N_{\mathscr{T}}(T))}. \tag{8}$$

This proves (7) for $n = 1$. The construction of $I_{\mathscr{T}}$ and ensuing estimate (7) are still valid for any $n > 1$ [11,48].

Proposition 1 (Quasi-interpolant without boundary values). *Let s, t be regularity indices with $0 \leq t \leq s \leq n + 1$, and $1 \leq p, q \leq \infty$ be integrability indices so that* $\operatorname{sob}(W^s_p) > \operatorname{sob}(W^t_q)$.

There exists a quasi-interpolation operator $I_{\mathscr{T}} : L^1(\Omega) \to \mathbb{S}^{n,0}(\mathscr{T})$, which is invariant in $S^{n,0}(\mathscr{T})$ and satisfies

$$\|D^t(v - I_{\mathscr{T}} v)\|_{L^q(T)} \lesssim h_T^{\operatorname{sob}(W^s_p) - \operatorname{sob}(W^t_q)} \|D^s v\|_{L^p(N_{\mathscr{T}}(T))} \quad \forall T \in \mathscr{T}. \tag{9}$$

The hidden constant in (9) *depends on the shape coefficient of $\mathscr{T}_0$ and d.*

To impose a vanishing trace on $I_{\mathscr{T}} v$ we may suitably modify the averaging process for boundary nodes. We thus define a set of dual functions with respect to an L^2-scalar product over $(d-1)$-subsimplices contained on $\partial\Omega$; see again [11,48] for details. This retains the invariance property of $I_{\mathscr{T}}$ on $\mathbb{S}^{n,0}(\mathscr{T})$ and guarantees that $I_{\mathscr{T}} v$ has a zero trace if $v \in W^1_1(\Omega)$ does. Hence, the above argument applies and (9) follows provided $s \geq 1$.

Proposition 2 (Quasi-interpolant with boundary values). *Let s, t, p, q be as in Proposition 1. There exists a quasi-interpolation operator $I_{\mathscr{T}} : W^1_1(\Omega) \to \mathbb{S}^{n,0}(\mathscr{T})$ invariant in $\mathbb{S}^{n,0}(\mathscr{T})$ which satisfies* (9) *for $s \geq 1$ and preserves the boundary values of v provided they are piecewise polynomial of degree $\leq n$. In particular, if $v \in W^1_1(\Omega)$ has a vanishing trace on $\partial\Omega$, then so does $I_{\mathscr{T}} v$.*

Remark 3 (Fractional regularity). We observe that (7) does not require the regularity indices t and s to be integer. The proof follows the same lines but replaces the polynomial degree $s-1$ by the greatest integer smaller that s; the generalization of (8) can be taken from [26].

Remark 4 (Local error estimate for lagrange Interpolant). Let the regularity index s and integrability index $1 \le p \le \infty$ satisfy $s - d/p > 0$. This implies that $\operatorname{sob}(W_p^s) > \operatorname{sob}(L^\infty)$, whence $W_p^s(\Omega) \subset C(\overline{\Omega})$ and the Lagrange interpolation operator $I_{\mathcal{T}} : W_p^s(\Omega) \to \mathbb{S}^{n,0}(\mathcal{T})$ is well defined and satisfies the *local error estimate*

$$\|D^t(v - I_{\mathcal{T}} v)\|_{L^q(T)} \lesssim h_T^{\operatorname{sob}(W_p^s)-\operatorname{sob}(W_q^t)} \|D^s v\|_{L^p(T)}, \tag{10}$$

provided $0 \le t \le s$, $1 \le q \le \infty$ are such that $\operatorname{sob}(W_p^s) > \operatorname{sob}(W_q^t)$. We point out that $N_{\mathcal{T}}(T)$ in (7) is now replaced by T in (10). We also remark that if v vanishes on $\partial\Omega$ so does $I_{\mathcal{T}} v$. The proof of (10) proceeds along the same lines as that of Proposition 1 except that the nodal evaluation does not extend beyond the element $T \in \mathcal{T}$ and the inverse and stability estimates over the reference element are replaced by

$$\|D^t I_{\mathcal{T}} v\|_{L^q(\widehat{T})} \lesssim \|I_{\mathcal{T}} v\|_{L^q(\widehat{T})} \lesssim \|v\|_{L^\infty(\widehat{T})} \lesssim \|v\|_{W_p^s(\widehat{T})}.$$

We are now in a position to derive a global interpolation error estimate. To this end, it is convenient to introduce the mesh-size function $h \in L^\infty(\Omega)$ given by

$$h_{|T} = h_T \quad \text{for all } T \in \mathcal{T}. \tag{11}$$

Notice that the following estimate encompasses the linear as well as the nonlinear Sobolev scales.

Theorem 2 (Global interpolation error estimate). *Let $1 \le s \le n+1$ and $1 \le p \le 2$ satisfy $r := \operatorname{sob}(W_p^s) - \operatorname{sob}(H^1) > 0$. If $v \in W_p^s(\Omega)$, then*

$$\|\nabla(v - I_{\mathcal{T}} v)\|_{L^2(\Omega)} \lesssim \|h^r D^s v\|_{L^p(\Omega)}. \tag{12}$$

Proof. Use Proposition 1 along with the elementary property of series $\sum_n a_n \le (\sum_n a_n^q)^{1/q}$ for $0 < q := p/2 \le 1$. □

Quasi-uniform meshes: We now apply Theorem 2 to quasi-uniform meshes, namely meshes $\mathcal{T} \in \mathbb{T}$ for which all its elements are of comparable size h regardless of the refinement level. In this case, we have

$$h \approx (\#\mathcal{T})^{-1/d}.$$

Corollary 1 (Quasi-uniform meshes). *Let $1 \le s \le n+1$ and $u \in H^s(\Omega)$. If $\mathcal{T} \in \mathbb{T}$ is quasi-uniform, then*

$$\|\nabla(v - I_{\mathcal{T}} v)\|_{L^2(\Omega)} \lesssim |v|_{H^s(\Omega)} (\#\mathcal{T})^{-(s-1)/d}. \tag{13}$$

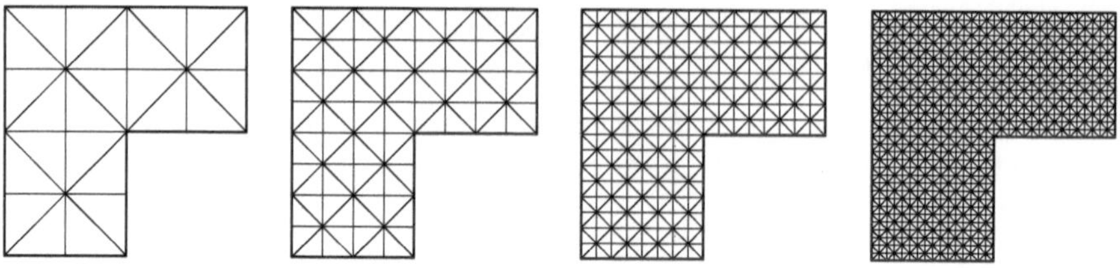

Fig. 8 Sequence of consecutive uniform meshes for L-shaped domain Ω created by two bisections

Table 1 The asymptotic rate of convergence in term of mesh-size h is about $h^{2/3}$, or equivalently $(\#\mathscr{T})^{-1/3}$, irrespective of the polynomial degree n. This provides a lower bound for $\|v - I_{\mathscr{T}} v\|_{L^2(\Omega)}$ and thus shows that (13) is sharp

h	Linear ($n = 1$)	Quadratic ($n = 2$)	Cubic ($n = 3$)
1/4	1.14	9.64	9.89
1/8	0.74	0.67	0.67
1/16	0.68	0.67	0.67
1/32	0.66	0.67	0.67
1/64	0.66	0.67	0.67
1/128	0.66	0.67	0.67

Remark 5 (Optimal Rate). If $s = n + 1$, and so v has the maximal regularity $v \in H^{n+1}(\Omega)$, then we obtain the optimal convergence rate in a linear Sobolev scale

$$\|\nabla(v - I_{\mathscr{T}} v)\|_{L^2(\Omega)} \lesssim |v|_{H^{n+1}(\Omega)} (\#\mathscr{T})^{-n/d}. \tag{14}$$

The order $-n/d$ is just dictated by the polynomial degree n and cannot be improved upon assuming either higher regularity that $H^{n+1}(\Omega)$ or a graded mesh $\mathscr{T}$.

Example (Corner singularity in $2d$). To explore the effect of a geometric singularity on (13), we let Ω be the L-shaped domain of Fig. 8 and $v \in H^1(\Omega)$ be

$$u(r, \vartheta) = r^{\frac{2}{3}} \sin(2\vartheta/3) - r^2/4.$$

This function $v \in H^1(\Omega)$ exhibits the typical corner singularity of the solution of $-\Delta v = f$ with suitable Dirichlet boundary condition: $v \in H^s(\Omega)$ for $s < 5/3$. Table 1 displays the best approximation error for polynomial degree $n = 1, 2, 3$ and the sequence of *uniform* refinements depicted in Fig. 8 in the seminorm $|\cdot|_{H^1(\Omega)}$. This gives a *lower* bound for the interpolation error in (13).
Even though s is fractional, the error estimate (13) is still valid as stated in Remark 3. In fact, for uniform refinement, (13) can be derived by space interpolation between $H^1(\Omega)$ and $H^{n+1}(\Omega)$. The asymptotic rate $(\#\mathscr{T})^{-1/3}$ reported in Table 1 is consistent with (13) and independent of the polynomial degree n; this shows that (13) is sharp. It is also suboptimal as compared with the optimal rate $(\#\mathscr{T})^{-n/2}$ of Remark 5.

The question arises whether the rate $(\#\mathscr{T})^{-1/3}$ in Table 1 is just a consequence of uniform refinement or unavoidable. It is important to realize that $v \notin H^s(\Omega)$ for

$s \geq 5/3$ and thus (13) is not applicable. However, the problem is not that second order derivatives of v do not exist but rather that they are not square-integrable. In particular, it is true that $v \in W_p^2(\Omega)$ if $1 \leq p < 3/2$. We therefore may apply Theorem 2 with, e.g., $n = 1$, $s = 2$, and $p \in [1, 3/2)$ and then ask whether the structure of (12) can be exploited, e.g., by compensating the local behavior of $D^s u$ with the local mesh-size h. This enterprise naturally leads to *graded* meshes adapted to v.

1.6 Adaptive Approximation

Principle of error equidistribution: We investigate the relation between local mesh-size and regularity for the design of graded meshes adapted to $v \in H^1(\Omega)$ for $d = 2$. We formulate this as an optimization problem:

> *Given a function $v \in C^2(\Omega) \cap W_p^2(\Omega)$ and an integer $N > 0$ find conditions for a shape regular mesh $\mathscr{T}$ to minimize the error $|v - I_{\mathscr{T}} v|_{H^1(\Omega)}$ subject to the constraint that the number of degrees of freedom $\#\mathscr{T} \leq N$.*

We first convert this *discrete* optimization problem into a *continuous model*, following Babuška and Rheinboldt [5]. Let

$$\#\mathscr{T} = \int_\Omega \frac{dx}{h(x)^2}$$

be the number of elements of $\mathscr{T}$ and let the Lagrange interpolation error

$$\|\nabla(v - I_{\mathscr{T}} v)\|_{L^2(\Omega)}^p = \int_\Omega h(x)^{2(p-1)} |D^2 v(x)|^p dx$$

be dictated by (12) with $s = 2$ and $1 < p \leq 2$; note that $r = \mathrm{sob}(W_p^2) - \mathrm{sob}(H^1) = 2 - 2/p$ whence $rp = 2(p-1)$ is the exponent of $h(x)$. We next propose the Lagrangian

$$\mathscr{L}[h, \lambda] = \int_\Omega \left(h(x)^{2(p-1)} |D^2 v(x)|^p - \frac{\lambda}{h(x)^2} \right) dx$$

with Lagrange multiplier $\lambda \in \mathbb{R}$. The optimality condition reads (Problem 4)

$$h(x)^{2(p-1)+2} |D^2 v(x)|^p = \Lambda, \tag{15}$$

where $\Lambda > 0$ is a constant. In order to interpret this expression, we compute the interpolation error E_T incurred in element $T \in \mathscr{T}$. According to (10), E_T is given by

$$E_T^p \approx h_T^{2(p-1)} \int_T |D^2 v(x)|^p \approx \Lambda$$

provided $D^2 v(x)$ is about constant in T. Therefore we reach the heuristic, but insightful, conclusion that E_T is about constant, or equivalently

A graded mesh is quasi-optimal if the local error is equidistributed. (16)

Corner singularities: Meshes satisfying (16) have been constructed by Babuška et al. [3] for corner singularities and $d = 2$; see also [30]. If the function v possess the typical behavior

$$v(x) \approx r(x)^\gamma, \quad 0 < \gamma < 1,$$

where $r(x)$ is the distance from $x \in \Omega$ to a reentrant corner of Ω, then (15) implies the mesh grading

$$h(x) = \Lambda^{\frac{1}{2p}} r(x)^{-\frac{1}{2}(\gamma-2)},$$

whence

$$\#\mathcal{T} = \int_\Omega h(x)^{-2} dx = \Lambda^{-\frac{1}{p}} \int_0^{\mathrm{diam}(\Omega)} r^{\gamma-1} dr \approx \Lambda^{-\frac{1}{p}}.$$

This crucial relation is valid for any $\gamma > 0$ and $p > 1$; in fact the only condition on p is that $r = 2 - 2/p > 0$, or equivalently $\mathrm{sob}(W_p^2) > \mathrm{sob}(H^1)$. Therefore,

$$\|\nabla(v - I_{\mathcal{T}} v)\|_{L^2(\Omega)}^2 = \sum_{T\in\mathcal{T}} E_T^2 = \Lambda^{\frac{2}{p}} (\#\mathcal{T}) = (\#\mathcal{T})^{-1} \tag{17}$$

gives the optimal decay rate for $d = 2, n = 1$, according to Remark 5. We explore the case $d \geq 2$ and $n \geq 1$ in Problem 6. What this argument does not address is whether such meshes $\mathcal{T}$ exist in general and, more importantly, whether they can actually be constructed upon bisecting the initial mesh $\mathcal{T}_0$ so that $\mathcal{T} \in \mathbb{T}$.

Thresholding: We now construct graded bisection meshes $\mathcal{T}$ for $n = 1, d = 2$ that achieve the optimal decay rate $(\#\mathcal{T})^{-1/2}$ of (14) and (17) under the global regularity assumption

$$v \in W_p^2(\Omega), \quad p > 1. \tag{18}$$

Following the work of Binev et al. [8], we use a thresholding algorithm that is based on the knowledge of the element errors and on bisection. The algorithm hinges on (16): if $\delta > 0$ is a given tolerance, the element error is equidistributed, that is $E_T \approx \delta^2$, and the global error decays with maximum rate $(\#\mathcal{T})^{-1/2}$, then

$$\delta^4 \#\mathcal{T} \approx \sum_{T\in\mathcal{T}} E_T^2 = |v - I_{\mathcal{T}} v|_{H^1(\Omega)}^2 \lesssim (\#\mathcal{T})^{-1}$$

that is $\#\mathscr{T} \lesssim \delta^{-2}$. With this in mind, we impose $E_T \leq \delta^2$ as a common threshold to stop refining and expect $\#\mathscr{T} \lesssim \delta^{-2}$. The following algorithm implements this idea.

Thresholding algorithm: Given a tolerance $\delta > 0$ and a conforming mesh $\mathscr{T}_0$, THRESHOLD finds a conforming refinement $\mathscr{T} \geq \mathscr{T}_0$ of $\mathscr{T}_0$ by bisection such that $E_T \leq \delta^2$ for all $T \in \mathscr{T}$: let $\mathscr{T} = \mathscr{T}_0$ and

```
THRESHOLD(𝒯, δ)
while ℳ := {T ∈ 𝒯 | E_T > δ²} ≠ ∅
   𝒯 := REFINE(𝒯, ℳ)
end while
return(𝒯)
```

We get $W_p^2(\Omega) \subset C^0(\overline{\Omega})$, because $p > 1$, and can use the Lagrange interpolant and local estimate (10) with $r = \text{sob}(W_p^2) - \text{sob}(H^1) = 2 - 2/p > 0$. We deduce that

$$E_T \lesssim h_T^r \, \|D^2 v\|_{L^p(T)}, \tag{19}$$

and that THRESHOLD *terminates* because h_T decreases monotonically to 0 with bisection. The quality of the resulting mesh is assessed next.

Theorem 3 (Thresholding). *If $v \in H_0^1(\Omega)$ verifies (18), then the output $\mathscr{T} \in \mathbb{T}$ of THRESHOLD satisfies*

$$|v - I_{\mathscr{T}} v|_{H^1(\Omega)} \leq \delta^2 (\#\mathscr{T})^{1/2}, \quad \#\mathscr{T} - \#\mathscr{T}_0 \lesssim \delta^{-2} \, |\Omega|^{1-1/p} \|D^2 v\|_{L^p(\Omega)}.$$

Proof. Let $k \geq 1$ be the number of iterations of THRESHOLD before termination. Let $\mathscr{M} = \mathscr{M}_0 \cup \cdots \cup \mathscr{M}_{k-1}$ be the set of marked elements. We organize the elements in $\mathscr{M}$ by size in such a way that allows for a counting argument. Let $\mathscr{P}_j$ be the set of elements T of $\mathscr{M}$ with size

$$2^{-(j+1)} \leq |T| < 2^{-j} \quad \Rightarrow \quad 2^{-(j+1)/2} \leq h_T < 2^{-j/2}.$$

We proceed in several steps.

[1] We first observe that all T's in $\mathscr{P}_j$ are *disjoint*. This is because if $T_1, T_2 \in \mathscr{P}_j$ and $\mathring{T}_1 \cap \mathring{T}_2 \neq \emptyset$, then one of them is contained in the other, say $T_1 \subset T_2$, due to the bisection procedure. Thus

$$|T_1| \leq \frac{1}{2} \, |T_2|$$

contradicting the definition of $\mathscr{P}_j$. This implies

$$2^{-(j+1)} \#\mathscr{P}_j \leq |\Omega| \quad \Rightarrow \quad \#\mathscr{P}_j \leq |\Omega| \, 2^{j+1}. \tag{20}$$

[2] In light of (19), we have for $T \in \mathscr{P}_j$

$$\delta^2 \leq E_T \lesssim 2^{-(j/2)r} \|D^2 v\|_{L^p(T)}.$$

Therefore

$$\delta^{2p}\,\#\mathscr{P}_j \lesssim 2^{-(j/2)rp} \sum_{T\in\mathscr{P}_j} \|D^2 v\|^p_{L^p(T)} \le 2^{-(j/2)rp}\,\|D^2 v\|^p_{L^p(\Omega)},$$

whence

$$\#\mathscr{P}_j \lesssim \delta^{-2p}\,2^{-(j/2)rp}\,\|D^2 v\|^p_{L^p(\Omega)}. \tag{21}$$

3 The two bounds for $\#\mathscr{P}$ in (20) and (21) are complementary. The first is good for j small whereas the second is suitable for j large (think of $\delta \ll 1$). The crossover takes place for j_0 such that

$$2^{j_0+1}|\Omega| = \delta^{-2p}\,2^{-j_0(rp/2)}\|D^2 v\|^p_{L^p(\Omega)} \quad\Rightarrow\quad 2^{j_0} \approx \delta^{-2}\,\frac{\|D^2 v\|_{L^p(\Omega)}}{|\Omega|^{1/p}}.$$

4 We now compute

$$\#\mathscr{M} = \sum_j \#\mathscr{P}_j \lesssim \sum_{j\le j_0} 2^j|\Omega| + \delta^{-2p}\,\|D^2 v\|^p_{L^p(\Omega)} \sum_{j>j_0} (2^{-rp/2})^j.$$

Since

$$\sum_{j\le j_0} 2^j \approx 2^{j_0}, \quad \sum_{j>j_0} (2^{-rp/2})^j \lesssim 2^{-(rp/2)j_0} = 2^{-(p-1)j_0},$$

we can write

$$\#\mathscr{M} \lesssim \left(\delta^{-2} + \delta^{-2p}\delta^{2(p-1)}\right)|\Omega|^{1-1/p}\,\|D^2 v\|_{L^p(\Omega)} \approx \delta^{-2}\,|\Omega|^{1-1/p}\,\|D^2 v\|_{L^p(\Omega)}.$$

We finally apply Theorem 1 to arrive at

$$\#\mathscr{T} - \#\mathscr{T}_0 \lesssim \#\mathscr{M} \lesssim \delta^{-2}\,|\Omega|^{1-1/p}\,\|D^2 v\|_{L^p(\Omega)}.$$

5 It remains to estimate the energy error. We have, upon termination of THRESHOLD, that $E_T \le \delta^2$ for all $T \in \mathscr{T}$. Then

$$|v - I_{\mathscr{T}} v|^2_{H^1(\Omega)} = \sum_{T\in\mathscr{T}} E_T^2 \le \delta^4\,\#\mathscr{T}.$$

This concludes the Theorem. □

By relating the threshold value δ and the number of refinements N, we obtain a result about the convergence rate.

Corollary 2 (Convergence rate). *Let $v \in H^1_0(\Omega)$ satisfy* (18). *Then for $N > \#\mathscr{T}_0$ integer there exists $\mathscr{T} \in \mathbb{T}$ such that*

$$|v - I_{\mathscr{T}} v|_{H^1(\Omega)} \lesssim |\Omega|^{1-1/p} \, \|D^2 v\|_{L^p(\Omega)} N^{-1/2}, \quad \#\mathscr{T} - \#\mathscr{T}_0 \lesssim N.$$

Proof. Choose $\delta^2 = |\Omega|^{1-1/p} \, \|D^2 v\|_{L^p(\Omega)} N^{-1}$ in Theorem 3. Then, there exists $\mathscr{T} \in \mathbb{T}$ such that $\#\mathscr{T} - \#\mathscr{T}_0 \lesssim N$ and

$$\begin{aligned} |v - I_{\mathscr{T}} v|_{H^1(\Omega)} &\lesssim |\Omega|^{1-1/p} \, \|D^2 v\|_{L^p(\Omega)} N^{-1} \big(N + \#\mathscr{T}_0\big)^{1/2} \\ &\lesssim |\Omega|^{1-1/p} \, \|D^2 v\|_{L^p(\Omega)} N^{-1/2} \end{aligned}$$

because $N > \#\mathscr{T}_0$. This finishes the Corollary. □

Remark 6 (Piecewise smoothness). The global regularity (18) can be weakened to *piecewise* W^2_p *regularity* over the initial mesh $\mathscr{T}_0$, namely $W^2_p(\Omega; \mathscr{T}_0)$, and global $H^1_0(\Omega)$. This is because $W^2_p(T) \hookrightarrow C^0(\overline{T})$ for all $T \in \mathscr{T}_0$, whence $I_{\mathscr{T}}$ can be taken to be the Lagrange interpolation operator.

Remark 7 (Case $p < 1$). We consider now polynomial degree $n \geq 1$. The integrability p corresponding to differentiability $n + 1$ results from equating Sobolev numbers:

$$n + 1 - \frac{d}{p} = \mathrm{sob}(H^1) = 1 - \frac{d}{2} \quad \Rightarrow \quad p = \frac{2d}{2n + d}.$$

Depending on $d \geq 2$ and $n \geq 1$, this may lead to $0 < p < 1$, in which case $W^{n+1}_p(\Omega)$ is to be replaced by the Besov space $B^{n+1}_{p,p}(\Omega)$ [22]; see Problem 6. The argument of Theorem 3 works provided we replace (19) by a modulus of regularity; in fact, $D^{n+1}v$ would not be locally integrable and so would fail to be a distribution.

Remark 8 (Isotropic vs anisotropic elements). Theorem 3 and Problem 5 show that isotropic graded meshes can always deal with geometric singularities for $d = 2$. This is no longer the case for $d > 2$ and is explored in Problem 6.

1.7 Nonconforming Meshes

More general subdivisions of Ω than those in Sect. 1.3 are used in practice. If the elements of $\mathscr{T}_0$ are quadrilaterals for $d = 2$, or their multidimensional variant for $d > 2$, then it is natural to allow for improper or *hanging nodes* for the resulting refinements $\mathscr{T}$ to be graded; see Fig. 9a. On the other hand, if $\mathscr{T}_0$ is made of triangles for $d = 2$, or simplices for $d > 2$, then red refinement without green completion also gives rise to graded meshes with hanging nodes; see Fig. 9b. In both cases, the presence of hanging nodes is inevitable to enforce mesh grading. Finally, bisection may produce meshes with hanging nodes, as depicted in Fig. 9c, if the completion process is incomplete. All three refinements maintain shape regularity,

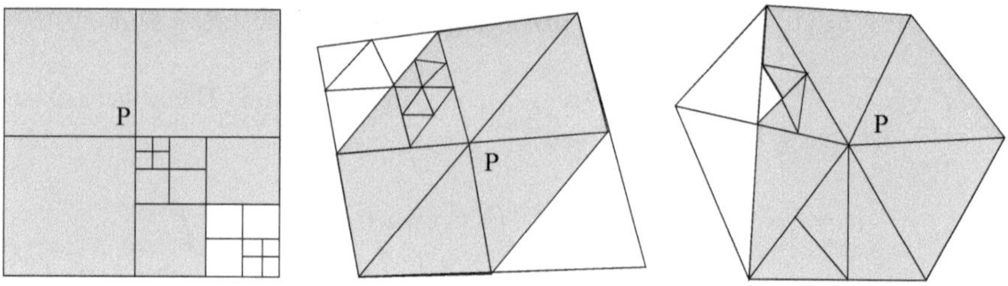

Fig. 9 Nonconforming meshes made of quadrilaterals (**a**), triangles with *red* refinement (**b**), and triangles with bisection (**c**). The shaded regions depict the domain of influence of a proper or conforming node P

but for both practice and theory, they cannot be arbitrary: we need to restrict the level of incompatibility; see Problem 10. We discuss this next.

We start with the notion of domain of influence of a proper node, introduced by Babuška and Miller in the context of K-meshes [4]; see Fig. 9. For simplicity, we restrict ourselves to polynomial degree $n = 1$. We say that a node P of $\mathscr{T}$ is a *proper* (or conforming) node if it is a vertex of all elements containing P; otherwise, we say that P is an *improper* (nonconforming or hanging) node. Since we only prescribe degrees of freedom at the proper nodes, it is natural to describe the canonical continuous piecewise linear basis functions φ_P associated with each proper node P.

We do this recursively. As in Sect. 1.3, the *generation* $g(T)$ of an element $T \in \mathscr{T}$ is the number of subdivisions needed to create T from its ancestor in the initial mesh $\mathscr{T}_0$, hereafter assumed to be conforming. We first rearrange the elements in $\mathscr{T} = \{T_i\}_{i=1}^{\#\mathscr{T}}$ by generation: $g(T_i) \le g(T_{i+1})$ for all $i \ge 0$. Suppose that φ_P has been already defined for each $T \in \mathscr{T}$ with $g(T) < i$. We proceed as follows to define φ_P at each vertex z of $T \in \mathscr{T}$ with $g(T) = i$:

- If z is a proper node, then we set $\varphi_P(z) = 1$ if $z = P$ and $\varphi_P(z) = 0$ otherwise
- If z is a hanging node, then z belongs to an edge of another element $T' \in \mathscr{T}$ with $g(T') < i$ and set $\varphi_P(z)|_T = \varphi_P(z)|_{T'}$

This definition is independent of the choice of T' since, by construction, φ_P is continuous across interelements of lower level. We also observe that $\{\varphi_P\}_{P\in\mathscr{N}}$ is a basis of the finite element space $\mathbb{V}(\mathscr{T})$ of *continuous* piecewise linear functions, thus

$$V = \sum_{P\in\mathscr{N}} V(P)\varphi_P \quad \forall\, V \in \mathbb{V}(\mathscr{T}).$$

The *domain of influence* of a proper node P is the support of φ_P:

$$\omega_{\mathscr{T}}(T) = \operatorname{supp}(\varphi_P).$$

We say that a sequence of nonconforming meshes $\{\mathscr{T}\}$ is *admissible* if there is a universal constant $\Lambda_* \le 1$, independent of the refinement level and $\mathscr{T}$, such that

$$\operatorname{diam}(\omega_{\mathscr{T}}(T)) \le \Lambda_* h_T \quad \forall\, T \in \mathscr{T}. \tag{22}$$

An important example is quadrilaterals with *one* hanging node per edge. We observe, however, that (22) can neither be guaranteed with more than one hanging node per edge for quadrilaterals, nor for triangles with one hanging node per edge (see Problem 10).

Given an admissible grid $\mathscr{T}$, a subset $\mathscr{M}$ of elements marked for refinement, and a desired number $\rho \ge 1$ of subdivisions to be performed in each marked element, the procedure

$$\mathscr{T}_* = \mathsf{REFINE}(\mathscr{T}, \mathscr{M})$$

creates a minimal admissible mesh $\mathscr{T}_* \ge \mathscr{T}$ such that all the elements of $\mathscr{M}$ are subdivided at least ρ times. In order for $\mathscr{T}_*$ to be admissible, perhaps other elements not in $\mathscr{M}$ must be partitioned. Despite the fact that admissibility is a constraint on the refinement procedure weaker than conformity, it cannot avoid the propagation of refinements beyond $\mathscr{M}$. The complexity of REFINE is again an issue which we discuss in Sect. 6.4: we show that Theorem 1 extends to this case.

Lemma 3 (REFINE for nonconforming meshes). *Let $\mathscr{T}_0$ be an arbitrary conforming partition of Ω, except for bisection in which case $\mathscr{T}_0$ satisfies the labeling* (6) *for $d = 2$ or its higher dimensional counterpart [53]. Then the estimate*

$$\#\mathscr{T}_k - \#\mathscr{T}_0 \le \Lambda_0 \sum_{j=0}^{k-1} \#\mathscr{M}_j \quad \forall k \ge 1$$

holds with a constant Λ_0 depending on $\mathscr{T}_0, d$ and ρ.

We conclude by emphasizing that the polynomial interpolation and adaptive approximation theories of Sects. 1.5 and 1.6 extend to nonconforming meshes with fixed level of incompatibility as well.

1.8 Notes

The use of Sobolev numbers is not so common in the finite element literature, but allows as to write compact error estimates and speak about nonlinear Sobolev scale. The latter concept is quite natural in nonlinear approximation theory [22].

The discussion of bisection for $d = 2$ follows Binev et al. [7]. Stevenson extended the theory to $d > 2$ [53]. We refer to the survey by Nochetto et al. [45] for a rather complete discussion for $d > 1$, and to Sect. 6 for a proof of Theorem 1 for $d = 2$, which easily extends to $d > 2$.

The discussion of finite element spaces [10, 11, 19] and polynomial interpolation [11, 26, 48] is rather classical. In contrast, the material of adaptive approximation is much less documented. The principle of equidistribution goes back to Babuška and

Rheimboldt [5] and the a priori design of optimal meshes for corner singularities for $d = 2$ is due to Babuška et al. [3]. The construction of optimal meshes via bisection using thresholding is extracted from Binev et al. [8].

Finally the discussion of nonconforming meshes follows Bonito and Nochetto [9], and continues in Sect. 6 with the proof of Lemma 3.

1.9 Problems

Problem 1 (Nonconforming element). Given a d-simplex T in R^d with vertices $z_0, \dots, z_d$, construct a basis $\bar{\lambda}_0, \dots, \bar{\lambda}_d$ of $\mathbb{P}_1(T)$ such that $\bar{\lambda}_i(\bar{z}_j) = \delta_{ij}$ for all $i, j \in \{1, \dots, d\}$, where $\bar{z}_j$ denotes the barycenter of the face opposite to the vertex z_j. Does this local basis also lead to a global one in $\mathbb{S}^{1,0}(\mathscr{T})$?

Problem 2 (Quadratic basis functions). Express the nodal basis of $\mathbb{P}_2(T)$ in terms of barycentric coordinates of $T \in \mathscr{T}$.

Problem 3 (Quadratic dual functions). Derive expressions for the dual functions of the quadratic local Lagrange basis of $\mathbb{P}_2(T)$ for each element $T \in \mathscr{T}$. Construct a global discontinuous dual basis $\varphi_z^* \in \mathbb{S}^{2,-1}(\mathscr{T})$ of the global Lagrange basis $\varphi_z \in \mathbb{S}^{2,0}(\mathscr{T})$ for all $z \in \mathscr{N}_2(\mathscr{T})$.

Problem 4 (Lagrangian). Let $h(x)$ be a smooth function locally equivalent to the mesh-size and $v \in C^2(\Omega) \cap W_p^2(\Omega)$. Prove that a stationary point of the Lagrangian

$$\mathscr{L}[h, \lambda] = \int_\Omega \left(h(x)^{2(p-1)} \, |D^2 v(x)|^p - \frac{\lambda}{h(x)^2} \right) dx$$

satisfies the optimality condition

$$h(x)^{2(p-1)+2} \, |D^2 v(x)|^p = \text{constant}.$$

Problem 5 (W_p^2-regularity). Consider the function $v(r, \vartheta) = r^\gamma \varphi(\vartheta)$ in polar coordinates (r, ϑ) for $d = 2$ with $\varphi(\vartheta)$ smooth. Show that $v \in W_p^2(\Omega) \setminus H^2(\Omega)$ for $1 \le p < 2/(2 - \gamma)$.

Problem 6 (Edge singularities). This problem explores *formally* the effect of edge singularities for dimension $d > 2$ and polynomial degree $n \ge 1$. Since edge (or line) singularities are two dimensional locally, away from corners, we assume the behavior $v(x) \approx r(x)^\gamma$ where $r(x)$ is the distance of $x \in \Omega$ to an edge of Ω and $\gamma > 0$.

(a) Use the Principle of Equidistribution with $p = 2$ to determine the mesh grading

$$h(x) \approx \Lambda^{\frac{1}{2n+d}} r(x)^{2d \frac{\gamma - (n+1)}{2n+d}}.$$

(b) Show the following relation between Λ and number of elements $\#\mathscr{T} = \int_\Omega h(x)^{-d}$

$$\gamma > \frac{(d-2)n}{d} \quad \Rightarrow \quad \#\mathscr{T} \approx \Lambda^{-\frac{d}{2n+d}}.$$

(c) If $\gamma > \frac{(d-2)n}{d}$, then deduce the optimal interpolation error decay

$$\|\nabla(v - I_{\mathscr{T}} v)\|_{L^2(\Omega)} \lesssim (\#\mathscr{T})^{-\frac{n}{d}}.$$

(d) Prove that $\gamma > \frac{(d-2)n}{d}$ is equivalent to the regularity $\int_\Omega |D^{n+1}v|^p < \infty$ for $p > \frac{2d}{2n+d}$. If $\tau := \frac{2d}{2n+d} \geq 1$, then this would mean $v \in W_p^{n+1}(\Omega)$. However, it is easy to find examples $d > 2$ or $n > 1$ for which $\tau < 1$, in which case the Sobolev space $W_p^{n+1}(\Omega)$ must be replaced by the Besov space $B_{p,p}^{n+1}(\Omega)$ [22]. Note that $p > \tau$ is precisely what yields the crucial relation between Sobolev numbers

$$\text{sob}(B_{p,p}^{n+1}) = n + 1 - \frac{d}{p} > \text{sob}(H^1) = 1 - \frac{d}{2}.$$

We observe that for $d = 2$ all singular exponents $\gamma > 0$ lead to optimal meshes, but this is not true for $d = 3$: $n = 1$ requires $\gamma > \frac{1}{3}$ whereas $n = 2$ needs $\gamma > \frac{2}{3}$. The latter corresponds to a dihedral angle $\omega > \frac{3\pi}{2}$ and can be easily checked computationally. We thus conclude that *isotropic* graded meshes are sufficient to deal with geometric singularities for $d = 2$ but not for $d > 2$, for which *anisotropic* graded meshes are the only ones which exhibit optimal behavior. Their adaptive construction is open.

Problem 7 (Local H^2-regularity). Consider the function $v(x) \approx r(x)^\gamma$ where $r(x)$ is the distance to the origin and $d = 2$.

(a) Examine the construction of a graded mesh via the thresholding algorithm.
(b) Repeat the proof of Theorem 3 replacing the W_p^2 regularity by the corresponding local H^2-regularity of v depending on the distance to the origin.

Problem 8 (Thresholding for $d > 2$). Let $d > 2$, $n = 1$, and $v \in W_p^2(\Omega)$ with $p > \frac{2d}{2+d}$. This implies that $v \in H^1(\Omega)$ but not necessarily in $C^0(\overline{\Omega})$. Use the quasi-interpolant $I_{\mathscr{T}}$ of Proposition 1 to define the local H^1-error E_T for each element $T \in \mathscr{T}$ and use the thresholding algorithm to show Theorem 3 and Corollary 2.

Problem 9 (Reduced rate). Let $d \geq 2$, $n = 1$, and $v \in W_p^s(\Omega)$ with $1 < s < 2$ and $\text{sob}(W_p^s) > \text{sob}(H^1)$, namely $s - \frac{d}{p} > 1 - \frac{d}{2}$. Use the quasi-interpolant $I_{\mathscr{T}}$ of Proposition 1 to define the local H^1-error E_T for each element $T \in \mathscr{T}$ and use the thresholding algorithm to show Corollary 2: given $N > \#\mathscr{T}_0$ there exists $\mathscr{T} \in \mathbb{T}$ with $\#\mathscr{T} - \#\mathscr{T}_0 \lesssim N$ such that

$$\|v - I_{\mathscr{T}} v\|_{H^1(\Omega)} \lesssim \|D^s v\|_{L^p(\Omega)} N^{-\frac{s-1}{d}}.$$

Problem 10 (Level of incompatibility). This problem shows that keeping the number of hanging nodes per side bounded does not guarantee a bounded level of incompatibility for $d = 2$. The situation is similar for $d > 2$.

(a) *Square elements*: construct a selfsimilar quad-refinement of the unit square with only two hanging nodes per side and unbounded level of incompatibility.
(b) *Triangular elements*: construct selfsimilar red-refinements and bisection refinements of the unit reference triangle with one hanging node per side and unbounded level of incompatibility.

Problem 11 (Quasi-interpolation of discontinuous functions). Let $\mathcal{T}$ be an admissible nonconforming mesh. Let $\mathbb{V}(\mathcal{T})$ denote the space of discontinuous piecewise polynomials of degree $\leq n$ over $\mathcal{T}$, and $\mathbb{V}^0(\mathcal{T})$ be the subspace of continuous functions. Construct a local quasi-interpolation operator $I_{\mathcal{T}} : \mathbb{V}(\mathcal{T}) \to \mathbb{V}^0(\mathcal{T})$ with the following approximation property for all $V \in \mathbb{V}(\mathcal{T})$ and $|\alpha| = 0, 1$

$$\|D^\alpha(V - I_{\mathcal{T}} V)\|_{L^2(T)} \lesssim h_T^{\frac{1-|\alpha|}{2}} \| [\![V]\!] \|_{L^2(\Sigma_{\mathcal{T}}(T))} \quad \forall T \in \mathcal{T},$$

where $\Sigma_{\mathcal{T}}(T)$ stands for all sides within $N_{\mathcal{T}}(T)$ and $[\![V]\!]$ denotes the jump of V across sides.

2 Error Bounds for Finite Element Solutions

In Sect. 1 we have seen that approximating a given known function with meshes which are adapted to that function can impressively outperform the approximation with quasi-uniform meshes. In view of the fact that the solution of a boundary value problem is given only implicitly, it is not all clear if this is also true for its adaptive numerical solution. Considering a simple model problem and discretization, we now derive two upper bounds for the error of the finite element solution: an a priori one and an a posteriori one. The a priori bound reveals that an adaptive variant of the finite element method has the potential of a similar performance. The a posteriori bound is a first step to design such a variant, which has to face the complication that the target function is given only implicitly.

2.1 *Model Boundary Value Problem*

In order to minimize technicalities in the presentation, let us consider the following simple boundary value problem as a model problem: find a scalar function $u = u(x)$ such that

$$\begin{aligned} -\operatorname{div}(\boldsymbol{A}\nabla u) &= f \quad \text{in } \Omega, \\ u &= 0 \quad \text{on } \partial\Omega, \end{aligned} \tag{23}$$

where $\Omega \subset \mathbb{R}^d$ is a bounded domain with Lipschitz boundary $\partial\Omega$, $\boldsymbol{A} = \boldsymbol{A}(x)$ a map into the positive definite $d \times d$ matrices, and $f = f(x)$ a scalar load term. Introducing the Hilbert space

$$\mathbb{V} := H_0^1(\Omega) := \{v \in H^1(\Omega) \mid v_{|\partial\Omega} = 0\}, \quad \|v\|_{\mathbb{V}} := \left(\int_\Omega |\nabla v|^2\right)^{1/2},$$

and the bilinear form

$$\mathscr{B}[v, w] := \int_\Omega \boldsymbol{A}\nabla v \cdot \nabla w, \quad v, w \in \mathbb{V},$$

the weak solution of (23) is characterized by

$$u \in \mathbb{V}: \quad \mathscr{B}[u, v] = \langle f, v \rangle \quad \text{for all } v \in \mathbb{V}. \tag{24}$$

Hereafter $\langle \cdot, \cdot \rangle$ stands for the $L^2(\Omega)$-scalar product and also for a duality paring. We assume that $f \in \mathbb{V}^* = H^{-1}(\Omega) := H_0^1(\Omega)^*$ and that there exist constants $0 < \alpha_1 \leq \alpha_2$ with

$$\forall x \in \Omega,\ \xi \in \mathbb{R}^d \quad \alpha_1|\xi|^2 \leq \boldsymbol{A}(x)\xi \cdot \xi \text{ and } |\boldsymbol{A}(x)\xi| \leq \alpha_2|\xi|. \tag{25}$$

The latter implies that the operator $-\operatorname{div}(\boldsymbol{A}\nabla\cdot)$ is uniformly elliptic. Moreover, the bilinear form $\mathscr{B}$ is coercive and continuous with constants α_1 and α_2, respectively. Lax–Milgram Theorem and Poincaré–Friedrichs Inequality

$$\|v\|_\Omega \leq \operatorname{diam}(\Omega)\|\nabla v\|_\Omega \quad \text{for all } v \in \mathbb{V} = H_0^1(\Omega) \tag{26}$$

thus ensure existence and uniqueness of the weak solution (24).

Note that $\boldsymbol{A}$ is not assumed to be symmetric and so the bilinear form $\mathscr{B}$ may be nonsymmetric. For the a posteriori upper bound, we will require some additional regularity on the data f and $\boldsymbol{A}$ in Sect. 2.4.

2.2 Galerkin Solutions

Since $\mathbb{V}$ has infinite dimension, problem (24) cannot be implemented on a computer and solved numerically. Given a subspace $\mathbb{S} \subset \mathbb{V}$, the corresponding Galerkin solution or approximation of (24) is given by

$$U \in \mathbb{S}: \quad \mathscr{B}[U, V] = \langle f, V \rangle \quad \text{for all } V \in \mathbb{S}. \tag{27}$$

We simply replaced each occurrence of $\mathbb{V}$ in (24) by $\mathbb{S}$. If $\mathbb{S}$ is finite-dimensional, we can choose a basis of $\mathbb{S}$ and the coefficients of the expansion of U can be determined by solving a square linear system.

Residual: Associate the functional $\mathscr{R} \in \mathbb{V}^*$ given by

$$\langle \mathscr{R}, v \rangle := \langle f, v \rangle - \mathscr{B}[U, v],$$

to $U \in \mathbb{S}$. The functional $\mathscr{R}$ is called the residual and depends only on the approximate solution U and data $\boldsymbol{A}$ and f. Moreover, it has the following properties:

- It relates to the typically unknown error function $u - U$ in the following manner:

$$\langle \mathscr{R}, v \rangle = \mathscr{B}[u - U, v] \quad \text{for all } v \in \mathbb{V}. \tag{28}$$

  This is a direct consequence of the characterization (24) of the exact solution.
- It vanishes for discrete test functions, which in the case of symmetric $\boldsymbol{A}$ corresponds to the so-called Galerkin orthogonality:

$$\mathscr{B}[u - U, V] = \langle \mathscr{R}, V \rangle = 0 \quad \text{for all } V \in \mathbb{S}. \tag{29}$$

  This immediately follows from (28) and the definition (27) of the Galerkin solution.

Quasi-best approximation: Property (25) of $\boldsymbol{A}$, Galerkin orthogonality (29) and Cauchy–Schwarz Inequality in $L^2(\Omega)$ imply

$$\begin{aligned} \alpha_1 \|u - U\|_{\mathbb{V}}^2 &\leq \mathscr{B}[u - U, u - U] = \mathscr{B}[u - U, u - V] \\ &\leq \alpha_2 \|u - U\|_{\mathbb{V}} \|u - V\|_{\mathbb{V}} \end{aligned}$$

for arbitrary $V \in \mathbb{S}$. This proves the famous

Theorem 4 (Céa Lemma). *The Galerkin solution is a quasi-best approximation from* $\mathbb{S}$ *with respect to the* $\mathbb{V}$*-norm:*

$$\|u - U\|_{\mathbb{V}} \leq \frac{\alpha_2}{\alpha_1} \inf_{V \in S} \|u - V\|_{\mathbb{V}}. \tag{30}$$

If the bilinear $\mathscr{B}$ is also symmetric and one considers the error with respect to the so-called energy norm $\mathscr{B}[\cdot, \cdot]^{1/2}$, the Galerkin solution is even the best approximation from $\mathbb{S}$; see Problem 12.

2.3 Finite Element Solutions and A Priori Bound

Problem (27) can be solved numerically on a computer, if we dispose of an implementable basis of $\mathbb{S}$. As an example of such space, let $\mathscr{T}$ be a conforming triangulation of Ω into d-simplices (this imposes further conditions on Ω) and consider

$$\mathbb{S} = \mathbb{V}(\mathscr{T}) := \{V \in \mathbb{S}^{n,0}(\mathscr{T}) \mid V_{|\partial\Omega} = 0\}, \tag{31}$$

where, as in Sect. 1.4, $\mathbb{S}^{n,0}(\mathscr{T})$ the space of continuous functions that are piecewise polynomial up to degree n. This is in fact a subspace of $\mathbb{V} = H_0^1(\Omega)$ thanks to the continuity requirement and boundary condition for the functions in $\mathbb{V}(\mathscr{T})$. Moreover, the basis $\{\varphi_z\}_{z \in \mathscr{N} \cap \Omega}$ from Sect. 1.4 can be easily constructed in the computer; see for example Siebert [50].

The space $\mathbb{V}(\mathscr{T})$ is a popular choice for $\mathbb{S}$ in (27) and their combination may be viewed as a model finite element discretization.

In Sect. 1.5 we studied the approximation properties of $\mathbb{S}^{n,0}(\mathscr{T})$ with the help of (quasi-)interpolation operators $I_{\mathscr{T}}$. Since the right-hand side of (30) is bounded in terms of $\|u - I_{\mathscr{T}} u\|_{\mathbb{V}} = \|\nabla(u - I_{\mathscr{T}} u)\|_{L^2(\Omega)}$ with $I_{\mathscr{T}}$ as in Proposition 2, the discussion of Sect. 1.5 applies also to the error of the Galerkin solution $U_{\mathscr{T}}$ in $\mathbb{V}(\mathscr{T})$. In particular, the combination of the Céa Lemma and Theorem 2 yields the following upper bound. Since it does not involve the discrete solution, it is also comes with the adjective "a priori".

Theorem 5 (A priori upper bound). *Assume that the exact solution u of (24) satisfies $u \in W_p^s(\Omega)$ with $1 \le s \le n+1$, $1 \le p \le 2$, and set*

$$r := \operatorname{sob}(W_p^s(\Omega)) - \operatorname{sob}(H^1(\Omega)) > 0.$$

Then the error of the finite element solution $U_{\mathscr{T}} \in \mathbb{S} = \mathbb{V}(\mathscr{T})$ of (27) satisfies the global a priori upper bound

$$\|u - U_{\mathscr{T}}\|_{\mathbb{V}} \lesssim \|h^r D^s u\|_{L^p(\Omega)}. \tag{32}$$

The discussion in Sect. 1.6 about adaptively graded meshes only partially carries over to the error of the finite element solution $U_{\mathscr{T}}$, from now on denoted U. In view of the Céa Lemma, Sect. 1.6 shows that there are sequences of meshes such that the error of U decays as $\#\mathscr{T}^{-1/2}$ if, for example, $d = 2$ and $u \in W^{2,p}(\Omega)$ with $p > 1$. Notice however that the thresholding algorithm utilizes the local errors $E_T = \|\nabla(u - I_{\mathscr{T}} u)\|_{L^2(T)}$, which are typically not computable. The construction of appropriate meshes when the target function is given only implicitly by a boundary value problem is much more subtle. A first step towards this goal is developed in the next section.

2.4 A Posteriori Upper Bound

The *a priori* upper bound (32) is not computable and essentially provides only asymptotic information, namely the asymptotic convergence rate. The goal of this section is to derive an alternative bound, so-called *a posteriori* bound, that provides information beyond asymptotics and is computable in terms of data and the approximate solution. It is worth noting that such bounds are useful not only for adaptivity but also for the quality assessment of the approximate solution.

Since in this section the grid $\mathscr{T}$ is (arbitrary but) fixed, we simplify the notation by suppressing the subscript indicating the dependence on the grid in case of the approximate solution and similar quantities.

Error and residual: Our starting point is the algebraic relationship (28) between the residual $\mathscr{R}$ and the error function $u - U$. It implies (Problem 13)

$$\|u - U\|_{\mathbb{V}} \leq \frac{1}{\alpha_1}\|\mathscr{R}\|_{\mathbb{V}^*} \leq \frac{\alpha_2}{\alpha_1}\|u - U\|_{\mathbb{V}}, \tag{33}$$

which means that the dual norm

$$\|\ell\|_{\mathbb{V}^*} := \sup\left\{\langle \ell, v\rangle \mid v \in \mathbb{V}, \|v\|_{\mathbb{V}} \leq 1\right\} \tag{34}$$

is a good measure for the residual $\mathscr{R}$ if we are interested in the error $\|u - U\|_{\mathbb{V}} = \|\nabla(u - U)\|_{L^2(\Omega)}$. However the evaluation of $\|\mathscr{R}\|_{\mathbb{V}^*} = \|\mathscr{R}\|_{H^{-1}(\Omega)}$ is impractical and, moreover, does not provide local information for guiding an adaptive mesh refinement. We therefore aim at a sharp upper bound of $\|\mathscr{R}\|_{H^{-1}(\Omega)}$ in terms of locally computable quantities.

Assumptions and structure of residual: For the derivation of a computable upper bound of the dual norm of the residual, we require that

$$f \in L^2(\Omega) \quad \text{and} \quad \boldsymbol{A} \in W^{1,\infty}(\Omega; \mathscr{T}), \tag{35}$$

where the latter means that $\boldsymbol{A}$ is Lipschitz in each element of $\mathscr{T}$. Under these assumptions, we can write $\langle \mathscr{R}, v\rangle$ as integrals over elements $T \in \mathscr{T}$ and elementwise integration by parts yields the representation:

$$\begin{aligned}\langle \mathscr{R}, v\rangle &= \int_\Omega fv - \boldsymbol{A}\nabla U \cdot \nabla v = \sum_{T\in\mathscr{T}} \int_T fv - \boldsymbol{A}\nabla U \cdot \nabla v \\ &= \sum_{T\in\mathscr{T}} \int_T rv + \sum_{S\in\mathscr{S}} \int_S jv,\end{aligned} \tag{36}$$

where

$$\begin{aligned}&r = f + \operatorname{div}(\boldsymbol{A}\nabla U) \quad \text{in any simplex } T \in \mathscr{T}, \\ &j = [\![\boldsymbol{A}\nabla U]\!] \cdot \boldsymbol{n} = \boldsymbol{n}^+ \cdot \boldsymbol{A}\nabla U_{|T^+} + \boldsymbol{n}^- \cdot \boldsymbol{A}\nabla U_{|T^-} \quad \text{on any internal side } S \in \mathscr{S}\end{aligned} \tag{37}$$

and $\boldsymbol{n}^+$, $\boldsymbol{n}^-$ are unit normals pointing towards T^+, $T^- \in \mathscr{T}$. We see that the distribution $\mathscr{R}$ consists of a regular part r, called *interior or element residual*, and a singular part j, called *jump or interelement residual*. The regular part is absolutely continuous w.r.t. the d-dimensional Lebesgue measure and is related to the strong form of the PDE. The singular part is supported on the skeleton $\Gamma = \bigcup_{S \in \mathcal{S}} S$ of $\mathcal{T}$ and is absolutely continuous w.r.t. the $(d-1)$-dimensional Hausdorff measure.

We point out that this structure of the residual is not special to the model problem and its discretization but rather arises from the weak formulation of the PDE and the piecewise construction of finite element spaces.

Scaled integral norms: In view of the structure of the residual $\mathcal{R}$, we make our goal precise as follows: we aim at a sharp upper bound for $\|\mathcal{R}\|_{H^{-1}(\Omega)}$ in terms of local Lebesgue norms of the element and interelement residuals r and j, which are considered to be computable because they can be easily approximated with numerical integration. This approach is usually called standard a posteriori error estimation.

The sharpness of these bounds crucially hinges on appropriate local scaling constants for the aforementioned Lebesgue norms, which depend on the local geometry of the mesh. For simplicity, we will explicitly trace only the dependence on the local mesh-size and write "$\lesssim$" instead of "$\leq C$", where the constant C is bounded in terms of the shape coefficient $\sigma(\mathcal{T}) = \max_{T \in \mathcal{T}} \overline{h}_T / \underline{h}_T$ of the triangulation $\mathcal{T}$ and the dimension d.

Localization: As a first step, we decompose the residual $\mathcal{R}$ into local contributions with the help of the Courant basis $\{\varphi_z\}_{z \in \mathcal{V}}$ from Sect. 1.4. Hereafter $\mathcal{V}$ stands for the set of vertices of $\mathcal{T}$, which coincide with the nodes of $\mathbb{S}^{1,0}(\mathcal{T})$. The Courant basis has the following properties:

- It provides a partition of unity:

$$\sum_{z \in \mathcal{V}} \varphi_z = 1 \quad \text{in } \Omega. \tag{38}$$

- For each interior vertex z, the corresponding basis function φ_z is contained in $\mathbb{V}(\mathcal{T})$ and so the residual is orthogonal to the interior contributions of the partition of unity:

$$\langle \mathcal{R}, \varphi_z \rangle = 0 \quad \text{for all } z \in \mathcal{V} \cap \Omega. \tag{39}$$

The second property corresponds to the Galerkin orthogonality. Notice that the first property involves all vertices, while in the second one the boundary vertices are excluded.

Given any $v \in H_0^1(\Omega)$, we apply (38) and then (39) to write

$$\langle \mathcal{R}, v \rangle = \sum_{z \in \mathcal{V}} \langle \mathcal{R}, v\varphi_z \rangle = \sum_{z \in \mathcal{V}} \langle \mathcal{R}, (v - c_z)\varphi_z \rangle, \tag{40}$$

where $c_z \in \mathbb{R}$ and $c_z = 0$ whenever $z \in \partial\Omega$. Exploiting representation (36), $0 \leq \varphi_z \leq 1$, and the fact that the φ_z are locally supported, we can bound each local contribution $\langle \mathcal{R}, (v - c_z)\varphi_z \rangle$ in the following manner:

$$|\langle \mathscr{R}, (v-c_z)\varphi_z\rangle| \le \left|\int_{\omega_z} r(v-c_z)\varphi_z\right| + \left|\int_{\gamma_z} j(v-c_z)\varphi_z\right|, \tag{41}$$

where $\omega_z := \cup_{T\ni z} T$ is the star (or patch) around a vertex $z \in \mathscr{V}$ in $\mathscr{T}$ and γ_z is the skeleton of ω_z, i.e. the union of all sides emanating from z; note that r in (41) is computed elementwise. We examine the two terms on the right-hand side separately.

Bounding the element residual: We first consider the terms associated with the element residual r. The key tool for a sharp bound is the following local Poincaré-type inequality. Let

$$h_z := |\omega_z|^{1/d}$$

and notice that this quantity is, up to the shape coefficient $\sigma(\mathscr{T})$, equivalent to the diameter of ω_z, to $h_T = |T|^{1/d}$ if T is a d-simplex of ω_z and to $h_S := |S|^{1/(d-1)}$ if S is a side of γ_z.

Lemma 4 (Local Poincaré-type inequality). *For any $v \in H_0^1(\Omega)$ and $z \in \mathscr{V}$ there exists $c_z \in \mathbb{R}$ such that*

$$\|v - c_z\|_{L^2(\omega_z)} \lesssim h_z \|\nabla v\|_{L^2(\omega_z)}. \tag{42}$$

If $z \in \partial\Omega$ is a boundary vertex, then we can take $c_z = 0$.

We postpone the proof of Lemma 4. Combining the Cauchy–Schwarz inequality in $L^2(\omega_z)$ and Lemma 4 readily yields

$$\left|\int_{\omega_z} r(v-c_z)\varphi_z\right| \le \|r\,\varphi_z^{1/2}\|_{L^2(\omega_z)}\|v-c_z\|_{L^2(\omega_z)} \lesssim h_z\|r\,\varphi_z^{1/2}\|_{L^2(\omega_z)}\|\nabla v\|_{L^2(\omega_z)}. \tag{43}$$

Notice that the right-hand side consists of two factors: a computable one in the desired form and one that involves the test function in a local variant of the norm of the test space.

Bounding the jump residual: Next, we consider the terms associated to the jump residual j. Recall that j is supported on sides and so proceeding similarly as for the element residual will bring up traces of the test function. The following trace inequality exactly meets our needs.

Lemma 5 (Scaled trace inequality). *For any side S of a d-simplex T the following inequality holds:*

$$\|w\|_{L^2(S)} \lesssim h_S^{-1/2}\|w\|_{L^2(T)} + h_S^{1/2}\|\nabla w\|_{L^2(T)} \quad \textit{for all } w \in H^1(T). \tag{44}$$

We again postpone the proof, now of Lemma 5. We apply first the Cauchy–Schwarz inequality in $L^2(\gamma_z)$, then Lemma 5 and finally Lemma 4 to obtain

$$\left|\int_{\gamma_z} j(v - c_z)\varphi_z\right| \leq \|j\,\varphi_z^{1/2}\|_{L^2(\gamma_z)}\|v - c_z\|_{L^2(\gamma_z)} \lesssim h_z^{1/2}\|j\,\varphi_z^{1/2}\|_{L^2(\gamma_z)}\|\nabla v\|_{L^2(\omega_z)}, \tag{45}$$

where the right-hand side has the same structure as that of the element residual.

Upper bound for residual norm: We collect the local estimates and sum them up in order to arrive at the desired bound for the dual norm of the residual. Inserting the estimates (43) and (45) for element and jump residuals into (41) gives

$$|\langle \mathscr{R}, v\,\varphi_z\rangle| \lesssim \left(h_z\|r\,\varphi_z^{1/2}\|_{L^2(\omega_z)} + h_z^{1/2}\|j\,\varphi_z^{1/2}\|_{L^2(\gamma_z)}\right)\|\nabla v\|_{L^2(\omega_z)}.$$

Recalling the decomposition (40), we sum over $z \in \mathscr{V}$ and use Cauchy–Schwarz in $\mathbb{R}^{\#\mathscr{T}}$ to arrive at

$$|\langle \mathscr{R}, v\rangle| \lesssim \left(\sum_{z\in\mathscr{V}} h_z^2\|r\,\varphi_z^{1/2}\|^2_{L^2(\omega_z)} + h_z\|j\,\varphi_z^{1/2}\|^2_{L^2(\gamma_z)}\right)^{1/2}\left(\sum_{z\in\mathscr{V}}\|\nabla v\|^2_{L^2(\omega_z)}\right)^{1/2}.$$

For bounding the second factor, we resort to the finite overlapping property of stars, namely

$$\sum_{z\in\mathscr{V}} \chi_{\omega_z}(x) \leq d + 1,$$

and infer that

$$\sum_{z\in\mathscr{V}} \|\nabla v\|^2_{L^2(\omega_z)} \lesssim \|\nabla v\|^2_{L^2(\Omega)}.$$

Since mesh refinement is typically based upon element subdivision, we regroup the terms within the first factor. To this end, denote by $h\colon \Omega \to \mathbb{R}^+$ the mesh-size function given by $h(x) := |S|^{1/k}$ if x belongs to the interior of the k-subsimplex S of $\mathscr{T}$ with $k \in \{1,\ldots,d\}$. Then for all $x \in \omega_z$ we have $h_z \lesssim h(x)$. Therefore employing (38) once more and recalling that Γ is the union of all interior sides of $\mathscr{T}$, we deduce

$$\begin{aligned}\sum_{z\in\mathscr{V}} h_z^2\|r\,\varphi_z^{1/2}\|^2_{L^2(\omega_z)} + h_z\|j\,\varphi_z^{1/2}\|^2_{L^2(\gamma_z)} &\lesssim \sum_{z\in\mathscr{V}} \|h\,r\,\varphi_z^{1/2}\|^2_{L^2(\Omega)} + \|h^{1/2}\,j\,\varphi_z^{1/2}\|^2_{L^2(\Gamma)}\\ &= \|h\,r\|^2_{L^2(\Omega)} + \|h^{1/2}\,j\|^2_{L^2(\Gamma)}.\end{aligned}$$

Thus, introducing the *element indicators*

$$\mathscr{E}^2_{\mathscr{T}}(U,T) := h_T^2\|r\|^2_{L^2(T)} + h_T\|j\|^2_{L^2(\partial T\setminus\partial\Omega)} \tag{46}$$

and the *error estimator*

$$\mathscr{E}^2_{\mathscr{T}}(U) = \sum_{T\in\mathscr{T}} \mathscr{E}^2_{\mathscr{T}}(U,T), \tag{47}$$

we arrive at the following upper bound for the dual norm of the residual:

$$\|\mathcal{R}\|_{H^{-1}(\Omega)} \lesssim \mathcal{E}_{\mathcal{T}}(U). \tag{48}$$

Hereafter, we write $\mathcal{E}_{\mathcal{T}}(U, \mathcal{M})$ to indicate that the estimator is computed over $\mathcal{M} \subset \mathcal{T}$, whereas $\mathcal{E}_{\mathcal{T}}(U, \mathcal{T}) = \mathcal{E}_{\mathcal{T}}(U)$ if no confusion arises.

Proofs of Poincaré-type and trace inequalities: We now prove Lemmas 4 and 5. We start with a formula for the mean value of a trace, which follows from the Divergence Theorem.

Lemma 6 (Trace identity). *Let T be a d-simplex, S a side of T, and z the vertex opposite to S. Defining the vector field $\boldsymbol{q}_S$ by*

$$\boldsymbol{q}_S(x) := x - z$$

the following equality holds:

$$\frac{1}{|S|}\int_S w = \frac{1}{|T|}\int_T w + \frac{1}{d|T|}\int_T \boldsymbol{q}_S \cdot \nabla w \quad \text{for all } w \in W_1^1(T).$$

Proof. We start with properties of the vector field $\boldsymbol{q}_S$. Let S' be an arbitrary side of T and fix some $y \in S'$. We then see $\boldsymbol{q}_S(x) \cdot \boldsymbol{n}_T = \boldsymbol{q}_S(y) \cdot \boldsymbol{n}_T + (x - y) \cdot \boldsymbol{n}_T = \boldsymbol{q}_S(y) \cdot \boldsymbol{n}_T$ for any $x \in S'$ since $x - y$ is a tangent vector to S'. Therefore, on each side of T, the associated normal flux $\boldsymbol{q}_S \cdot \boldsymbol{n}_T$ is constant. In particular, we see $\boldsymbol{q}_S \cdot \boldsymbol{n}_T$ vanishes on $\partial T \setminus S$ by choosing $y = z$ for sides emanating from z. Moreover, $\operatorname{div} \boldsymbol{q}_S = d$. Thus, if $w \in C^1(\overline{T})$, the Divergence Theorem yields

$$\int_T \boldsymbol{q}_S \cdot \nabla w = -d\int_T w + (\boldsymbol{q}_S \cdot \boldsymbol{n}_T)_{|S}\int_S w.$$

Take $w = 1$ to show $(\boldsymbol{q}_S \cdot \boldsymbol{n}_T)_{|S} = d|T|/|S|$ and extend the result to $w \in W_1^1(T)$ by density. □

Proof of Lemma 5. Apply Lemma 6 to $|w|^2$; for the details see Problem 17. □

Proof of Lemma 4. [1] For any $z \in \mathcal{V}$ the value

$$\bar{c}_z = \frac{1}{|\omega_z|}\int_{\omega_z} v$$

is an optimal choice and (42) follows from (8) with $c_z = \bar{c}_z$.
[2] If $z \in \partial\Omega$, then we observe that there exists a side $S \subset \partial\omega_z \cap \partial\Omega$ such that $v = 0$ on S. We therefore can write

$$v = v - \frac{1}{|S|}\int_S v = (v - \bar{c}_z) - \frac{1}{|S|}\int_S (v - \bar{c}_z)$$

whence, using Lemma 5 and Step 1 for the second term,

$$\|v\|_{L^2(\omega_z)} \lesssim \|v - \bar{c}_z\|_{L^2(\omega_z)} + h_z \|\nabla v\|_{L^2(\omega_z)} \lesssim h_z \|\nabla v\|_{L^2(\omega_z)},$$

which establishes the supplement for boundary vertices. □

Upper bound for error: Inserting the bound (48) for the dual norm of the residual in the first bound of (33), we obtain the main result of this section.

Theorem 6 (A posteriori upper bound). *Let u be the exact solution of the model problem* (24) *satisfying* (25) *and* (35). *The error of the finite element solution $U \in \mathbb{S} = \mathbb{V}(\mathscr{T})$ of* (27) *is bounded in terms of the estimator* (47) *as follows:*

$$\|u - U\|_{\mathbb{V}} \lesssim \frac{1}{\alpha_1} \mathscr{E}_{\mathscr{T}}(U), \tag{49}$$

where the hidden constant depends only on the shape coefficient $\sigma(\mathscr{T})$ of the triangulation $\mathscr{T}$ and on the dimension d.

Notice that the a posteriori bound in Theorem 6 does not require additional regularity on the exact solution u as the a priori one in Theorem 5. On the other hand, the dependence of the estimator on the approximate solution prevents us from directly extracting information such as asymptotic decay rate of the error. The question thus arises how sharp the a posteriori bound in Theorem 6 is.

In this context it is worth noticing that if we did not exploit orthogonality and used a global Poincaré-type inequality instead of the local ones, the resulting scalings of the element and jump residuals would be, respectively, 1 and $h_T^{-1/2}$ and the corresponding upper bound would have a lower asymptotic decay rate. We will show in the next Sect. 3 that the upper bound in Theorem 6 is sharp in an asymptotic sense.

2.5 Notes

The discussion of the quasi-best approximation and the a priori upper bound of the error of the finite element solution are classical; see Braess [10], Brenner and Scott [11], and Ciarlet [19]. The core of the a posteriori upper bound is a bound of the dual norm of the residual in terms of scaled Lebesgue norms. This approach is usually called *standard a posteriori error estimation* and has been successfully used for a variety of problems and discretizations. For alternative approaches we refer to the monographs of Ainsworth and Oden [2] and Verfürth [58] on a posteriori error estimation.

Typically standard a posteriori error estimation is carried out with the help of error estimates for quasi-interpolation as in Sect. 1.5, which in turn rely on local Bramble-Hilbert lemmas. The above presentation invokes only the special case of Poincaré-type inequalities. It is a simplified version of derivation in Veeser and Verfürth [56], which has been influenced by Babuška and Rheinboldt [5], Carstensen and Funken [12], and Morin et al. [42], and provides in particular constants that are explicit in terms of local Poincaré constants. It is worth mentioning that the ensuing constants are found in [56] for sample meshes and have values close to 1.

The setting and assumption of the model problem and discretization in this section avoids the following complications: numerical integration, approximation of boundary values, approximation of the domain, and inexact solution of the discrete system. While all these issues have been analyzed in an a priori context, only some of them have been considered in a posteriori error estimation; see Ainsworth and Kelly [1], Dörfler and Rumpf [25], Morin et al. [42], Nochetto et al. [46], and Sacchi and Veeser [47].

2.6 Problems

Problem 12 (Best approximation for symmetric problems). Consider the model problem (24), assume in addition to (25) that A is symmetric and denote the energy norm associated with the differential operator $-\operatorname{div}(A\nabla\cdot)$ by

$$|\!|\!| v |\!|\!|_\Omega := \left(\int_\Omega A\nabla v \cdot \nabla v\right)^{1/2}.$$

Prove that the Galerkin solution is the best approximation from $\mathbb{S} = \mathbb{V}(\mathscr{T})$ with respect to the energy norm:

$$|\!|\!| u - U |\!|\!|_\Omega = \min_{V \in \mathbb{S}} |\!|\!| u - V |\!|\!|_\Omega . \tag{50}$$

Derive from this that in this case (30) improves to

$$\|u - U\|_{\mathbb{V}} \le \sqrt{\frac{\alpha_2}{\alpha_1}} \inf_{V \in \mathbb{S}} \|u - V\|_{\mathbb{V}}.$$

Problem 13 (Equivalence of error and residual norm). Prove the equivalence (33) between error and dual norm of the residual. Consider the model problem also with a symmetric A and derive a similar relationship for the energy norm error.

Problem 14 (Dominance of jump residual). Considering the model problem (24) and its discretization (27) with (31) and $n = 1$, show that, up to higher order terms, the jump residual

$$\eta_{\mathscr{T}}(U) = \left(\sum_{S \in \mathscr{S}} \|h^{1/2} j\|^2_{L^2(S)}\right)^{1/2}$$

bounds $\|\mathscr{R}\|_{H^{-1}(\Omega)}$, which entails that the estimator $\mathscr{E}_{\mathscr{T}}(U)$ is dominated by $\eta_{\mathscr{T}}(U)$. To this end, revise the proof of the upper bound for $\|\mathscr{R}\|_{H^{-1}(\Omega)}$, use

$$c_z = \frac{1}{\int_{\omega_z} \varphi_z} \int_{\omega_z} f \varphi_z$$

and rewrite $\int_{\omega_z} f\,(v - c_z)\,\varphi_z$ by exploiting this weighted L^2-orthogonality.

Problem 15 (A posteriori upper bound with quasi-interpolation). Consider the model problem (24) and its discretization (27) with space $\mathbb{S} = \mathbb{V}(\mathscr{T})$, and derive the upper a posteriori error bound without using the discrete partition of unity. To this end, use (36) and combine the scaled trace inequality (44) with the local interpolation error estimate (7). Show as an intermediate step the upper bound

$$|\langle \mathscr{R}, v\rangle| \lesssim \sum_{T\in\mathscr{T}} \mathscr{E}_{\mathscr{T}}(U,T)\|\nabla v\|_{L^2(N_{\mathscr{T}}(T))} \tag{51}$$

with $N_{\mathscr{T}}(T)$ from Sect. 1.5. This bound will be useful in Sect. 4.

Problem 16 (Upper bound for certain singular loads). Revising the proof of Theorem 6, derive an a posteriori upper bound in the case of right-hand sides of the form

$$\langle f, v\rangle = \int_\Omega g_0 v + \int_\Gamma g_1 v, \quad v \in \mathbb{V} = H_0^1(\Omega),$$

where $g_0 \in L^2(\Omega)$, $g_1 \in L^2(\Gamma)$, and Γ stands for the skeleton of the mesh $\mathscr{T}$.

Problem 17 (Scaled trace inequality). Work out the details of the proof of Lemma 5, taking into account that $h_T \approx |T| / |S| \approx h_S$.

Problem 18 (A posteriori upper bound for L^2-error). Assuming that Ω is convex and applying a duality argument, establish a variant of (33) between the L^2-error $\|u - U\|_{L^2(\Omega)}$ and a suitable dual norm of the residual. Use this to derive the a posteriori upper bound

$$\|u - U\|_{L^2(\Omega)} \lesssim \left(\sum_{T\in\mathscr{T}} h_T^2 \mathscr{E}_{\mathscr{T}}(U,T)^2\right)^{1/2},$$

where the hidden constant depends in addition on the domain Ω.

3 Lower A Posteriori Bounds

The goal of this section is to assess the sharpness of the a posteriori upper bound for the model problem and discretization. We show not only that it is sharp in an asymptotic sense like the a priori bound but also in a local sense and, for certain data, in a non-asymptotic sense. Moreover, we verify that the latter cannot be true for all data and argue that this is the price to pay for the upper bound to be computable.

As in Sect. 2.4, "$\lesssim$" stands for "$\leq C$", where the constant C is bounded in terms of the shape coefficient $\sigma(\mathscr{T})$ of the triangulation $\mathscr{T}$ and the dimension d and, often, we do not indicate the dependence on the arbitrary but fixed triangulation.

3.1 Local Lower Bounds

The first step in the derivation of the upper bound (49) is that the error is bounded in terms of an appropriate dual norm of the residual. In the case of the model problem (24) this relies on the continuity of $[-\operatorname{div}(\boldsymbol{A}\nabla\cdot)]^{-1} : H^{-1}(\Omega) \to H_0^1(\Omega)$. Notice that the inverse is a global operator, while $-\operatorname{div}(\boldsymbol{A}\nabla\cdot)$ in the classical sense is a local one. One thus may suspect that an appropriate local dual norm of the residual is bounded in terms of the local error. Let us verify this for the model problem (24).

Local dual norms: Let ω be a subdomain of Ω and notice that $H^{-1}(\omega)$ is a good candidate for the local counterpart of $H^{-1}(\Omega)$. Given $v \in H_0^1(\omega)$ (and extending it by zero on $\Omega \setminus \omega$), the algebraic relationship (28), the Cauchy–Schwarz inequality in $L^2(\omega)$, and (25) readily yield

$$\langle \mathscr{R}, v\rangle = \mathscr{B}[u-U, v] = \int_\omega \boldsymbol{A}\nabla(u-U)\cdot\nabla v \le \alpha_2 \|\nabla(u-U)\|_{L^2(\omega)}\|\nabla v\|_{L^2(\omega)}.$$

Consequently,

$$\|\mathscr{R}\|_{H^{-1}(\omega)} \le \alpha_2 \|\nabla(u-U)\|_{L^2(\omega)}, \tag{52}$$

entailing that lower bounds for the local error $\|\nabla(u-U)\|_{L^2(\omega)}$ may be shown by bounding the local dual norm $\|\mathscr{R}\|_{H^{-1}(\omega)}$ from below.

Local dual and scaled integral norms: As for the a posteriori upper bound, we assume (35). If we take $\omega = T \in \mathscr{T}$ in the preceding paragraph, then there holds

$$\|\mathscr{R}\|_{H^{-1}(T)} = \sup_{\|\nabla v\|_{L^2(T)}\le 1} \langle \mathscr{R}, v\rangle = \sup_{\|\nabla v\|_{L^2(T)}\le 1} \int_T rv = \|r\|_{H^{-1}(T)} \tag{53}$$

thanks to the representation (36). Recall that the corresponding indicator $\mathscr{E}_{\mathscr{T}}(U,T)$ contains the term $h_T\|r\|_{L^2(T)}$ and therefore we wonder about the relationship of $\|r\|_{H^{-1}(T)}$ and $h_T\|r\|_{L^2(T)}$. Mimicking the local part in the derivation of the a posteriori upper bound in Sect. 2.4, we obtain

$$\int_T rv \le \|r\|_{L^2(T)}\|v\|_{L^2(T)} \lesssim h_T\|r\|_{L^2(T)}\|\nabla v\|_{L^2(T)}$$

with the help of the Poincaré–Friedrichs inequality (26). Hence there holds

$$\|r\|_{H^{-1}(T)} \lesssim h_T\|r\|_{L^2(T)}. \tag{54}$$

Since $L^2(\Omega)$ is a proper subspace of $H^{-1}(\Omega)$, the inverse inequality cannot hold for arbitrary r. Consequently, $h_T\|r\|_{L^2(T)}$ may overestimate $\|r\|_{H^{-1}(T)}$. On the other hand, if $r \in \mathbb{R}$ is *constant* and $\eta = \eta_T$ denotes a non-negative function with properties

$$|T| \lesssim \int_T \eta, \quad \operatorname{supp}\eta = T, \quad \|\nabla\eta\|_{L^\infty(T)} \lesssim h_T^{-1} \tag{55}$$

(postpone the question of existence until (59) below), we deduce

$$\begin{aligned}\|r\|_{L^2(T)}^2 &\lesssim \int_T r(r\eta) \le \|r\|_{H^{-1}(T)}\|\nabla(r\eta)\|_{L^2(T)}\\ &\le \|r\|_{H^{-1}(T)}\|r\|_{L^2(T)}\|\nabla\eta\|_{L^\infty(T)} \lesssim h_T^{-1}\|r\|_{H^{-1}(T)}\|r\|_{L^2(T)},\end{aligned}$$

whence

$$h_T\|r\|_{L^2(T)} \lesssim \|r\|_{H^{-1}(T)}. \tag{56}$$

This shows that overestimation in (54) is caused by *oscillation* of r at a scale finer than the mesh-size. Notice that (56) is a so-called inverse estimate, where one norm is a dual norm. It is also valid for $r \in \mathbb{P}_n(T)$, but the constant deteriorates with the degree n; see Problem 22.

Local lower bound with element residual: Motivated by the observations of the preceding paragraph, we expect that $h_T\|r\|_{L^2(T)}$ bounds asymptotically $\|\mathscr{R}\|_{H^{-1}(T)}$ from below and introduce the *oscillation of the interior residual* in T defined by

$$h_T\|r - \overline{r}_T\|_{L^2(T)},$$

where $\overline{r}_T$ denotes the mean value of r in T. Replacing r by $\overline{r}_T$ in (56) and by $r - \overline{r}_T$ in (54), as well as recalling (53), we derive

$$\begin{aligned}h_T\|r\|_{L^2(T)} &\le h_T\|\overline{r}_T\|_{L^2(T)} + h_T\|r-\overline{r}_T\|_{L^2(T)}\\ &\lesssim \|\overline{r}_T\|_{H^{-1}(T)} + h_T\|r-\overline{r}_T\|_{L^2(T)}\\ &\lesssim \|r\|_{H^{-1}(T)} + \|r-\overline{r}_T\|_{H^{-1}(T)} + h_T\|r-\overline{r}_T\|_{L^2(T)}\\ &\lesssim \|\mathscr{R}\|_{H^{-1}(T)} + h_T\|r-\overline{r}_T\|_{L^2(T)}.\end{aligned} \tag{57}$$

This is the desired statement because the oscillation $h_T\|r - \overline{r}_T\|_{L^2(T)}$ is expected to convergence faster than $h_T\|r\|_{L^2(T)}$ under refinement. In particular, if $n = 1$, then $r = f$ and the oscillation of the interior residual becomes *data oscillation*:

$$\operatorname{osc}_{\mathscr{T}}(f,T) := \|h(f - \bar{f}_T)\|_{L^2(T)} \quad \text{for all } T \in \mathscr{T}. \tag{58}$$

Note that in this case there is one additional order of convergence if $f \in H^1(\Omega)$.

The inequality (57) holds also with $\overline{r}_T$ chosen from $\mathbb{P}_{n_1}(T)$, with $n_1 \ge 1$, at the price of a larger constant hidden in $\lesssim$. We postpone the discussion of the higher order nature of the oscillation in this case after Theorem 7 below.

We conclude this paragraph by commenting on the choice of the cut-off function $\eta_T \in W^1_\infty(T)$ with (55). For example, we may take

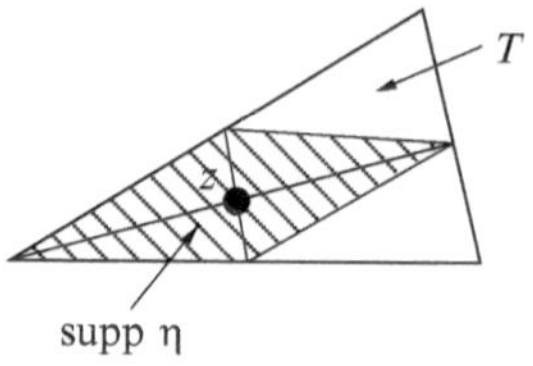

Fig. 10 Virtual refinement of a triangle for the Dörfler cut-off function

$$\eta_T = (d+1)^{d+1} \prod_{z \in \mathscr{V} \cap T} \lambda_z, \tag{59}$$

where λ_z, $z \in \mathscr{V} \cap T$, are the barycentric coordinates of T from Sect. 1.4. This choice is due to Verfürth [57, 58]. Another choice, due to Dörfler [24], can be defined as follows: refine T such that there appears an interior node and take the corresponding Courant basis function on the virtual triangulation of T; see Fig. 10 for the 2-dimensional case.

The Dörfler cut-off function has the additional property that it is an element of a refined finite element space. This is not important here but useful when proving lower bounds for the differences of two discrete solutions; see e.g. Problem 23. Such estimates are therefore called *discrete lower bound* whereas the bound for the true error is called *continuous lower bound.*

Local lower bound with jump residual: We next strive for a local lower bound for the error in terms of the jump residual $h_S^{1/2} \|j\|_{L^2(S)}$, $S \in \mathscr{S}$, and use the local lower bound in terms of the element residual as guideline.

We first notice that $j = [\![\boldsymbol{A} \nabla U]\!]$ may not be constant on an interior side $S \in \mathscr{S}$ due to the presence of $\boldsymbol{A}$. We therefore introduce the *oscillation of the jump residual* in S,

$$h_S^{1/2} \|j - \bar{j}_S\|_{L^2(S)},$$

where $\bar{j}_S$ stands for the mean value of j on S, and write

$$h_S^{1/2} \|j\|_{L^2(S)} \le h_S^{1/2} \|\bar{j}_S\|_{L^2(S)} + h_S^{1/2} \|j - \bar{j}_S\|_{L^2(S)}. \tag{60}$$

Notice that here the important question about the order of this oscillation is not obvious because, in contrast to the oscillation of the element residual in the case $n = 1$, the approximate solution U is involved. We postpone the corresponding discussion until after Theorem 7 below.

To choose a counterpart of η_T, let ω_S denote the patch composed of the two elements of $\mathscr{T}$ sharing S; see Fig. 11, left for the 2-dimensional case. Obviously ω_S has a nonempty interior. Let $\eta_S \in W^1_\infty(\omega_S)$ be a cut-off function with the properties

$$|S| \lesssim \int_S \eta_S, \quad \operatorname{supp} \eta_S = \omega_S, \quad \|\eta_S\|_{L^\infty(\omega_S)} = 1, \quad \|\nabla \eta_S\|_{L^\infty(\omega_S)} \lesssim h_S^{-1}. \tag{61}$$

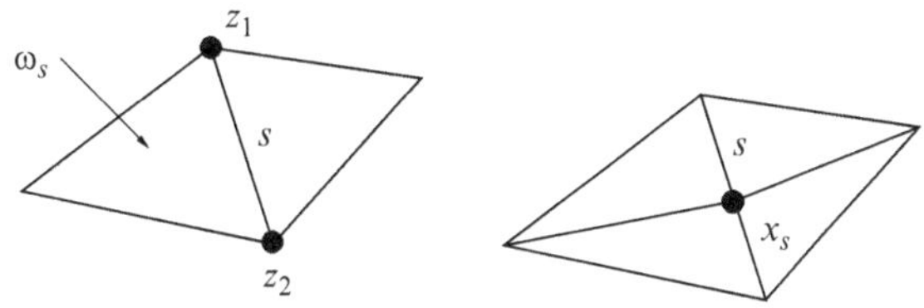

Fig. 11 Patch ω_S of triangles associated to interior side (*left*) and its refinement for Dörfler cut-off function (*right*)

Following Verfürth [57, 58] we may take η_S given by

$$\eta_{S|T} = d^d \prod_{z \in \mathscr{V} \cap S} \lambda_z^T, \tag{62}$$

where $T \subset \omega_S$ and λ_z^T, $z \in \mathscr{V} \cap T$, are the barycentric coordinates of T. Also here Dörfler [24] proposed the following alternative: refine ω_S such that there appears an interior node of S and take the corresponding Courant basis function on the virtual triangulation of ω_S; see Fig. 11, right for the 2-dimensional case.

After these preparations we are ready to derive a counterpart of (57). In view of the properties of η_S, we have

$$\|\bar{j}_S\|_{L^2(S)}^2 \lesssim \int_S \bar{j}_S (\bar{j}_S \eta_S) = \int_S j\, v_S + \int_S (\bar{j}_S - j) v_S \tag{63}$$

with $v_S = \bar{j}_S \eta_S$. We rewrite the first term on the right-hand side with the representation formula (36) as follows:

$$\int_S j\, v_S = -\int_{\omega_S} r\, v_S + \langle \mathscr{R}, v_S \rangle;$$

in contrast to (53), the jump residual couples with the element residual. Hence

$$\left| \int_S j\, v_S \right| \le \|r\|_{L^2(\omega_S)} \|v_S\|_{L^2(\omega_S)} + \|\mathscr{R}\|_{H^{-1}(\omega_S)} \|\nabla v_S\|_{L^2(\omega_S)}.$$

In view of the Poincaré–Friedrichs inequality (26), $|\omega_S| \lesssim h_S |S|$ and (61), we have

$$\|v_S\|_{L^2(\omega_S)} \lesssim h_S \|\nabla v_S\|_{L^2(\omega_S)} \le h_S \|\bar{j}_S\|_{L^2(\omega_S)} \|\nabla \eta_S\|_{L^\infty(\omega_S)} \lesssim h_S^{1/2} \|\bar{j}_S\|_{L^2(S)}.$$

We thus infer that

$$\left| \int_S j\, v_S \right| \lesssim \left(h_S^{1/2} \|r\|_{L^2(\omega_S)} + h_S^{-1/2} \|\mathscr{R}\|_{H^{-1}(\omega_S)} \right) \|\bar{j}_S\|_{L^2(S)}$$

and, using (61)

$$\left| \int_S (\bar{j}_S - j) v_S \right| \le \|\bar{j}_S - j\|_{L^2(S)} \|v_S\|_{L^2(S)} \lesssim \|\bar{j}_S - j\|_{L^2(S)} \|\bar{j}_S\|_{L^2(S)}.$$

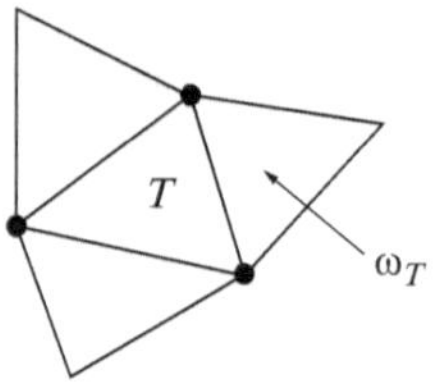

Fig. 12 Patch associated to a triangle in the local lower bound

Inserting these estimates into (63) yields

$$\|\overline{j}_S\|_{L^2(S)}^2 \lesssim \left(h_S^{1/2}\|r\|_{L^2(\omega_S)} + h_S^{-1/2}\|\mathscr{R}\|_{H^{-1}(\omega_S)} + \|\overline{j}_S - j\|_{L^2(S)}\right)\|\overline{j}_S\|_{L^2(S)}$$

whence, recalling (60),

$$h_S^{1/2}\|j\|_{L^2(S)} \lesssim \|\mathscr{R}\|_{H^{-1}(\omega_S)} + h_S\|r\|_{L^2(\omega_S)} + h_S^{1/2}\|\overline{j}_S - j\|_{L^2(S)}. \qquad (64)$$

This estimate also holds if $\overline{j}_S \in \mathbb{P}_{n_2}(S)$ is a polynomial of degree $\leq n_2$ (Problem 27).

Local lower bound with indicator: We combine the two results on interior and jump residual and exploit also the local relationship between residual and error in order to obtain a local lower bound in terms of a single indicator.

To this end, we introduce the following notation for the oscillations. Recall the mesh-size function h from Sect. 1.5 and let

$$\overline{r} = P_{2n-2}r, \quad \overline{j} = P_{2n-1}j,$$

where $P_{2n-2}r_{|T}$ and $P_{2n-1}j_{|S}$ are the L^2-orthogonal projections of r and j onto $\mathbb{P}_{2n-2}(T)$ and $\mathbb{P}_{2n-1}(S)$, respectively. The choice of the polynomial degrees arise from the desire that the oscillations are of higher order. Details are discussed after Theorem 7. Moreover, we associate with each simplex $T \in \mathscr{T}$ the patch

$$\omega_T := \bigcup_{S\subset\partial T\setminus\partial\Omega} \omega_S$$

(see Fig. 12 for the 2-dimensional case), and define the oscillation in ω_T by

$$\operatorname{osc}_{\mathscr{T}}(U,\omega_T) = \|h(r-\overline{r})\|_{L^2(\omega_T)} + \|h^{1/2}(j-\overline{j})\|_{L^2(\partial T\setminus\partial\Omega)}. \qquad (65)$$

In general, as indicated by the notation, the oscillation depends on the approximation U. However, in certain situations, it may be independent of the approximation U and then becomes *data* oscillation (58); see also Problem 19.

Theorem 7 (Local lower bound). *Let u be the exact solution of the model problem* (24) *satisfying* (25) *and* (35). *Each element indicator of* (46) *bounds, up to* *oscillation, the local error of an approximation* $U \in \mathbb{V}(\mathcal{T})$ *from below:*

$$\mathcal{E}_{\mathcal{T}}(U,T) \lesssim \alpha_2 \|\nabla(u-U)\|_{L^2(\omega_T)} + \operatorname{osc}_{\mathcal{T}}(U,\omega_T) \quad \text{for all } T \in \mathcal{T}, \tag{66}$$

where the hidden constant depends only on the shape coefficients of the simplices in ω_T*, the dimension* d *and the polynomial degree* n*.*

Proof. We combine (57) and (64), where $\overline{r}$ and $\overline{j}$ are piecewise polynomial of degree $2n-2$ and $2n-1$, respectively. Noting $h_S \approx h_T$ for all interior sides $S \in \mathcal{S}$ with $S \subset \partial T$ and $\|\mathcal{R}\|_{H^{-1}(T')} \leq \|\mathcal{R}\|_{H^{-1}(\omega_T)}$ for $T' \subset T$, we thus derive

$$\mathcal{E}_{\mathcal{T}}(U,T) \lesssim \|\mathcal{R}\|_{H^{-1}(\omega_T)} + \|h(r-\overline{r})\|_{L^2(\omega_T)} + \|h^{1/2}(\overline{j}-j)\|_{L^2(\partial T\setminus\partial\Omega)}.$$

Thus, the special case

$$\|\mathcal{R}\|_{H^{-1}(\omega_T)} \leq \alpha_2 \|\nabla(u-U)\|_{L^2(\omega_T)}$$

of (52) finishes the proof. □

A discussion of the significance of local lower bound in Theorem 7 is in order. To this end, we first consider the decay properties of the oscillation terms, which are crucial for the relevance of the aforementioned bound. Then we remark about the importance of the fact that the bound in Theorem 7 is local. Finally, in the next section, we provide a global lower bound as corollary and discuss its relationship with the upper bound in Theorem 6.

Higher order nature of oscillation: In some sense the oscillation pollutes the local lower bound in Theorem 7. It is therefore important that the oscillation is or gets small relative to the local error. We therefore compare the convergence order of the oscillation (65) with that of the local error.

To this end, let us first observe that the choices of the polynomial degrees in the oscillation allow us to derive the following upper bound of the oscillation (see Problem 29):

$$\begin{aligned}\operatorname{osc}_{\mathcal{T}}(U,\omega_T) \lesssim{}& \|h(f-P_{2n-2}f)\|_{L^2(\omega_T)}\\ &+\Big(\|h(\operatorname{div}\boldsymbol{A}-P_{n-1}(\operatorname{div}\boldsymbol{A}))\|_{L^\infty(\omega_T)} + \|\boldsymbol{A}-P_n\boldsymbol{A}\|_{L^\infty(\omega_T)}\Big)\\ &\times \|\nabla U\|_{L^2(\omega_T)}.\end{aligned} \tag{67}$$

If f and $\boldsymbol{A}$ are smooth, one expects that the local error vanishes like

$$\|\nabla(u-U)\|_{L^2(T)} = \mathcal{O}(h_T^{d/2+n})$$

and, in view of (67), oscillation like

$$\operatorname{osc}_{\mathcal{T}}(U,\omega_T) = \mathcal{O}(h_T^{d/2+n+1}).$$

See also Problem 30 for a stronger result for the jump residual.

The oscillation $\mathrm{osc}_{\mathcal{T}}(U,\omega_T)$ is therefore expected to be of higher order as $h_T \downarrow 0$. However, as Problem 32 below illustrates, it may be relevant on relatively coarse triangulations $\mathcal{T}$.

Local lower bound and marking: In contrast to the upper bound in Theorem 6, the lower bound in Theorem 7 is local. This is very welcome in a context of adaptivity. In fact, if $\mathrm{osc}_{\mathcal{T}}(U,\omega_T) \ll \|\nabla(u-U)\|_{L^2(\omega_T)}$, as we expect asymptotically, then (66) translates into

$$\mathcal{E}_{\mathcal{T}}(U,T) \lesssim \alpha_2 \, \|\nabla(u-U)\|_{L^2(\omega_T)}. \tag{68}$$

This means that an element T with relatively large error indicator contains a large portion of the error. To improve the solution U effectively, such T must be split giving rise to a procedure that tries to equidistribute errors. This is consistent with the discussion of adaptive approximation of Sect. 1.1 for $d = 1$ and of Sect. 1.6 for $d > 1$.

3.2 Global Lower Bound

We derive a global lower bound from the local one in Theorem 7 and discuss its relationship with the global upper bound in Theorem 6.

The global counterpart of $\mathrm{osc}_{\mathcal{T}}(U,\omega_T)$ from (65) is given by

$$\mathrm{osc}_{\mathcal{T}}(U) = \|h(r-\overline{r})\|_{L^2(\Omega)} + \|h^{1/2}(j-\overline{j})\|_{L^2(\Gamma)}, \tag{69}$$

where r is computed elementwise over $\mathcal{T}$ and Γ is the interior skeleton of $\mathcal{T}$.

Corollary 3 (Global lower bound). *Let u be the exact solution of the model problem* (24) *satisfying* (25) *and* (35)*. The estimator* (47) *bounds, up to oscillation, the error of an approximation $U \in \mathbb{V}(\mathcal{T})$ from below:*

$$\mathcal{E}_{\mathcal{T}}(U) \lesssim \alpha_2 \|u-U\|_{\mathbb{V}} + \mathrm{osc}_{\mathcal{T}}(U), \tag{70}$$

where the hidden constant depends on the shape coefficient of $\mathcal{T}$, the dimension d, and the polynomial degree n.

Proof. Sum (66) over all $T \in \mathcal{T}$ and take into account that each element is contained in at most by $d+2$ patches ω_T. □

Supposing that the approximation U is the Galerkin solution (27) with (31), the upper and lower a posteriori bounds in Theorem 6 and Corollary 3 imply

$$\|u-U\|_{\mathbb{V}} \lesssim \frac{1}{\alpha_1}\mathcal{E}_{\mathcal{T}}(U) \lesssim \frac{\alpha_2}{\alpha_1}\|u-U\|_{\mathbb{V}} + \frac{1}{\alpha_1}\,\mathrm{osc}_{\mathcal{T}}(U). \tag{71}$$

In other words, the error and estimator are equivalent up to oscillation.

In Problem 32 we present an example for which the ratio $\|u - U\|_{\mathbb{V}}/\mathscr{E}_{\mathscr{T}}(U)$ can be made arbitrarily small. Consequently, a lower bound without pollution and a perfect equivalence of error and estimator cannot be true in general. Moreover, for that example there holds $\mathscr{E}_{\mathscr{T}}(U) = \operatorname{osc}_{\mathscr{T}}(U)$, indicating that $\operatorname{osc}_{\mathscr{T}}(U)$ is a good measure to account for the discrepancy.

We see that $\operatorname{osc}_{\mathscr{T}}(U)$ intervenes in the relationship of error and estimator and, therefore, cannot be ignored in an analysis of an adaptive algorithm using the estimator $\mathscr{E}_{\mathscr{T}}(U)$; we will come back to this in Sect. 7. The case of data oscillation will be simpler than the general case in which $\operatorname{osc}_{\mathscr{T}}(U)$ depends on the approximation U; the latter dependence creates a nonlinear interaction in the adaptive algorithm.

The presence of oscillation is also consistent with our previous comparison of local dual norms and scaled integral norms. Since we invoked scaled integral norm in order to have an (almost) computable upper bound, this suggests that, at least for standard a posteriori error estimation, oscillation is a price that we have to pay for computability.

Fortunately, as we have illustrated in Sect. 3.1, oscillation is typically of higher order and then the a posteriori upper bound in Theorem 6 is asymptotically sharp in that its decay rate coincides with the one of the error, as the a priori bound of Theorem 5. Notice however the lower bound in Corollary 3 provides information beyond asymptotics: for example, if we consider the linear finite element method, that is $n = 1$, then $\operatorname{osc}_{\mathscr{T}}(U)$ vanishes for all triangulations on which f and A are piecewise constant and in this class of meshes error and estimator are thus equivalent:

$$\|\nabla(u - U)\|_{L^2(\Omega)} \approx \mathscr{E}_{\mathscr{T}}(U).$$

In summary: the estimator $\mathscr{E}_{\mathscr{T}}(U)$ from (47) is computable, it may be used to quantify the error and, in view of the local properties in Sect. 3.1, its indicators may be employed to provide the problem-specific information for local refinement.

3.3 Notes

Local lower bounds first appear in the work of Babuška and Miller [4]. Their derivation with continuous bubble functions is due to Verfürth [57], while the discrete lower bounds are due to Dörfler [24].

The discussion of the relationship between local dual norms and scaled integral norms as the reason for oscillation is an elaborated version of Sacchi–Veeser's one [47, Remark 3.1]. It is worth mentioning that there the indicators associated with the approximation of the Dirichlet boundary values do not need to invoke scaled Lebesgue norms and are overestimation-free. Binev et al. [7] and Stevenson [52] arrange the a posteriori analysis such that oscillation is measured in $H^{-1}(\Omega)$. This avoids overestimation but brings back the question how to (approximately) evaluate the $H^{-1}(\Omega)$-norm at acceptable cost. This question is open.

One may think that the issue of oscillation is specific to standard a posteriori error estimation. However all estimators we are aware of suffer from oscillations of the data that are finer than the mesh-size. For example, in the case of hierarchical estimators $\eta_{\mathscr{T}}(U)$ [2, 55, 58], as well as those based upon local discrete problems [2, 12, 42] or on gradient recovery [2, 27], the oscillation arises in the upper but not in the lower bounds and so creates a similar gap as that discussed here, namely

$$\eta_{\mathscr{T}}(U) \lesssim \|\nabla(u-U)\|_{L^2(\Omega)} \lesssim \eta_{\mathscr{T}}(U) + \mathrm{osc}_{\mathscr{T}}(U). \tag{72}$$

3.4 Problems

Problem 19 (Data oscillation). Check that $\mathrm{osc}_{\mathscr{T}}(U, \omega_T)$ in (65) does not depend on the approximation U if U is piecewise affine and $\boldsymbol{A}$ is piecewise constant, and is given by

$$\mathrm{osc}(f, \omega_T) = \|h(f - \bar{f})\|_{L^2(\omega_T)},$$

which corresponds to element data oscillation in (58).

Problem 20 (Energy norm case). Consider model problem (24) and discretization (27) with $\mathbb{S} = \mathbb{V}(\mathscr{T})$ and $\boldsymbol{A}$ symmetric. Derive the counterparts of (66) and (71) for the energy norm and discuss the difference to the case presented here.

Problem 21 (Cut-off functions for simplices). Verify that a suitable multiple of the Verfürth cut-off function (59) satisfies the properties (55). To this end, exploit affine equivalence of T to a fixed reference simplex and shape regularity. Repeat for the corresponding Dörfler cut-off function.

Problem 22 (Inverse estimate for general polynomials). (Try this problem after Problem 21.) Show that the choice (59) for η_T verifies, for all $p \in \mathbb{P}_n(T)$,

$$\int_T p^2 \lesssim \int_T p^2 \eta_T, \quad \|\nabla(p\eta_T)\|_{L^2(T)} \lesssim h_T^{-1} \|p\|_{L^2(T)}$$

with constants depending on n and the shape coefficient of T. To this end, recall the equivalence of norms in finite-dimensional spaces. Derive the estimate

$$h_T \|r\|_{L^2(T)} \lesssim \|r\|_{H^{-1}(T)}$$

for $r \in \mathbb{P}_n(T)$.

Problem 23 (Lower bound for correction). Consider the model problem and its discretization for $d = 2$ and $n = 1$. Let U_1 be the solution over a triangulation $\mathscr{T}_1$ and U_2 the solution over $\mathscr{T}_2$, where $\mathscr{T}_2$ has been obtained by applying at least three bisections to every triangle of $\mathscr{T}_1$. Moreover, suppose that f is piecewise constant over $\mathscr{T}_1$. Show

$$\|\nabla(U_2 - U_1)\|_{L^2(\Omega)} \geq \|h_1 f\|_{L^2(\Omega)},$$

where h_1 is the mesh-size function of $\mathscr{T}_1$.

Problem 24 (Cut-off functions for sides). Verify that a suitable multiple of the Verfürth cut-off function (62) satisfies the properties (61). Repeat for the corresponding Dörfler cut-off function.

Problem 25 (Polynomial extension). Let S be a side of a simplex T. Show that for each $q \in \mathbb{P}_n(S)$ there exists a $p \in \mathbb{P}_n(T)$ such that

$$p = q \text{ on } S \quad \text{and} \quad \|p\|_{L^2(T)} \lesssim h_T^{1/2} \|q\|_{L^2(S)}.$$

Problem 26 (Norm equivalences with cut-off functions of sides). Let S be a side of a simplex T. Show that the choice (62) for η_S verifies, for all $q \in \mathbb{P}_n(S)$ and all $p \in \mathbb{P}_m(T)$,

$$\int_S q^2 \lesssim \int_S q^2 \eta_S, \quad \|\nabla(p\eta_S)\|_{L^2(T)} \lesssim h_T^{-1} \|p\|_{L^2(T)}$$

with constants depending on m, n, and the shape coefficient of T.

Problem 27 (Lower bound with jump residual and general oscillation). Exploit the claims in Problems 25 and 26, to rederive the estimate (64) but this time with $\bar{r}$ and $\bar{j}$ piecewise polynomials of degree $\leq n_1$ and n_2.

Problem 28 (Best approximation of a product). Let K be either a d or a $(d-1)$-simplex. For $\ell \in \mathbb{N}$ denote by $P_m^p \colon L^p(K, \mathbb{R}^\ell) \to \mathbb{P}_m(K, \mathbb{R}^\ell)$ the operator of best L^p-approximation in K. Prove that, for all $v \in L^\infty(K, \mathbb{R}^\ell)$, $V \in \mathbb{P}_n(K, \mathbb{R}^\ell)$ and $m \geq n$,

$$\|vV - P_m^2(vV)\|_{L^2(K)} \leq \|v - P_{m-n}^\infty v\|_{L^\infty(K)} \|V\|_{L^2(K)}.$$

Problem 29 (Upper bound for oscillation). Verify the upper bound (67) for the oscillation by exploiting Problem 28.

Problem 30 (Superconvergence of jump residual oscillation). Show that if A is smooth across interelement boundaries, then the oscillation of the jump residual is superconvergent in that

$$\|j - \bar{j}_S\|_{L^2(S)} = \mathscr{O}(h_S^n) \|j\|_{L^2(S)} \quad \text{as } h_S \searrow 0.$$

Problem 31 (Simplified bound of oscillation). Using (67), show that (35) implies

$$\operatorname{osc}_{\mathscr{T}}(U, \omega_T) \lesssim h_T \Big(\|f\|_{L^2(\omega_T)} + \|\nabla U\|_{L^2(\omega_T)} \Big), \tag{73}$$

where the hidden constant depends also on $\boldsymbol{A}$.

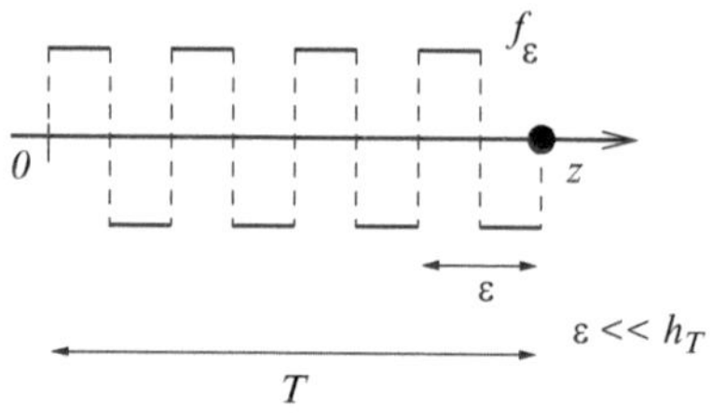

Fig. 13 A strongly oscillating forcing function

Problem 32 (Necessity of oscillation). Let $\epsilon = 2^{-K}$ for K integer and extend the function $\frac{1}{2}x(\epsilon - |x|)$ defined on $(-\epsilon, \epsilon)$ to a 2ϵ-periodic C^1 function u_ϵ on $\Omega = (-1, 1)$. Moreover, let the forcing function be $f_\epsilon = -u''$, which is 2ϵ-periodic and piecewise constant with values ± 1 that change at multiples of ϵ; see Fig. 13. Let $\mathscr{T}_\varepsilon$ be a uniform mesh with mesh-size $h = 2^{-k}$, with $k \ll K$. We consider piecewise linear finite elements $\mathbb{V}(\mathscr{T}_\varepsilon)$ and corresponding Galerkin solution $U_\varepsilon \in \mathbb{V}(\mathscr{T}_\varepsilon)$. Observing that f_ε is $L^2(\Omega)$-orthogonal to both the space of piecewise constants and linears over $\mathscr{T}_\varepsilon$, show that

$$\|u_\varepsilon' - U_\varepsilon'\|_{L^2(\Omega)} = \|u_\varepsilon'\|_{L^2(\Omega)} = \frac{\epsilon}{\sqrt{6}} = \frac{2^{-K}}{\sqrt{6}}$$
$$\ll 2^{-k} = h = \|hf_\varepsilon\|_{L^2(\Omega)} = \operatorname{osc}_{\mathscr{T}_\varepsilon}(U_\varepsilon) = \mathscr{E}_{\mathscr{T}_\varepsilon}(U_\varepsilon).$$

Extend this 1d example via a checkerboard pattern to any dimension.

4 Convergence of AFEM

The purpose of this section is to formulate an adaptive finite element method (AFEM) and to prove that it generates a sequence of approximate solutions converging to the exact one. The method consists in the following main steps:

$$\mathsf{SOLVE} \rightarrow \mathsf{ESTIMATE} \rightarrow \mathsf{MARK} \rightarrow \mathsf{REFINE}.$$

By their nature, adaptive algorithms define the sequence of approximate solutions as well as associated meshes and spaces only implicitly. This fact requires an approach that differs from "classical" convergence proofs. In particular, a proof of convergence will hinge on results of an a posteriori analysis as in Sects. 2.4 and 3.

The approach presented in this section covers wide classes of problems, discrete spaces, estimators and marking strategies. Here we do not strive for such generality but instead, in order to minimize technicalities, illustrate the main arguments only in a model case and then hint on possible generalizations.

It is worth noticing that, conceptually, the following convergence proof does not suppose any additional regularity of the exact solution. Consequently, it does not (and cannot) provide any information about the convergence speed. This important issue will be the concern of Sect. 7 for smaller classes of problems and algorithms.

4.1 A Model Adaptive Algorithm

We first present an AFEM for the model problem (24), which is an example of a *standard* iterative process that is often used in practice. In Sect. 4.2 we then prove its convergence and, finally, in Sect. 4.3 we comment on generalizations still covered by the given approach.

AFEM: The main structure of the adaptive finite element method is as follows: given an initial grid $\mathscr{T}_0$, set $k = 0$ and iterate

1. $U_k = \mathsf{SOLVE}(\mathscr{T}_k)$;
2. $\{\mathscr{E}_k(U_k, T)\}_{T\in\mathscr{T}_k} = \mathsf{ESTIMATE}(U_k, \mathscr{T}_k)$;
3. $\mathscr{M}_k = \mathsf{MARK}\big(\{\mathscr{E}_k(U_k, T)\}_{T\in\mathscr{T}_k}, \mathscr{T}_k\big)$;
4. $\mathscr{T}_{k+1} = \mathsf{REFINE}(\mathscr{M}_k, \mathscr{T}_k)$; $k \leftarrow k + 1$.

Thus, the algorithm produces sequences $(\mathscr{T}_k)_{k=0}^{\infty}$ of meshes, $(U_k)_{k=0}^{\infty}$ of approximate solutions, and, implicitly, $(\mathbb{V}_k)_{k=0}^{\infty}$ of discrete spaces.

We next state our main assumptions and define the aforementioned modules for the problem at hand in detail.

Assumptions on continuous problem: We assume that the model problem (24) satisfies (25) and (35) so that the a posteriori error bounds of Sects. 2 and 3 are available.

Initial grid: Assume that $\mathscr{T}_0$ is some initial triangulation of Ω such that $\boldsymbol{A}$ is piecewise Lipschitz over $\mathscr{T}_0$.

Solve: Let

$$\mathbb{V}_k := \mathbb{V}(\mathscr{T}_k) := \big\{V \in \mathbb{S}^{n,0}(\mathscr{T}_k) \mid V_{|\partial\Omega} = 0\big\}$$

be the space of continuous functions that are piecewise polynomial of degree $\leq n$ over $\mathscr{T}_k$, and compute the Galerkin solution U_k in $\mathbb{V}_k$ given by

$$U_k \in \mathbb{V}_k : \quad \int_\Omega \boldsymbol{A}\nabla U_k \cdot \nabla V = \int_\Omega f V \quad \text{for all } V \in \mathbb{V}_k.$$

Estimate: Compute the error estimator $\{\mathscr{E}_k(U_k, T)\}_{T\in\mathscr{T}_k}$ given by

$$\mathscr{E}_k(U_k, T) := \Big(h_T^2 \|r\|_{L^2(T)}^2 + h_T \|j\|_{L^2(\partial T\setminus\partial\Omega)}^2\Big)^{1/2},$$

where $h_T = |T|^{1/d}$, r and j are the element and jump residuals from (37) associated to the approximate solution U_k.

Mark: Collect a subset $\mathcal{M}_k \subset \mathcal{T}_k$ of marked elements with the following property:

$$\forall T \in \mathcal{T}_k \quad \mathcal{E}_k(U_k, T) = \mathcal{E}_{k,\max} > 0 \quad \Longrightarrow \quad T \in \mathcal{M}_k \tag{74}$$

with $\mathcal{E}_{k,\max} := \max_{T \in \mathcal{T}_k} \mathcal{E}_k(U_k, T)$.

Refine: Refine $\mathcal{T}_k$ into $\mathcal{T}_{k+1}$ using bisection, as explained in Sect. 1.3, in such a way that each element in $\mathcal{M}_k$ is bisected at least once and, finally, increment k.

Classical convergence proofs consider the case of uniform, or "non-adaptive", refinement, which is included in the above class of algorithms by choosing $\mathcal{M}_k = \mathcal{T}_k$, thereby ignoring the information provided by the estimator. These convergence proofs rely on the fact that the maximum mesh-size decreases to 0 and therefore $\overline{\bigcup_{k=0}^{\infty} \mathbb{V}_k} = H_0^1(\Omega)$. The above algorithm does not require this property, neither explicitly nor implicitly in general. In fact this property is not desirable in an adaptive context, since (30) reveals that it is sufficient to approximate only one function of $H_0^1(\Omega)$, namely the exact solution u of (24). In the next section we see that the above algorithm ensures just convergence to u by a subtle combination of properties of estimator and marking strategy.

4.2 Convergence

The goal of this section is to prove the convergence of the AFEM in Sect. 4.1. More precisely, we show that the sequence $(U_k)_{k=0}^{\infty}$ of approximate solutions converges to the exact solution u of the model problem (24).

Throughout this section "$\lesssim$" stands for "$\leq C$", where the constant is independent of the iteration number k in the adaptive algorithm.

Convergence to some function: We expect the Galerkin solutions $(U_k)_{k=0}^{\infty}$ to approximate the exact solution u in $\mathbb{V} = H_0^1(\Omega)$. In any event, we may regard them as approximations to the Galerkin solution U_∞ in the limit

$$\mathbb{V}_\infty := \overline{\bigcup_{k=0}^{\infty} \mathbb{V}_k}$$

of the discrete spaces. Notice that $\mathbb{V}_\infty$ is a subspace of $\mathbb{V}$, which may not coincide with $\mathbb{V}$ (see below). In the next lemma we adopt this viewpoint and show that $(U_k)_{k=0}^{\infty}$ converges to U_∞.

Lemma 7 (Limit of approximate solutions). *The finite element solutions $(U_k)_{k=0}^{\infty}$ converge in $\mathbb{V}$ to the Galerkin solution $U_\infty \in \mathbb{V}_\infty$ given by*

$$\int_\Omega \boldsymbol{A} \nabla U_\infty \cdot \nabla V = \int_\Omega f V \quad \text{for all } V \in \mathbb{V}_\infty.$$

Proof. Since the sequence of $(\mathbb{V}_k)_{k=0}^{\infty}$ is nested (see Problem 33), the set $\mathbb{V}_\infty$ is a closed linear subspace of $\mathbb{V}$. Hence $\mathbb{V}_\infty$ is a Hilbert space and the bilinear form $\mathscr{B}$ is coercive and continuous on $\mathbb{V}_\infty$. The Lax–Milgram Theorem therefore ensures existence and uniqueness of U_∞.

Let $k \in \mathbb{N}_0$ and note that $\mathbb{V}_k \subset \mathbb{V}_\infty$. We can therefore replace $\mathbb{V}$ by $\mathbb{V}_\infty$ in Theorem 4 and obtain

$$\|U_\infty - U_k\|_{\mathbb{V}} \le \frac{\alpha_2}{\alpha_1} \inf_{V \in \mathbb{V}_k} \|U_\infty - V\|_{\mathbb{V}}.$$

Sending $k \to \infty$ then finishes the proof, because the right-hand side decreases to 0 by the very definition of $\mathbb{V}_\infty$. □

Lemma 7 reduces our task to showing that $U_\infty = u$. Notice that this is equivalent to the condition $u \in \mathbb{V}_\infty$, illustrating that in general there is no need for $\mathbb{V}_\infty = \mathbb{V}$.

The identity $U_\infty = u$ hinges on the design of the adaptive algorithm. To illustrate this point, let us consider two extreme examples:

- It may happen that all indicators vanish in iteration k^*. Then $\mathscr{E}_{k^*,\max} = 0$ and (74) is compatible with $\mathscr{M}_k = \emptyset$ and $\mathbb{V}_\infty = \mathbb{V}_k$ for all $k \ge k^*$. In this case, $U_\infty = U_{k^*}$ and convergence is only ensured if a vanishing estimator implies a vanishing error. The latter is given in particular if the estimator bounds the error from above.
- It may happen that only the simplices containing a fixed point are bisected in each iteration, but the exact solution u has a more complex structure so that $u \notin \mathbb{V}_\infty$. Since $u \neq U_\infty$, and uniform refinement is not enforced, the adaptive procedure must depend on the unknown function u.

Convergence therefore will require that the module ESTIMATE extracts enough relevant information about the error, the module MARK uses this information correctly, and the module REFINE reduces the mesh-size where requested.

Mesh-size before bisection: The module REFINE bisects elements and so halves their volume. This implies the following useful property of elements to be bisected, which include the marked elements.

Lemma 8 (Sequences of elements to be bisected). *For any sequence $(T_k)_{k=0}^{\infty}$ of elements with $T_k \in \mathscr{T}_k \setminus \mathscr{T}_{k+1}$ there holds*

$$\lim_{k\to\infty} |T_k| = 0.$$

Proof. Suppose that $\limsup_{k\to\infty} |T_k| \ge c > 0$, that is there exists a infinite subsequence $(T_{k_\ell})_\ell$ such that $\lim_{\ell\to\infty} |T_{k_\ell}| \ge c$. Recall that the children of a bisection have half the volume of the parent. Consequently, only a finite number of children of any generation of each T_{k_ℓ} can appear in the sequence $(T_{k_\ell})_\ell$. Eliminating inductively these children, we obtain an infinite sequence of simplices with disjoint interiors and volume greater than $c > 0$. This however contradicts the boundedness of Ω, whence $\limsup_{k\to\infty} |T_k| \le 0$, which is equivalent to the assertion. □

It is instructive and convenient to reformulate Lemma 8 in terms of mesh-size functions.

Lemma 9 (Mesh-size of elements to be bisected). *If* χ_k *denotes the characteristic function of the union* $\cup_{T\in\mathscr{T}_k\setminus\mathscr{T}_{k+1}} T$ *of elements to be bisected and* h_k *is the mesh-size function of* $\mathscr{T}_k$*, then*

$$\lim_{k\to\infty} \|h_k\chi_k\|_{L^\infty(\Omega)} = 0.$$

Proof. We may assume that $\mathscr{T}_k\setminus\mathscr{T}_{k+1} \neq \emptyset$ for all $k \in \mathbb{N}_0$ without loss of generality. Choose $(T_k)_{k=0}^{\infty}$ such that $T_k \in \mathscr{T}_k \setminus \mathscr{T}_{k+1}$ and $h_{T_k} = \max_{T\in\mathscr{T}_k\setminus\mathscr{T}_{k+1}} h_T$ and, recalling that $h_T = |T|^{1/d}$, use Lemma 8 to deduce the assertion. □

Lemma 9 may be viewed as a generalization of $\lim_{k\to\infty}\|h_k\|_{L^\infty(\Omega)} = 0$ in the case of uniform refinement. It may be proven also by invoking the limiting mesh-size h_∞; see Problems 34 and 35. The limiting mesh-size describes the local structure of $\mathbb{V}_\infty$ and may differ from the zero function.

Convergence to exact solution: In order to achieve $U_\infty = u$, we may investigate the residual of U_∞, which is related to the residuals of the finite element solutions U_k. The latter are in turn controlled by the element indicators $\mathscr{E}_k(U_k, T)$, $T \in \mathscr{T}_k$, which are employed in the step MARK. The fact that indicators with maximum value are marked yields the following property of the largest element indicator $\mathscr{E}_{k,\max}$.

Lemma 10 (Convergence of maximum indicator). *There holds*

$$\lim_{k\to\infty} \mathscr{E}_{k,max} = 0.$$

Proof. We may assume that $\mathscr{M}_k \neq \emptyset$ for all $k \in \mathbb{N}_0$ without loss of generality. Choose a sequence $(T_k)_{k=0}^{\infty}$ of elements such that $T_k \in \mathscr{T}_k$ and $\mathscr{E}_k(U_k, T_k) = \mathscr{E}_{k,\max}$. Thanks to (74), we have $T_k \in \mathscr{M}_k$ and so Lemma 8 and module REFINE yield $\lim_{k\to\infty}|T_k| = 0$. Exploiting the local lower bound in Theorem 7 and the simplified upper bound for the local oscillation (73), we derive the following estimate for any indicator for $T \in \mathscr{T}_k$:

$$\begin{aligned}\mathscr{E}_k(U_k, T) &\lesssim \|\nabla(U_k - U_\infty)\|_{L^2(\omega_T)} + \|\nabla(U_\infty - u)\|_{L^2(\omega_T)} \\ &\quad + h_T\Big(\|f\|_{L^2(\omega_T)} + \|\nabla U_k\|_{L^2(\omega_T)}\Big).\end{aligned} \tag{75}$$

Taking $T = T_k$, we obtain

$$\begin{aligned}\mathscr{E}_{k,\max} = \mathscr{E}_k(U_k, T_k) &\lesssim \|U_k - U_\infty\|_{\mathbb{V}} + \|\nabla(U_\infty - u)\|_{L^2(\omega_k)} \\ &\quad + |T_k|^{1/d}\Big(\|f\|_{L^2(\omega_k)} + \|U_k\|_{\mathbb{V}}\Big)\end{aligned}$$

with $\omega_k := \omega_{T_k}$. Consequently, Lemma 7 and $\lim_{k\to\infty} |T_k| = 0$, which also entails $\lim_{k\to\infty} |\omega_k| = 0$, prove the assertion. □

With these preparations we are ready for the first main result of this section.

Theorem 8 (Convergence of approximate solutions). *Let u be the exact solution of the model problem (24) satisfying (25) and (35). The finite element solutions $(U_k)_{k=0}^{\infty}$ of the AFEM of Sect. 4.1 converge to the exact one in $\mathbb{V}$:*

$$U_k \to u \text{ in } \mathbb{V} \text{ as } k \to \infty.$$

Proof. [1] In view of Lemma 7, it remains to show that $U_\infty = u$. This is equivalent to

$$0 = \langle \mathscr{R}_\infty, v\rangle := \int_\Omega fv - \int_\Omega A\nabla U_\infty \cdot \nabla v \quad \text{for all } v \in \mathbb{V} = H_0^1(\Omega). \tag{76}$$

Here we can take the test functions from $C_0^\infty(\Omega)$, because $C_0^\infty(\Omega)$ is a dense subset of the Hilbert space $H_0^1(\Omega)$. Lemma 7 therefore ensures that (76) follows from

$$0 = \lim_{k\to\infty} \langle \mathscr{R}_k, \varphi\rangle \quad \forall \varphi \in C_0^\infty(\Omega), \tag{77}$$

where $\mathscr{R}_k \in \mathbb{V}^*$ is the residual of U_k given by

$$\langle \mathscr{R}_k, v\rangle := \int_\Omega fv - \int_\Omega A\nabla U_k \cdot \nabla v.$$

[2] In order to show (77), let $\varphi \in C_0^\infty(\Omega)$ and introduce the set

$$\mathscr{T}_\ell^* := \bigcap_{m\geq \ell} \mathscr{T}_m$$

of elements in $\mathscr{T}_\ell$ that will no longer be bisected; note that if $\mathscr{T}_\ell^* \neq \emptyset$, then $\mathbb{V} \neq \mathbb{V}_\infty$. Given $\ell \leq k$, $\mathbb{V}_\ell \subset \mathbb{V}_k$ and (51) imply

$$\langle \mathscr{R}_k, \varphi\rangle = \langle \mathscr{R}_k, \varphi - I_\ell\varphi\rangle \lesssim S_{\ell,k} + S_{\ell,k}^*, \tag{78}$$

where we expect that

$$S_{\ell,k} := \sum_{T\in\mathscr{T}_k\setminus\mathscr{T}_\ell^*} \mathscr{E}_k(U_k, T)\|\nabla(\varphi - I_\ell\varphi)\|_{L^2(N_k(T))}$$

gets small because of decreasing mesh-size, whereas

$$S^*_{\ell,k} := \sum_{T\in\mathscr{T}^*_\ell} \mathscr{E}_k(U_k,T)\|\nabla(\varphi - I_\ell\varphi)\|_{L^2(N_k(T))}$$

gets small because of properties of the adaptive algorithm.

3 We first deal with $S_{\ell,k}$. The Cauchy–Schwarz inequality in some $\mathbb{R}^N$ yields

$$S_{\ell,k} \le \mathscr{E}_k(U_k, \mathscr{T}_k \setminus \mathscr{T}^*_\ell)\Big(\sum_{T\in\mathscr{T}_k\setminus\mathscr{T}^*_\ell} \|\nabla(\varphi - I_\ell\varphi)\|^2_{L^2(N_k(T))}\Big)^{1/2},$$

where the first factor

$$\begin{aligned}\mathscr{E}_k(U_k, \mathscr{T}_k \setminus \mathscr{T}^*_\ell) &\lesssim \|U_k - U_\infty\|_{\mathbb{V}} + \|U_\infty - u\|_{\mathbb{V}} \\ &\quad + \|h_k\chi_\ell\|_{L^\infty(\Omega)}\big(\|f\|_{L^2(\Omega)} + \|U_k\|_{\mathbb{V}}\big) \lesssim 1\end{aligned} \tag{79}$$

is uniformly bounded thanks to (75) and the second factor satisfies

$$\begin{aligned}\Big(\sum_{T\in\mathscr{T}_k\setminus\mathscr{T}^*_\ell} \|\nabla(\varphi - I_\ell\varphi)\|^2_{L^2(N_k(T))}\Big)^{1/2} &\lesssim \Big(\sum_{T\in\mathscr{T}_\ell\setminus\mathscr{T}^*_\ell} \|\nabla(\varphi - I_\ell\varphi)\|^2_{L^2(N_l(T))}\Big)^{1/2} \\ &\lesssim \|h_\ell\chi_\ell\|^n_{L^\infty(\Omega)}\|D^{n+1}\varphi\|_{L^2(\Omega)}\end{aligned}$$

because of $\mathscr{T}_k \ge \mathscr{T}_\ell$ and Proposition 2. Hence Lemma 9 implies

$$S_{\ell,k} \to 0 \quad \text{as } \ell \to \infty \text{ uniformly in } k. \tag{80}$$

4 Next, we deal with $S^*_{\ell,k}$. Here the Cauchy–Schwarz inequality yields

$$S^*_{\ell,k} \le \mathscr{E}_k(U_k, \mathscr{T}^*_\ell)\Big(\sum_{T\in\mathscr{T}^*_\ell} \|\nabla(\varphi - I_\ell\varphi)\|^2_{L^2(N_k(T))}\Big)^{1/2},$$

where the first factor satisfies

$$\mathscr{E}_k(U_k, \mathscr{T}^*_\ell) \le \#\mathscr{T}_\ell\, \mathscr{E}_{k,\max} \tag{81}$$

and the second factor

$$\Big(\sum_{T\in\mathscr{T}^*_\ell} \|\nabla(\varphi - I_\ell\varphi)\|^2_{L^2(N_k(T))}\Big)^{1/2} \lesssim \|h_\ell\|^n_{L^\infty(\Omega)}\|D^{n+1}\varphi\|_{L^2(\Omega)} \lesssim 1$$

is uniformly bounded. Lemma 10 therefore implies

$$S^*_{\ell,k} \to 0 \quad \text{as } k \to \infty \text{ for any fixed } \ell. \tag{82}$$

5 Given $\varepsilon > 0$, we exploit (80) and (82) by first choosing ℓ so that $S_{\ell,k} \le \varepsilon/2$ and next $k \ge \ell$ so that $S^*_{\ell,k} \le \varepsilon/2$. Inserting this into (78) yields the desired convergence (77) and finishes the proof. □

Convergence of estimator: Theorem 8 ensures convergence of the finite element solutions U_k but says nothing about the behavior of the estimators

$$\mathscr{E}_k(U_k) = \Big(\sum_{T\in\mathscr{T}_k} \mathscr{E}_k(U_k,T)^2\Big)^{1/2},$$

which enables one to monitor that convergence. The convergence of the estimators is ensured by the following theorem. Notice that this is not a simple consequence of Theorem 8 and Corollary 3 because of the presence of the oscillation $\operatorname{osc}_k(U_k)$ in the global lower bound; see also Problem 36.

Corollary 4 (Estimator convergence). *Assume again that the model problem* (24) *satisfies* (25) *and* (35). *The estimators* $(\mathscr{E}_k(U_k))_{k=0}^{\infty}$ *of AFEM in Sect. 4.1 converge to* 0*:*

$$\lim_{k\to\infty} \mathscr{E}_k(U_k) = 0.$$

Proof. Theorem 8 implies $U_\infty = u$. Using this and $h_k \le h_\ell$ for $\ell \le k$, along with $\|U_k - U_\infty\|_{\mathbb{V}} \lesssim \|U_\ell - U_\infty\|_{\mathbb{V}}$, after the first inequality of (79) yields

$$\mathscr{E}_k(U_k, \mathscr{T}_k \setminus \mathscr{T}_\ell^*) \to 0 \quad \text{as } \ell\to\infty \text{ uniformly in } k \tag{83}$$

with the help of Lemmas 7 and 9. In view of

$$\mathscr{E}_k^2(U_k) = \mathscr{E}_k^2(U_k, \mathscr{T}_k \setminus \mathscr{T}_\ell^*) + \mathscr{E}_k^2(U_k, \mathscr{T}_\ell^*),$$

we realize that (81), (83), and Lemma 10 complete the proof. □

We conclude this section with a few remarks about variants of Theorem 8 and Corollary 4 for general estimators. Theorem 8 holds for any estimator that provides an upper bound of the form

$$|\langle \mathscr{R}_k, v\rangle| \lesssim \sum_{T\in\mathscr{T}_k} \mathscr{E}_k(U_k,T)\|\nabla v\|_{L^2(N_k(T))} \quad \text{for all } v\in\mathbb{V}, \tag{84}$$

which is locally stable in the sense

$$\mathscr{E}_k(U_k,T) \lesssim h_T\|f\|_{L^2(\omega_T)} + \|\nabla U_k\|_{L^2(\omega_T)} \quad \text{for all } T\in\mathscr{T}_k; \tag{85}$$

see Problem 37. While the first assumption (84) appears natural, and is in fact crucial in view of the first example after Lemma 7, the second assumption (85) may appear artificial. However, Problem 38 reveals that is also crucial and, thus, the suggested variant of Theorem 8 is "sharp". Problem 39 proposes the construction of an estimator verifying the two assumptions (84) and (85) which, however, does not decrease to 0. On the other hand, Corollary 4 hinges on the local lower bound (75), which is a sort of minimal requirement of efficiency if the finite element solutions U_k converge. Roughly speaking, convergence of U_k relies on reliability and stability of the estimator, while the convergence of the estimator depends on the efficiency of the estimator. This shows that the assumptions on the estimator for Theorem 8 and Corollary 4 are of different nature. In particular, we see that convergence of U_k can be achieved even with estimators that are too poor to quantify the error.

4.3 Notes

The convergence proof in Sect. 4.2 is a simplified version of Siebert [49], which unifies the work of Morin et al. [44] with the standard a priori convergence theory based on (global) density. In order to further discuss the underlying assumptions of the approach in Sect. 4.2, we now compare these two works in more detail.

Solve: Both works [44] and [49] consider well-posed linear problems and invoke a generalization of Lemma 7 that follows from a discrete inf-sup condition on the discretization. The latter assumption appears natural since it is necessary for convergence in the particular case of uniform refinement; see [10, Problem 3.9]. In the case of a problem with potential or "energy", the explicit construction of U_∞ can be replaced by a convergent sequence of approximate energy minima. Examples are the convergence analyses for the p-Laplacian by Veeser [55] and for the obstacle problem by Siebert and Veeser [51], which are the first steps in the terrain of nonlinear and nonsmooth problems and are predecessors of [44,49].

Estimate and mark: Paper [44] differs from [49] on the assumptions on estimators and marking strategy. More precisely, [44] assumes that the estimator provides a discrete local lower bound and that the marking strategy essentially ensures

$$\mathscr{E}_k(U_k,T) \leq \Big(\sum_{T' \in \mathscr{M}_k} \mathscr{E}_k(U_k,T')^2 \Big)^{1/2} \quad \text{for all } T \in \mathscr{T}_k \setminus \mathscr{M}_k, \tag{86}$$

whereas [49] essentially assumes (84), (85), and (74). Thus, the assumptions on the estimator are weaker in [49], while those on the marking strategy are weaker in [44]; see also Problem 40. Since both works verify that their assumptions on the marking strategy are necessary, this shows that (minimal) assumptions on the estimator and marking strategy are coupled.

Refine: Both [44,49] consider the same framework for REFINE. This does not only include bisection for conforming meshes (see Sect. 1.3), but also nonconforming meshes (see Sect. 1.7) and other manners of subdividing elements. Moreover,

[44, 49] assume the minimal requirement of subdividing the marked elements, as in Sect. 4.1.

Further variants and generalizations: These approaches can be further developed in several directions:

- Morin et al. [43] give a proof of convergence of a variant of the AFEM in Sect. 4.1 when the estimator provides upper and local lower bounds for the error in "weak" norms, e.g. similar to the L^2-norm in Problem 18.
- Demlow [20] proves convergence of a variant of the AFEM in Sect. 4.1 with estimators for local energy norm errors.
- Garau et al. [28] show convergence of a variant of the AFEM in Sect. 4.1 for symmetric eigenvalue problems.
- Holst et al. [31] extend [44] to nonlinear partial differential equations, the linearization of which are well-posed.

4.4 Problems

Problem 33 (Nesting of spaces). Let $\mathscr{T}_1$ and $\mathscr{T}_2$ be triangulations such that $\mathscr{T}_1 \leq \mathscr{T}_2$, that is $\mathscr{T}_2$ is a refinement by bisection of $\mathscr{T}_1$. Show that the corresponding Lagrange finite element spaces from (31) are nested, i.e., $\mathbb{V}(\mathscr{T}_1) \subset \mathbb{V}(\mathscr{T}_2)$.

Problem 34 (Limiting mesh-size function). Prove that there exists a limiting mesh-size function $h_\infty \in L^\infty(\Omega)$ such that

$$\|h_k - h_\infty\|_{L^\infty(\Omega)} \to 0 \quad \text{as } k \to \infty.$$

Can you construct an example with $h_\infty \neq 0$?

Problem 35 (Alternative proof of Lemma 9). For any iteration k, let χ_k be the characteristic function of the union $\cup_{T \in \mathscr{T}_k \setminus \mathscr{T}_{k+1}} T$ of elements to be bisected and h_k the mesh-size function of $\mathscr{T}_k$. Show

$$\lim_{k \to \infty} \|h_k \chi_k\|_{L^\infty(\Omega)} = 0$$

by means of Problem 34 and the fact that bisection reduces the mesh-size.

Problem 36 (Persistence of oscillation). Choosing appropriately the data of the model problem (24), provide an example where the exact solution is (locally) piecewise affine and the (local) oscillation does not vanish.

Problem 37 (Convergence for general estimators). Check that Lemma 10 and Theorem 8 hold for any estimator $\{\mathscr{E}_k(U_k, T)\}_{T \in \mathscr{T}_k}$ that is reliable in the sense of (84) and locally stable in the sense of (85).

Problem 38 ("Necessity" of local estimator stability). Construct an estimator that satisfies (84) and its indicators are always largest around a fixed point, entailing that (74) is compatible with refinement only around that fixed point, irrespective of the exact solution u.

Problem 39 (No estimator convergence). Assuming that the exact solution u of the model problem (24) does not vanish, construct an estimator satisfying (84) and (85) which does not decrease to 0.

Problem 40 (Assumptions for marking strategies). Check that (86) is weaker that (74) by considering the bulk-chasing strategy (90).

5 Contraction Property of AFEM

This section discusses the contraction property for a special AFEM for the *model problem* (23), which we rewrite for convenience:

$$-\operatorname{div}(\boldsymbol{A}(x)\,\nabla u) = f \quad \text{in } \Omega, \qquad u = 0 \quad \text{on } \partial\Omega, \tag{87}$$

with piecewise smooth coefficient matrix $\boldsymbol{A}$ on $\mathscr{T}_0$. The matrix $\boldsymbol{A}$ is assumed to be (uniformly) SPD so that the problem is *coercive* and *symmetric*. We consider a loop of the form

$$\textsf{SOLVE} \rightarrow \textsf{ESTIMATE} \rightarrow \textsf{MARK} \rightarrow \textsf{REFINE}$$

that produces a sequence $(\mathscr{T}_k, \mathbb{V}_k, U_k)_{k=0}^{\infty}$ of conforming meshes $\mathscr{T}_k$, spaces of conforming elements $\mathbb{V}_k$ (typically C^0 piecewise linears $n = 1$), and Galerkin solutions $U_k \in \mathbb{V}_k$.

The desired contraction property hinges on *error monotonicity*. Since this is closely related to a minimization principle, it is natural to consider the coercive problem (87). We cannot expect a similar theory for problems governed by an inf-sup condition; this is an important open problem.

We next follow Cascón et al. [14]. We refer to [7, 9, 16, 17, 23, 24, 37, 40–42] for other approaches and to Sect. 5.6 for a discussion.

5.1 *Modules of AFEM for the Model Problem*

We present further properties of the four basic modules of AFEM for (87). The main additional restrictions with respect to Sect. 4 are symmetry and coercivity of the bilinear form and the marking strategy.

Module ***SOLVE*** : If $\mathscr{T} \in \mathbb{T}$ is a conforming refinement of $\mathscr{T}_0$ and $\mathbb{V} = \mathbb{V}(\mathscr{T})$ is the finite element space of C^0 piecewise polynomials of degree $\leq n$, then

$$U = \mathsf{SOLVE}(\mathcal{T})$$

determines the Galerkin solution *exactly*, namely,

$$U \in \mathbb{V}: \quad \int_\Omega A\nabla U \cdot \nabla V = \int_\Omega f V \quad \text{for all } V \in \mathbb{V}. \tag{88}$$

Module ***ESTIMATE*** : Given a conforming mesh $\mathcal{T} \in \mathbb{T}$ and the Galerkin solution $U \in \mathbb{V}(\mathcal{T})$, the output of

$$\{\mathcal{E}_{\mathcal{T}}(U,T)\}_{T\in\mathcal{T}} = \mathsf{ESTIMATE}(U,\mathcal{T}).$$

are the element indicators defined in (46). For convenience, we recall the definitions (37) of *interior* and *jump residuals*

$$\begin{aligned} r(V)|_T &= f + \operatorname{div}(A\nabla V) && \text{for all } T \in \mathcal{T}, \\ j(V)|_S &= [\![A\nabla V]\!] \cdot \boldsymbol{n}\,|_S && \text{for all } S \in \mathcal{S} \quad \text{(internal sides of } \mathcal{T}), \end{aligned}$$

and $j(V)|_S = 0$ for boundary sides $S \in \mathcal{S}$, as well as the element indicator

$$\mathcal{E}^2_{\mathcal{T}}(V,T) = h_T^2 \, \|r(V)\|^2_{L^2(T)} + h_T \, \|j(V)\|^2_{L^2(\partial T)} \quad \text{for all } T \in \mathcal{T}. \tag{89}$$

We observe that we now write explicitly the argument V in both r and j because this dependence is relevant for the present discussion.

Module ***MARK***: Given $\mathcal{T} \in \mathbb{T}$, the Galerkin solution $U \in \mathbb{V}(\mathcal{T})$, and element indicators $\{\mathcal{E}_{\mathcal{T}}(U,T)\}_{T\in\mathcal{T}}$, the module **MARK** selects elements for refinement using *Dörfler Marking* (or bulk chasing), i.e., using a fixed parameter $\vartheta \in (0,1]$ the output

$$\mathcal{M} = \mathsf{MARK}\big(\{\mathcal{E}_{\mathcal{T}}(U,T)\}_{T\in\mathcal{T}}, \mathcal{T}\big)$$

satisfies

$$\mathcal{E}_{\mathcal{T}}(U,\mathcal{M}) \geq \vartheta\, \mathcal{E}_{\mathcal{T}}(U,\mathcal{T}). \tag{90}$$

This marking guarantees that $\mathcal{M}$ contains a substantial part of the total (or bulk), thus its name. This is a crucial property in our arguments. The choice of $\mathcal{M}$ does not have to be minimal at this stage, that is, the marked elements $T \in \mathcal{M}$ do not necessarily must be those with largest indicators. However, minimality of $\mathcal{M}$ will be crucial to derive rates of convergence in Sect. 7.

Module ***REFINE***: Let $b \in \mathbb{N}$ be the number of desired bisections per marked element. Given $\mathcal{T} \in \mathbb{T}$ and a subset $\mathcal{M}$ of marked elements, the output $\mathcal{T}_* \in \mathbb{T}$ of

$$\mathcal{T}_* = \mathsf{REFINE}\big(\mathcal{T}, \mathcal{M}\big)$$

is the smallest refinement $\mathscr{T}_*$ of $\mathscr{T}$ so that all elements of $\mathscr{M}$ are at least bisected b times. Therefore, we have $h_{\mathscr{T}_*} \le h_{\mathscr{T}}$ and the strict reduction property

$$h_{\mathscr{T}_*}|_T \le 2^{-b/d} h_{\mathscr{T}}|_T \quad \text{for all } T \in \mathscr{M}. \tag{91}$$

We finally let $\mathscr{R}_{\mathscr{T}\to\mathscr{T}_*}$ be the subset of refined elements of $\mathscr{T}$ and note that

$$\mathscr{M} \subset \mathscr{R}_{\mathscr{T}\to\mathscr{T}_*}.$$

AFEM: The following procedure is identical to that of Sect. 4.1 except for the module MARK, which uses Dörfler marking with parameter $0 < \vartheta \le 1$: given an initial grid $\mathscr{T}_0$, set $k = 0$ and iterate

1. $U_k = \mathsf{SOLVE}(\mathscr{T}_k)$;
2. $\{\mathscr{E}_k(U_k, T)\}_{T\in\mathscr{T}_k} = \mathsf{ESTIMATE}(U_k, \mathscr{T}_k)$;
3. $\mathscr{M}_k = \mathsf{MARK}\big(\{\mathscr{E}_k(U_k, T)\}_{T\in\mathscr{T}_k}, \mathscr{T}_k\big)$;
4. $\mathscr{T}_{k+1} = \mathsf{REFINE}(\mathscr{T}_k, \mathscr{M}_k)$; $k \leftarrow k + 1$.

5.2 Basic Properties of AFEM

We next summarize some basic properties of AFEM that emanate from the symmetry of the differential operator (i.e. of $\boldsymbol{A}$) and features of the modules. In doing this, any explicit constant or hidden constant in $\lesssim$ will only depend on the uniform shape-regularity of $\mathbb{T}$, the dimension d, the polynomial degree n, and the (global) eigenvalues of $\boldsymbol{A}$, but not on a specific grid $\mathscr{T} \in \mathbb{T}$, except if explicitly stated. Furthermore, u will always be the weak solution of (24).

The following property relies on the fact that the bilinear form $\mathscr{B}$ is coercive and symmetric, and so induces a scalar product in $\mathbb{V}$ equivalent to the H_0^1-scalar product.

Lemma 11 (Pythagoras). *Let $\mathscr{T}, \mathscr{T}_* \in \mathbb{T}$ such that $\mathscr{T} \le \mathscr{T}_*$. The corresponding Galerkin solutions $U \in \mathbb{V}(\mathscr{T})$ and $U_* \in \mathbb{V}(\mathscr{T}_*)$ satisfy the following orthogonality property in the energy norm $|\!|\!|\cdot|\!|\!|_\Omega$*

$$|\!|\!|u - U|\!|\!|_\Omega^2 = |\!|\!|u - U_*|\!|\!|_\Omega^2 + |\!|\!|U_* - U|\!|\!|_\Omega^2. \tag{92}$$

Proof. See Problem 41. □

Property (92) is valid for (87) for the energy norm exclusively. This restricts the subsequent analysis to the energy norm, or equivalent norms, but does not extend to other, perhaps more practical, norms such as the maximum norm. This is an important open problem and a serious limitation of this theory.

We now recall the concept of oscillation from Sect. 3.1. In view of (65), we denote by $\operatorname{osc}_{\mathscr{T}}(V, T)$ the *element oscillation* for any $V \in \mathbb{V}$

$$\operatorname{osc}_{\mathcal{T}}(V,T) = \|h(r(V) - \overline{r(V)})\|_{L^2(T)} + \|h^{1/2}(j(V) - \overline{j(V)})\|_{L^2(\partial T \cap \Omega)}, \quad (93)$$

where $\overline{r(V)} = P_{2n-2}r(V)$ and $\overline{j(V)} = P_{2n-1}j(V)$ stand for L^2-projections of the residuals $r(V)$ and $j(V)$ onto the polynomials $\mathbb{P}_{2n-2}(T)$ and $\mathbb{P}_{2n-1}(S)$ defined on the element T or side $S \subset \partial T$, respectively. For variable $\boldsymbol{A}$, $\operatorname{osc}_{\mathcal{T}}(V,T)$ depends on the discrete function $V \in \mathbb{V}$, and its study is more involved than for piecewise constant $\boldsymbol{A}$. In the latter case, $\operatorname{osc}_{\mathcal{T}}(V,T)$ is given by (58) and is called *data oscillation*; see also Problem 19.

We now rewrite the a posteriori error estimates of Theorems 6 and 7 in the energy norm.

Lemma 12 (A posteriori error estimates). *There exist constants $0 < C_2 \le C_1$, such that for any $\mathcal{T} \in \mathbb{T}$ and the corresponding Galerkin solution $U \in \mathbb{V}(\mathcal{T})$ there holds*

$$|||u - U|||_{\Omega}^2 \le C_1\, \mathcal{E}_{\mathcal{T}}^2(U), \quad (94a)$$

$$C_2\, \mathcal{E}_{\mathcal{T}}^2(U) \le |||u - U|||_{\Omega}^2 + \operatorname{osc}_{\mathcal{T}}^2(U). \quad (94b)$$

The constants C_1 and C_2 depend on the smallest and largest global eigenvalues of $\boldsymbol{A}$. This dependence can be improved if the a posteriori analysis is carried out directly in the energy norm instead of the H_0^1-norm; see Problem 20. The definitions of $\overline{r(V)}$ and $\overline{j(V)}$, as well as the lower bound (94b), are immaterial for deriving a contraction property. However, they will be important for proving convergence rates in Sect. 7.

One serious difficulty in dealing with AFEM is that one has access to the energy error $|||u - U|||_{\Omega}$ only through the estimator $\mathcal{E}_{\mathcal{T}}(U)$. The latter, however, fails to be monotone because it depends on the discrete solution $U \in \mathbb{V}(\mathcal{T})$ that changes with the mesh. We first show that $\mathcal{E}_{\mathcal{T}}(V)$ decreases strictly provided V does not change (Lemma 13) and next we account for the effect of changing V but keeping the mesh (Lemma 14). Combining these two lemmas we get Proposition 3. In formulating these results we rely on the following notation: given $\mathcal{T} \in \mathbb{T}$ let $\mathcal{M} \subset \mathcal{T}$ denote a set of elements that are bisected $b \ge 1$ times at least, let $\mathcal{T}_* \ge \mathcal{T}$ be a conforming refinement of $\mathcal{T}$ that contains the bisected elements of $\mathcal{M}$, and let

$$\lambda = 1 - 2^{-b/d}.$$

Lemma 13 (Reduction of $\mathcal{E}_{\mathcal{T}}(V)$ wrt $\mathcal{T}$). *For any $V \in \mathbb{V}(\mathcal{T})$, we have*

$$\mathcal{E}_{\mathcal{T}_*}^2(V, \mathcal{T}_*) \le \mathcal{E}_{\mathcal{T}}^2(V, \mathcal{T}) - \lambda \mathcal{E}_{\mathcal{T}}^2(V, \mathcal{M}). \quad (95)$$

Proof. We decompose $\mathcal{E}_{\mathcal{T}_*}^2(V, \mathcal{T}_*)$ over elements $T \in \mathcal{T}$, and distinguish whether or not $T \in \mathcal{M}$. If $T \in \mathcal{M}$, then T is bisected at least b times and so T can be written as the union of elements $T' \in \mathcal{T}_*$ We denote this set of elements $\mathcal{T}_*(T)$ and observe that, according with (91), $h_{T'} \le 2^{-b/d}\, h_T$ for all $T' \in \mathcal{T}_*(T)$. Therefore

$$\sum_{T' \in \mathscr{T}_*(T)} h_{T'}^2 \|r(V)\|_{L^2(T')}^2 \le 2^{-2b/d} h_T^2 \|r(V)\|_{L^2(T)}^2$$

and

$$\sum_{T' \in \mathscr{T}_*(T)} h_{T'} \|j(V)\|_{L^2(\partial T' \cap \Omega)}^2 \le 2^{-b/d} h_T \|j(V)\|_{L^2(\partial T \cap \Omega)}^2,$$

because $V \in \mathbb{V}(\mathscr{T})$ only jumps across the boundary of T. This implies

$$\mathscr{E}_{\mathscr{T}_*}^2(V,T) \le 2^{-b/d} \mathscr{E}_{\mathscr{T}}^2(V,T) \quad \text{for all } T \in \mathscr{M}.$$

For the remaining elements $T \in \mathscr{T} \setminus \mathscr{M}$ we only know that mesh-size does not increase because $\mathscr{T} \le \mathscr{T}_*$, whence

$$\mathscr{E}_{\mathscr{T}_*}^2(V,T) \le \mathscr{E}_{\mathscr{T}}^2(V,T) \quad \text{for all } T \in \mathscr{T} \setminus \mathscr{M}.$$

Combining the two estimates we see that

$$\begin{aligned}\mathscr{E}_{\mathscr{T}_*}^2(V,\mathscr{T}_*) &\le 2^{-b/d} \mathscr{E}_{\mathscr{T}}^2(V,\mathscr{M}) + \mathscr{E}_{\mathscr{T}}^2(V,\mathscr{T} \setminus \mathscr{M}) \\ &= \mathscr{E}_{\mathscr{T}}^2(V,\mathscr{T}) - \left(1 - 2^{-b/d}\right) \mathscr{E}_{\mathscr{T}}^2(V,\mathscr{M}),\end{aligned}$$

which, in light of the definition of λ, is the desired estimate. □

Lemma 14 (Lipschitz property of $\mathscr{E}_{\mathscr{T}}(V,T)$ wrt V). *For all $T \in \mathscr{T}$, let ω_T denote the union of elements sharing a side with T, $\operatorname{div} \boldsymbol{A} \in L^\infty(\Omega;\mathbb{R}^d)$ be the divergence of $\boldsymbol{A}$ computed by rows, and*

$$\eta_{\mathscr{T}}(\boldsymbol{A},T) := h_T \|\operatorname{div} \boldsymbol{A}\|_{L^\infty(T)} + \|\boldsymbol{A}\|_{L^\infty(\omega_T)}.$$

Then the following estimate is valid

$$|\mathscr{E}_{\mathscr{T}}(V,T) - \mathscr{E}_{\mathscr{T}}(W,T)| \lesssim \eta_{\mathscr{T}}(\boldsymbol{A},T) \|\nabla(V-W)\|_{L^2(\omega_T)} \quad \textit{for all } V,W \in \mathbb{V}(\mathscr{T}).$$

Proof. Recalling the definition of the indicators, the triangle inequality yields

$$|\mathscr{E}_{\mathscr{T}}(V,T) - \mathscr{E}_{\mathscr{T}}(W,T)| \le h_T \|r(V) - r(W)\|_{L^2(T)} + h_T^{1/2} \|j(V) - j(W)\|_{L^2(\partial T)}.$$

We set $E := V - W \in \mathbb{V}(\mathscr{T})$, and observe that

$$r(V) - r(W) = \operatorname{div}(\boldsymbol{A} \nabla E) = \operatorname{div} \boldsymbol{A} \cdot \nabla E + \boldsymbol{A} : D^2 E,$$

where $D^2 E$ is the Hessian of E. Since E is a polynomial of degree $\le n$ in T, applying the inverse estimate $\|D^2 E\|_{L^2(T)} \lesssim h_T^{-1} \|\nabla E\|_{L^2(T)}$, we deduce

$$h_T \|r(V) - r(W)\|_{L^2(T)} \lesssim \eta_{\mathscr{T}}(\boldsymbol{A},T) \|\nabla E\|_{L^2(T)}.$$

On the other hand, for any $S \subset \partial T$ applying the inverse estimate of Problem 43 gives

$$\|j(V) - j(W)\|_{L^2(S)} = \|j(E)\|_{L^2(S)} = \| [\![A\nabla E]\!] \|_{L^2(S)} \lesssim h_T^{-1/2} \|\nabla E\|_{L^2(\omega_T)},$$

where the hidden constant is proportional to $\eta_{\mathscr{T}}(\boldsymbol{A}, T)$. This finishes the proof. □

Proposition 3 (Estimator reduction). *Given $\mathscr{T} \in \mathbb{T}$ and a subset $\mathscr{M} \subset \mathscr{T}$ of marked elements, let $\mathscr{T}_* = \mathsf{REFINE}(\mathscr{T}, \mathscr{M})$. Then there exists a constant $\Lambda > 0$, such that for all $V \in \mathbb{V}(\mathscr{T})$, $V_* \in \mathbb{V}_*(\mathscr{T}_*)$ and any $\delta > 0$ we have*

$$\begin{aligned}\mathscr{E}^2_{\mathscr{T}_*}(V_*, \mathscr{T}_*) &\leq (1+\delta)\big(\mathscr{E}^2_{\mathscr{T}}(V, \mathscr{T}) - \lambda\, \mathscr{E}^2_{\mathscr{T}}(V, \mathscr{M})\big) \\ &\quad + (1+\delta^{-1})\, \Lambda\, \eta^2_{\mathscr{T}}(\boldsymbol{A}, \mathscr{T})\, \|\!|V_* - V|\!\|^2_{\Omega}.\end{aligned}$$

Proof. Apply Lemma 14 to $V, V_* \in \mathbb{V}(\mathscr{T}_*)$ in conjunction with Lemma 13 for V (see Problem 44). □

5.3 Contraction Property of AFEM

A key question to ask is what is (are) the quantity(ies) that AFEM may contract. In light of (92), an obvious candidate is the energy error $\|\!|u - U_k|\!\|_{\Omega}$. We first show, in the simplest scenario of piecewise constant data $\boldsymbol{A}$ and f, that this is in fact the case provided an interior node property holds; see Lemma 15. However, the energy error may not contract in general unless REFINE enforces several levels of refinement; see Example 1. We then present a more general approach that eliminates the interior node property at the expense of a more complicated contractive quantity, the quasi error; see Theorem 9.

Piecewise constant data: We now assume that both f and $\boldsymbol{A}$ are piecewise constant in the initial mesh $\mathscr{T}_0$, so that $\mathrm{osc}_k(U_k) = 0$ for all $k \geq 0$. The following property was introduced by Morin et al. [40].

Definition 1 (Interior node property). The refinement $\mathscr{T}_{k+1} \geq \mathscr{T}_k$ satisfies an interior node property with respect to $\mathscr{T}_k$ if each element $T \in \mathscr{M}_k$ contains at least one node of $\mathscr{T}_{k+1}$ in the interiors of T and of each side of T.

This property is valid upon enforcing a fixed number b_* of bisections ($b_* = 3, 6$ for $d = 2, 3$). An immediate consequence of this property, proved in [40,41], is the following *discrete* lower a posteriori bound:

$$C_2 \mathscr{E}^2_k(U_k, \mathscr{M}_k) \leq \|\!|U_k - U_{k+1}|\!\|^2_{\Omega} + \mathrm{osc}^2_k(U_k); \tag{96}$$

see also Problem 23 for a related result.

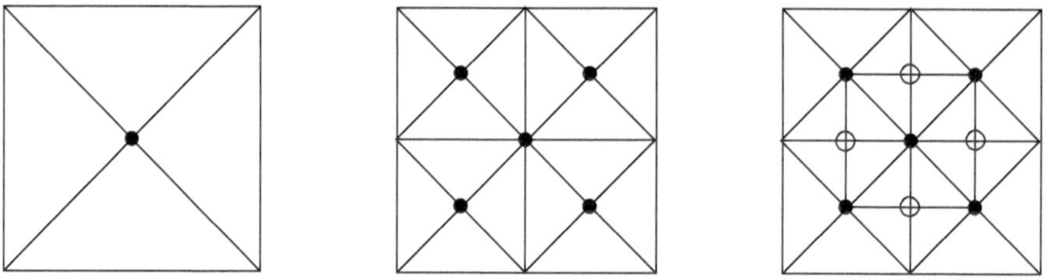

Fig. 14 Grids $\mathscr{T}_0$, $\mathscr{T}_1$, and $\mathscr{T}_2$ of Example 1. The mesh $\mathscr{T}_1$ has nodes in the middle of sides of $\mathscr{T}_0$, but only $\mathscr{T}_2$ has nodes in the interior of elements of $\mathscr{T}_0$. Hence, $\mathscr{T}_2$ satisfies the interior node property of Definition 1 with respect to $\mathscr{T}_0$

Lemma 15 (Contraction property for piecewise constant data). *Let A, f be piecewise constant in the initial mesh $\mathscr{T}_0$. If $\mathscr{T}_{k+1}$ satisfies an interior node property with respect to $\mathscr{T}_k$, then for $\alpha := (1 - \vartheta^2 \frac{C_2}{C_1})^{1/2} < 1$ there holds*

$$|||u - U_{k+1}|||_\Omega \le \alpha \, |||u - U_k|||_\Omega \,, \tag{97}$$

where $0 < \vartheta < 1$ is the parameter in (90) *and $C_1 \ge C_2$ are the constants in* (94).

Proof. For convenience, we use the notation

$$e_k = |||u - U_k|||_\Omega \,, E_k = |||U_{k+1} - U_k|||_\Omega \,, \mathscr{E}_k = \mathscr{E}_k(U_k, \mathscr{T}_k), \mathscr{E}_k(\mathscr{M}_k) = \mathscr{E}_k(U_k, \mathscr{M}_k).$$

The key idea is to use the Pythagoras equality (92)

$$e_{k+1}^2 = e_k^2 - E_k^2,$$

and show that E_k is a significant portion of e_k. Since (96) with $\operatorname{osc}_k(U_k) = 0$ implies

$$C_2 \mathscr{E}_k^2(\mathscr{M}_k) \le E_k^2, \tag{98}$$

applying Dörfler marking (90) and the upper bound (94a), we deduce

$$E_k^2 \ge C_2 \vartheta^2 \mathscr{E}_k^2 \ge \frac{C_2}{C_1} \vartheta^2 e_k^2.$$

This is the desired property of E_k and leads to (97). □

Example 1 (Strict monotoniticity). Let $\Omega = (0,1)^2$, $\boldsymbol{A} = \boldsymbol{I}$, $f = 1$ (constant data), and consider the following sequences of meshes depicted in Fig. 14. If φ_0 denotes the basis function associated with the only interior node of the initial mesh $\mathscr{T}_0$, then

$$U_0 = U_1 = \frac{1}{12} \varphi_0, \quad U_2 \ne U_1.$$

The mesh $\mathscr{T}_1 \geq \mathscr{T}_0$ is produced by a standard 2-step bisection ($b = 2$) in $2d$. Since $U_0 = U_1$ we conclude that the energy error may not change

$$|||u - U_0|||_\Omega = |||u - U_1|||_\Omega$$

between two consecutive steps of AFEM for $b = d = 2$. This is no longer true provided an interior node in each marked element is created, as in Definition 1, because then Lemma 15 holds. This example appeared first in [40, 41], and was used to justify the interior node property.

General data: If $\mathrm{osc}_k(U_k) \neq 0$, then the contraction property of AFEM becomes trickier because the energy error and estimator are no longer equivalent regardless of the interior node property. The first question to ask is what quantity replaces the energy error in the analysis. We explore this next and remove the interior node property.

Heuristics: According to (92), the energy error is monotone

$$|||u - U_{k+1}|||_\Omega \leq |||u - U_k|||_\Omega ,$$

but the previous example shows that strict inequality may fail. However, if $U_{k+1} = U_k$, estimate (95) reveals a strict estimator reduction $\mathscr{E}_{k+1}(U_k) < \mathscr{E}_k(U_k)$. We thus expect that, for a suitable scaling factor $\gamma > 0$, the so-called *quasi error*

$$|||u - U_k|||_\Omega^2 + \gamma\, \mathscr{E}_k^2(U_k) \tag{99}$$

may be contractive. This heuristics illustrates a distinct aspect of AFEM theory, the interplay between continuous quantities such the energy error $|||u - U_k|||_\Omega$ and discrete ones such as the estimator $\mathscr{E}_k(U_k)$: no one alone has the requisite properties to yield a contraction between consecutive adaptive steps.

Theorem 9 (Contraction property). *Let $\vartheta \in (0, 1]$ be the Dörfler Marking parameter, and $\{\mathscr{T}_k, \mathbb{V}_k, U_k\}_{k=0}^\infty$ be a sequence of conforming meshes, finite element spaces and discrete solutions created by AFEM for the model problem* (87).

Then there exist constants $\gamma > 0$ and $0 < \alpha < 1$, additionally depending on the number $b \geq 1$ of bisections and ϑ, such that for all $k \geq 0$

$$|||u - U_{k+1}|||_\Omega^2 + \gamma\, \mathscr{E}_{k+1}^2(U_{k+1}) \leq \alpha^2 \Big(|||u - U_k|||_\Omega^2 + \gamma\, \mathscr{E}_k^2(U_k)\Big). \tag{100}$$

Proof. We split the proof into four steps and use the notation in Lemma 15.
$\boxed{1}$ The error orthogonality (92) reads

$$e_{k+1}^2 = e_k^2 - E_k^2. \tag{101}$$

Employing Proposition 3 with $\mathscr{T} = \mathscr{T}_k$, $\mathscr{T}_* = \mathscr{T}_{k+1}$, $V = U_k$ and $V_* = U_{k+1}$ gives

$$\mathscr{E}_{k+1}^2 \leq (1+\delta)\big(\mathscr{E}_k^2 - \lambda\, \mathscr{E}_k^2(\mathscr{M}_k)\big) + (1+\delta^{-1})\, \Lambda_0\, E_k^2, \tag{102}$$

where $\Lambda_0 = \Lambda \eta^2_{\mathcal{T}_0}(\boldsymbol{A}, \mathcal{T}_0) \geq \Lambda \eta^2_{\mathcal{T}_k}(\boldsymbol{A}, \mathcal{T}_k)$. After multiplying (102) by $\gamma > 0$, to be determined later, we add (101) and (102) to obtain

$$e_{k+1}^2 + \gamma\,\mathcal{E}_{k+1}^2 \leq e_k^2 + \big(\gamma\,(1+\delta^{-1})\,\Lambda_0 - 1\big)\,E_k^2 + \gamma\,(1+\delta)\,\big(\mathcal{E}_k^2 - \lambda\,\mathcal{E}_k^2(\mathcal{M}_k)\big).$$

2 We now choose the parameters δ, γ, the former so that

$$(1+\delta)\big(1-\lambda\vartheta^2\big) = 1 - \frac{\lambda\vartheta^2}{2},$$

and the latter to verify

$$\gamma\,(1+\delta^{-1})\,\Lambda_0 = 1.$$

Note that this choice of γ yields

$$e_{k+1}^2 + \gamma\,\mathcal{E}_{k+1}^2 \leq e_k^2 + \gamma\,(1+\delta)\,\big(\mathcal{E}_k^2 - \lambda\,\mathcal{E}_k^2(\mathcal{M}_k)\big).$$

3 We next employ Dörfler Marking, namely $\mathcal{E}_k(\mathcal{M}_k) \geq \vartheta\,\mathcal{E}_k$, to deduce

$$e_{k+1}^2 + \gamma\,\mathcal{E}_{k+1}^2 \leq e_k^2 + \gamma(1+\delta)(1-\lambda\vartheta^2)\mathcal{E}_k^2$$

which, in conjunction with the choice of δ, gives

$$e_{k+1}^2 + \gamma\,\mathcal{E}_{k+1}^2 \leq e_k^2 + \gamma\left(1 - \frac{\lambda\vartheta^2}{2}\right)\mathcal{E}_k^2 = e_k^2 - \frac{\gamma\lambda\vartheta^2}{4}\mathcal{E}_k^2 + \gamma\left(1 - \frac{\lambda\vartheta^2}{4}\right)\mathcal{E}_k^2.$$

4 Finally, the upper bound (94a), namely $e_k^2 \leq C_1\,\mathcal{E}_k^2$, implies that

$$e_{k+1}^2 + \gamma\,\mathcal{E}_{k+1}^2 \leq \left(1 - \frac{\gamma\lambda\vartheta^2}{4C_1}\right)e_k^2 + \gamma\left(1 - \frac{\lambda\vartheta^2}{4}\right)\mathcal{E}_k^2.$$

This in turn leads to

$$e_{k+1}^2 + \gamma\,\mathcal{E}_{k+1}^2 \leq \alpha^2\big(e_k^2 + \gamma\,\mathcal{E}_k^2\big)$$

with

$$\alpha^2 := \max\left\{1 - \frac{\gamma\lambda\vartheta^2}{4C_1}, 1 - \frac{\lambda\vartheta^2}{4}\right\},$$

and proves the theorem because $\alpha^2 < 1$. □

Remark 9 (Ingredients). The basic ingredients of this proof are: Dörfler marking; coercivity and symmetry of $\mathcal{B}$ and nesting of spaces, which imply the Pythagoras identity (Lemma 11); the a posteriori upper bound (Lemma 12); and the estimator reduction property (Proposition 3). It does not use the lower bound (94b) and does not require marking by oscillation, as previous proofs do [17, 37, 40–42].

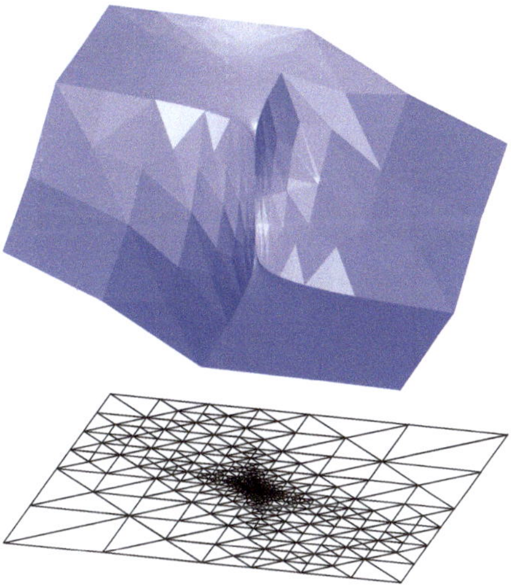

Fig. 15 Discontinuous coefficients in checkerboard pattern: Graph of the discrete solution, which is $u \approx r^{0.1}$, and underlying strongly graded grid. Notice the singularity of u at the origin

Remark 10 (Separate marking). MARK is driven by $\mathcal{E}_k$ exclusively, as it happens in all practical AFEM. Previous proofs in [17, 37, 40–42] require separate marking by estimator and oscillation. It is shown in [14] that separate marking may lead to suboptimal convergence rates. On the other hand, we will prove in Sect. 7 that the present AFEM yields quasi-optimal convergence rates.

5.4 Example: Discontinuous Coefficients

We invoke the formulas derived by Kellogg [34] to construct an exact solution of an elliptic problem with piecewise constant coefficients and vanishing right-hand side f; data oscillation is thus immaterial. We now write these formulas in the particular case $\Omega = (-1,1)^2$, $\boldsymbol{A} = a_1 \boldsymbol{I}$ in the first and third quadrants, and $\boldsymbol{A} = a_2 \boldsymbol{I}$ in the second and fourth quadrants. An exact weak solution u of the model problem (87) for $f \equiv 0$ is given in polar coordinates by $u(r,\vartheta) = r^\gamma \mu(\vartheta)$ (see Fig. 15), where

$$\mu(\vartheta) = \begin{cases} \cos((\pi/2-\sigma)\gamma)\cdot\cos((\vartheta-\pi/2+\rho)\gamma) & \text{if } 0 \le \vartheta \le \pi/2,\\ \cos(\rho\gamma)\cdot\cos((\vartheta-\pi+\sigma)\gamma) & \text{if } \pi/2 \le \vartheta \le \pi,\\ \cos(\sigma\gamma)\cdot\cos((\vartheta-\pi-\rho)\gamma) & \text{if } \pi \le \vartheta < 3\pi/2,\\ \cos((\pi/2-\rho)\gamma)\cdot\cos((\vartheta-3\pi/2-\sigma)\gamma) & \text{if } 3\pi/2 \le \vartheta \le 2\pi, \end{cases}$$

and the numbers γ, ρ, σ satisfy the nonlinear relations

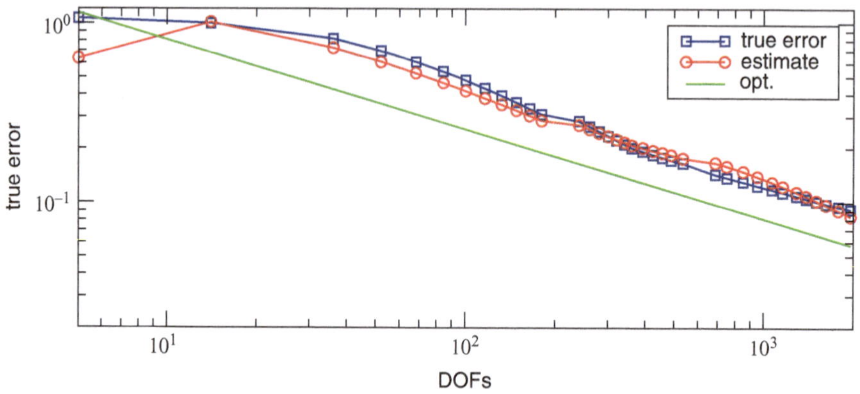

Fig. 16 Quasi-optimality of AFEM for discontinuous coefficients: estimate and true error. The optimal decay for piecewise linear elements in 2d is indicated by the line with slope $-1/2$

$$\begin{cases} R := a_1/a_2 = -\tan((\pi/2-\sigma)\gamma)\cdot\cot(\rho\gamma), \\ 1/R = -\tan(\rho\gamma)\cdot\cot(\sigma\gamma), \\ R = -\tan(\sigma\gamma)\cdot\cot((\pi/2-\rho)\gamma), \\ 0<\gamma<2, \\ \max\{0,\pi\gamma-\pi\} < 2\gamma\rho < \min\{\pi\gamma,\pi\}, \\ \max\{0,\pi-\pi\gamma\} < -2\gamma\sigma < \min\{\pi,2\pi-\pi\gamma\}. \end{cases} \tag{103}$$

Since we want to test the algorithm AFEM in a worst case scenario, we choose $\gamma = 0.1$, which produces a very singular solution u that is barely in H^1; in fact $u \in H^s(\Omega)$ for $s < 1.1$ and piecewise in $W^2_p(\Omega)$ for $p > 1$. We then solve (103) for R, ρ, and σ using Newton's method to obtain

$$R = a_1/a_2 \cong 161.4476387975881, \quad \rho = \pi/4, \quad \sigma \cong -14.92256510455152,$$

and finally choose $a_1 = R$ and $a_2 = 1$. A smaller γ would lead to a larger ratio R, but in principle γ may be as close to 0 as desired.

We realize from Fig. 16 that AFEM attains optimal decay rate for the energy norm. This is consistent with adaptive approximation for functions piecewise in $W^2_p(\Omega)$ (see Sect. 1.6), but nonobvious for AFEM which does not have direct access to u; this is the topic of Sect. 7. We also notice from Fig. 17 that a graded mesh with mesh-size of order 10^{-10} at the origin is achieved with about 2×10^3 elements. To reach a similar resolution with a uniform mesh we would need $N \approx 10^{20}$ elements! This example clearly reveals the advantages and potentials of adaptivity for the FEM even with modest computational resources.

What is missing is an explanation of the recovery of optimal error decay $N^{-1/2}$ through mesh grading. This is the subject of Sect. 7, where we have to deal with

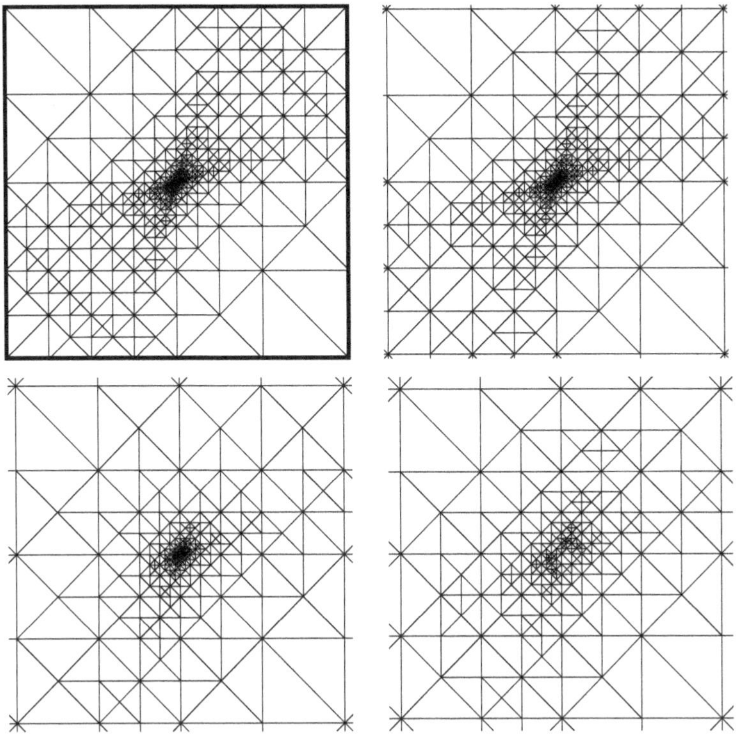

Fig. 17 Discontinuous coefficients in checkerboard pattern: Final grid (full grid with $< 2{,}000$ nodes) (*top left*), zooms to $(-10^{-3}, 10^{-3})^2$ (*top right*), $(-10^{-6}, 10^{-6})^2$ (*bottom left*), and $(-10^{-9}, 10^{-9})^2$ (*bottom right*). The grid is highly graded towards the origin. For a similar resolution, a uniform grid would require $N \approx 10^{20}$ elements

the interplay between continuous and discrete quantities as already alluded to in the heuristics.

5.5 Extensions and Restrictions

It is important to take a critical look at the theory just developed and wonder about its applicability. Below we list a few extensions of the theory and acknowledge some restrictions.

Nonconforming meshes: Theorem 9 easily extends to non-conforming meshes since conformity plays no role. This is reported in Bonito and Nochetto [9].

Non-residual estimators: The contraction property (100) has been derived for residual estimators $\mathscr{E}_k(U_k)$. This is because the estimator reduction property (95) is not known to hold for other estimators, such as hierarchical, Zienkiewicz–Zhu, and Braess–Schoerbel estimators, as well as those based on the solution of local problems. A common feature of these estimators $\eta_{\mathcal{T}}(U)$ is the lack of reliability in the preasymptotic regime, in which oscillation $\operatorname{osc}_{\mathcal{T}}(U)$ may dominate. In fact, we recall the upper a posteriori bound from (72)

$$|||u - U|||_{\Omega}^2 \leq C_1 \Big(\eta_{\mathcal{T}}^2(U) + \operatorname{osc}_{\mathcal{T}}^2(U) \Big) =: \mathcal{E}_{\mathcal{T}}^2(U),$$

which gives rise to Dörfler marking for the total estimator $\mathcal{E}_{\mathcal{T}}(U)$. Cascón and Nochetto [15] have recently extended Theorem 9 for $n = 1$ upon allowing an interior node property after a fixed number of adaptive loops and combining Lemma 15 with Theorem 9; this is easy to implement within ALBERTA [50]. At the same time, using the local equivalence of the above estimators with the residual one, Kreuzer and Siebert have proved an error reduction property after several adaptive loops [35].

Elliptic PDE on manifolds: Meckhay, Morin and Nochetto extended this theory to the Laplace–Beltrami operator [38]. In this case, an additional geometric error due to piecewise polynomial approximation of the surface must be accounted for.

Discontinuous Galerkin methods (dG): The convergence results available in the literature are for the *interior penalty* method [9, 32, 33]. The simplest contraction property (97) for a right-hand side f in the finite element space and the Laplace operator was first derived by Karakashian and Pascal [33], and later improved by Hoppe et al. [32] for $f \in L^2$ and just one bisection per marked element. In both cases, the theory is developed for $d = 2$. The most general result, valid for $d \geq 2$, operators with discontinuous variable coefficients, and L^2 data, has been developed by Bonito and Nochetto [9]. The theory in [9] deals with nonconforming meshes made of quadrilaterals or triangles, or their multidimensional generalizations, which are natural in the dG context. A key theoretical issue is the control of the jump term, which is not monotone with refinement [9, 33].

Saddle point problems: The contraction properties (97) and (100) rely crucially on the Pythagoras orthogonality property (92) and does not extend to saddle point problems. However, a modified AFEM based on an inexact Uzawa iteration and separate marking was shown to converge by Bänsch, Morin, and Nochetto for the Stokes equation [6]. The situation is somewhat simpler for mixed FEM for scalar second order elliptic PDE, and has been tackled directly for $d = 2$ by Carstensen and Hoppe for the lowest order Raviart–Thomas element [13], and by Chen et al. for any order [18]. They exploit the underlying special structure: the flux error is L^2-orthogonal to the discrete divergence free subspace, whereas the nonvanishing divergence component of the flux error can be bounded by data oscillation. This is not valid for the Stokes system, which remains open.

Beyond the energy framework: The contraction properties (97) and (100) may fail also for other norms of practical interest. An example is the maximum norm, for which there is no convergence result known yet of AFEM. Demlow proved a contraction property for local energy errors [20], and Demlow and Stevenson [21] showed a contraction property for the L^2 norm provided that the mesh grading is sufficiently mild.

5.6 Notes

The theory for conforming meshes in dimension $d > 1$ started with Dörfler [24], who introduced the crucial marking (90), the so-called *Dörfler marking*, and proved strict energy error reduction for the Laplacian provided the initial mesh $\mathscr{T}_0$ satisfies a fineness assumption. This marking plays an essential role in the present discussion, which does not seem to extend to other marking strategies such as those in Sect. 4. Morin et al. [40, 41] showed that such strict energy error reduction does not hold in general even for the Laplacian. They introduced the concept of data oscillation and the interior node property, and proved convergence of the AFEM without restrictions on $\mathscr{T}_0$. The latter result, however, is valid only for $\boldsymbol{A}$ in (23) piecewise constant on $\mathscr{T}_0$. Inspired by the work of Chen and Feng [17], Mekchay and Nochetto [37] proved a contraction property for the *total error*, namely the sum of the energy error plus oscillation for $\boldsymbol{A}$ piecewise smooth. The total error will reappear in the study of convergence rates in Sect. 7.

Diening and Kreuzer proved a similar contraction property for the p-Laplacian replacing the energy norm by a so-called quasi-norm [23]. They were able to avoid marking for oscillation by using the fact that oscillation is dominated by the estimator. Most results for nonlinear problems utilize the equivalence of the energy error and error in the associated (nonlinear) energy; compare with Problem 42. This equivalence was first used by Veeser in a convergence analysis for the p-Laplacian [55] and later on by Siebert and Veeser for the obstacle problem [51].

The result of Diening and Kreuzer inspired the work by Cascón et al. [14]. This approach hinges solely on a strict reduction of the mesh-size within refined elements, the upper a posteriori error bound, an orthogonality property natural for (87) in nested approximation spaces, and Dörfler marking. This appears to be the simplest approach currently available.

5.7 Problems

Problem 41 (Pythagoras). Let $\mathbb{V}_1 \subset \mathbb{V}_2 \subset \mathbb{V} = H_0^1(\Omega)$ be nested, conforming and closed subspaces. Let $u \in \mathbb{V}$ be the weak solution to (87), $U_1 \in \mathbb{V}_1$ and $U_2 \in \mathbb{V}_2$ the respective Ritz–Galerkin approximations to u. Prove the orthogonality property

$$|||u - U_1|||_\Omega^2 = |||u - U_2|||_\Omega^2 + |||U_2 - U_1|||_\Omega^2 . \tag{104}$$

Problem 42 (Error in energy). Let $\mathbb{V}_1 \subset \mathbb{V}_2 \subset \mathbb{V}$ and U_1, U_2, u be as in Problem 41. Recall that u, U_1, U_2 are the unique minimizers of the quadratic energy

$$I[v] := \tfrac{1}{2}\mathscr{B}[v,v] - \langle f, v\rangle$$

in $\mathbb{V}$, $\mathbb{V}_1$, $\mathbb{V}_2$ respectively. Show that (104) is equivalent to the identity

$$I[U_1] - I[u] = (I[U_2] - I[u]) + (I[U_1] - I[U_2]).$$

To this end prove

$$I[U_i] - I[u] = \tfrac{1}{2}\, |||U_i - u|||_{\Omega}^2 \quad \text{and} \quad I[U_1] - I[U_2] = \tfrac{1}{2}\, |||U_1 - U_2|||_{\Omega}^2 .$$

Problem 43 (Inverse estimate). Let $S \in \mathscr{S}$ be an interior side of $T \in \mathscr{T}$, and let $A \in L^\infty(S)$. Make use of a scaling argument to the reference element to show

$$\|A \nabla V\|_S \lesssim h_S^{-1/2} \|\nabla V\|_T \quad \text{for all } V \in \mathbb{V}(\mathscr{T}),$$

where the hidden constant depends on the shape coefficient of $\mathscr{T}$, the dimension d, and $\|A\|_{L^\infty(S)}$.

Problem 44 (Proposition 3). Complete the proof of Proposition 3 upon using Young inequality

$$(a+b)^2 \le (1+\delta)a^2 + (1+\delta^{-1})b^2 \quad \text{for all } a, b \in \mathbb{R}.$$

Problem 45 (Quasi-local Lipschitz property). Let $A \in W^1_\infty(T)$ for all $T \in \mathscr{T}$. Prove

$$|\operatorname{osc}_{\mathscr{T}}(V,T) - \operatorname{osc}_{\mathscr{T}}(W,T)| \lesssim \operatorname{osc}_{\mathscr{T}}(A,T)\, \|V - W\|_{H^1(\omega_T)} \quad \text{for all } V, W \in \mathbb{V},$$

where $\operatorname{osc}_{\mathscr{T}}(A,T) = h_T \|\operatorname{div} A - P^\infty_{n-1}(\operatorname{div} A)\|_{L^\infty(T)} + \|A - P^\infty_n A\|_{L^\infty(\omega_T)}$. Proceed as in the proof of Lemma 14 and use Problem 28.

Problem 46 (Perturbation). Let $\mathscr{T}, \mathscr{T}_* \in \mathbb{T}$, with $\mathscr{T} \le \mathscr{T}_*$. Use Problem 45 to prove that, for all $V \in \mathbb{V}(\mathscr{T})$ and $V_* \in \mathbb{V}(\mathscr{T}_*)$, there is a constant $\Lambda_1 > 0$ such that

$$\operatorname{osc}^2_{\mathscr{T}}(V, \mathscr{T} \cap \mathscr{T}_*) \le 2 \operatorname{osc}^2_{\mathscr{T}_*}(V_*, \mathscr{T} \cap \mathscr{T}_*) + \Lambda_1 \operatorname{osc}_{\mathscr{T}_0}(A, \mathscr{T}_0)^2\, |||V - V_*|||_{\Omega}^2 .$$

6 Complexity of Refinement

This section is devoted to proving Theorem 1 for conforming meshes and Lemma 3 for nonconforming meshes. The results of Sects. 6.1 and 6.2 are valid for $d = 2$ but the proofs of Theorem 1 in Sect. 6.3 and Lemma 3 in Sect. 6.4 extend easily to $d > 2$. We refer to the survey [45] for a full discussion for $d \ge 2$.

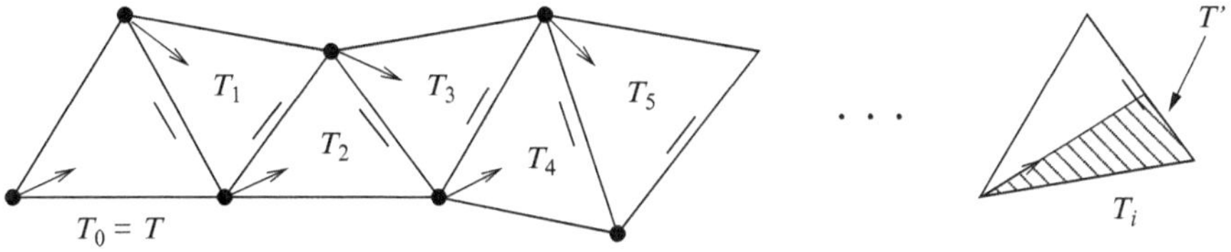

Fig. 18 Typical chain $\mathscr{C}(T, \mathscr{T}) = \{T_j\}_{j=0}^{i}$ emanating from $T = T_0 \in \mathscr{T}$ with $T_j = F(T_{j-1})$, $j \geq 1$

6.1 Chains and Labeling for $d = 2$

In order to study nonlocal effects of bisection for $d = 2$ we introduce now the concept of chain [7]; this concept is not adequate for $d > 2$ [45, 53]. Recall that $E(T)$ denotes the edge of T assigned for refinement. To each $T \in \mathscr{T}$ we associate the element $F(T) \in \mathscr{T}$ sharing the edge $E(T)$ if $E(T)$ is interior and $F(T) = \emptyset$ if $E(T)$ is on $\partial\Omega$. A *chain* $\mathscr{C}(T, \mathscr{T})$, with starting element $T \in \mathscr{T}$, is a sequence $\{T, F(T), \ldots, F^m(T)\}$ with no repetitions of elements and with

$$F^{m+1}(T) = F^k(T) \text{ for } k \in \{0, \ldots, m-1\} \quad \text{or} \quad F^{m+1}(T) = \emptyset;$$

see Fig. 18. We observe that if an element T belongs to two different grids, then the corresponding chains may be different as well. Two adjacent elements $T, T' = F(T)$ are *compatibly divisible* (or equivalently T, T' form a *compatible bisection patch*) if $F(T') = T$. Hence, $\mathscr{C}(T, \mathscr{T}) = \{T, T'\}$ and a bisection of either T or T' does not propagate outside the patch.

Example (Chains): Let $\mathscr{F} = \{T_i\}_{i=1}^{12}$ be the forest of Fig. 3. Then $\mathscr{C}(T_6, \mathscr{T}) = \{T_6, T_7\}, \mathscr{C}(T_9, \mathscr{T}) = \{T_9\}$, and $\mathscr{C}(T_{10}, \mathscr{T}) = \{T_{10}, T_8, T_2\}$ are chains, but only $\mathscr{C}(T_6, \mathscr{T})$ is a compatible bisection patch.

To study the structure of chains we rely on the initial labeling (6) and the bisection rule of Sect. 1.3 (see Fig. 5):

> *Every triangle $T \in \mathscr{T}$ with generation $g(T) = i$ receives the label $(i+1, i+1, i)$ with i corresponding to the refinement edge $E(T)$, its side i is bisected and both new sides as well as the bisector are labeled $i+2$ whereas the remaining labels do not change.* (105)

We first show that once the initial labeling and bisection rule are set, the resulting master forest $\mathbb{F}$ is uniquely determined: the label of an edge is independent of the elements sharing this edge and no ambiguity arises in the recursion process.

Lemma 16 (Labeling). *Let the initial labeling* (6) *for $\mathscr{T}_0$ and above bisection rule be enforced. If $\mathscr{T}_0 \leq \mathscr{T}_1 \leq \cdots \leq \mathscr{T}_n$ are generated according to* (105)*, then each side in $\mathscr{T}_k$ has a unique label independent of the two triangles sharing this edge.*

Proof. We argue by induction over $\mathscr{T}_k$. For $k = 0$ the assertion is valid due to the initial labeling. Suppose the statement is true for $\mathscr{T}_k$. An edge S in $\mathscr{T}_{k+1}$ can be obtained in two ways. The first is that S is a bisector, and so a new edge, in which case there is nothing to prove about its label being unique. The second possibility is that S was obtained by bisecting an edge $S' \in \mathscr{S}_k$. Let T, $T' \in \mathscr{T}_k$ be the elements sharing S', and let us assume that $E(T') = S'$. Let $(i+1, i+1, i)$ be the label of T', which means that S is assigned the label $i+2$. By induction assumption over $\mathscr{T}_k$, the label of S' as an edge of T is also i. There are two possible cases for the label of T:

- Label $(i+1, i+1, i)$: this situation is symmetric, $E(T) = S'$, and S' is bisected with both halves getting label $i+2$. This is depicted in the figure below.

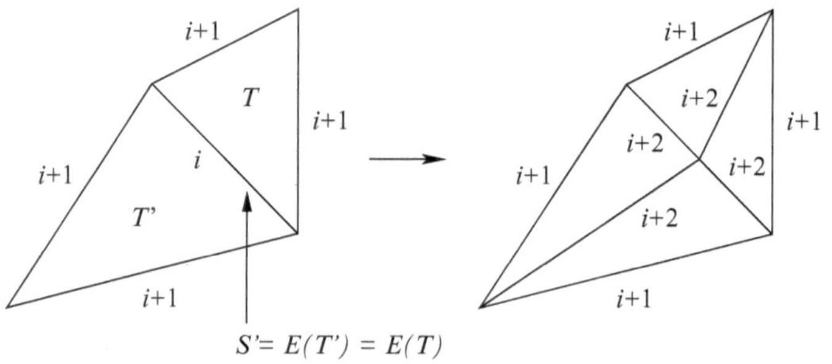

- Label $(i, i, i-1)$: a bisection of side $E(T)$ with label $i-1$ creates a children T'' with label $(i+1, i+1, i)$ that is compatibly divisible with T'. Joining the new node of T with the midpoint of S' creates a conforming partition with level $i+2$ assigned to S. This is depicted in the figure below.

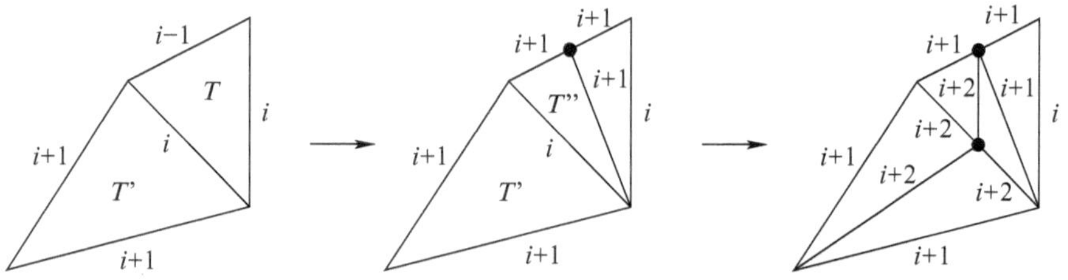

Therefore, in both cases the label $i+2$ assigned to S is the same from both sides, as asserted. □

The two possible configurations displayed in the two figures above lead readily to the following statement about generations.

Corollary 5 (Generation of consecutive elements). *For any* $\mathscr{T} \in \mathbb{T}$ *and* T, $T' = F(T) \in \mathscr{T}$ *we either have:*

(a) $g(T) = g(T')$ *and* T, T' *are compatibly divisible, or*
(b) $g(T') = g(T) - 1$ *and* T *is compatibly divisible with a child of* T'.

Corollary 6 (Generations within a chain). *For all* $\mathscr{T} \in \mathbb{T}$ *and* $T \in \mathscr{T}$, *its chain* $\mathscr{C}(T, \mathscr{T}) = \{T_k\}_{k=0}^m$ *with* $T_k = F^k(T)$ *have the property*

$$g(T_k) = g(T) - k \quad 0 \leq k \leq m-1,$$

and $T_m = F^m(T)$ *has generation* $g(T_m) = g(T_{m-1})$ *or it is a boundary element with lowest labeled edge on* $\partial\Omega$. *In the first case,* T_{m-1} *and* T_m *are compatibly divisible.*

Proof. Apply Corollary 5 repeatedly to consecutive elements of $\mathscr{C}(T, \mathscr{T})$. □

6.2 Recursive Bisection

Given an element $T \in \mathscr{M}$ to be refined, the routine REFINE_RECURSIVE $(\mathscr{T}, T)$ recursively refines the chain $\mathscr{C}(T, \mathscr{T})$ of T, from the end back to T, and creates a minimal conforming partition $\mathscr{T}_* \geq \mathscr{T}$ such that T is bisected once. This procedure reads as follows:

```
REFINE_RECURSIVE (𝒯, T)
if g(F(T)) < g(T)
    𝒯 := REFINE_RECURSIVE (𝒯, F(T));
else
    bisect the compatible bisection patch 𝒞(T, 𝒯);
    update 𝒯;
end if
return (𝒯)
```

We denote by $\mathscr{C}_*(T, \mathscr{T}) \subset \mathscr{T}_*$ the recursive refinement of $\mathscr{C}(T, \mathscr{T})$ (or completion of $\mathscr{C}(T, \mathscr{T})$) caused by bisection of T. Since REFINE_RECURSIVE refines solely compatible bisection patches, intermediate meshes are always conforming.

We refer to Fig. 19 for an example of recursive bisection $\mathscr{C}_*(T_{10}, \mathscr{T})$ of $\mathscr{C}(T_{10}, \mathscr{T}) = \{T_{10}, T_8, T_2\}$ in Fig. 2: REFINE_RECURSIVE starts bisecting from the end of $\mathscr{C}(T_{10}, \mathscr{T})$, namely T_2, which is a boundary element, and goes back the chain bisecting elements twice until it gets to T_{10}.

We now establish a fundamental property of REFINE_RECURSIVE $(\mathscr{T}, T)$ relating the generation of elements within $\mathscr{C}_*(T, \mathscr{T})$.

Lemma 17 (Recursive refinement). *Let* $\mathscr{T}_0$ *satisfy the labeling* (6), *and let* $\mathscr{T} \in \mathbb{T}$ *be a conforming refinement of* $\mathscr{T}_0$. *A call to REFINE_RECURSIVE* $(\mathscr{T}, T)$ *terminates, for all* $T \in \mathscr{M}$, *and outputs the smallest conforming refinement* $\mathscr{T}_*$ *of* $\mathscr{T}$ *such that* T *is bisected. In addition, all newly created* $T' \in \mathscr{C}_*(T, \mathscr{T})$ *satisfy*

$$g(T') \leq g(T) + 1. \tag{106}$$

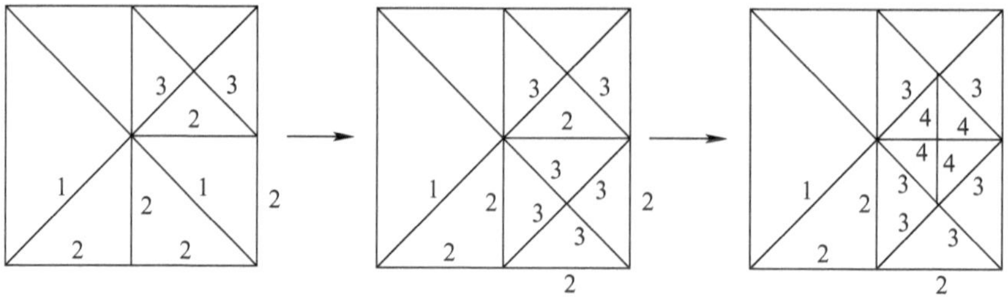

Fig. 19 Recursive refinement of $T_{10} \in \mathscr{T}$ in Fig. 2 by REFINE_RECURSIVE. This entails refining the chain $\mathscr{C}(T_{10}, \mathscr{T}) = \{T_{10}, T_8, T_2\}$, starting from the last element $T_2 \in \mathscr{T}$, which form alone a compatible bisection patch because its refinement edge is on the boundary, and continuing with $T_8 \in \mathscr{T}$ and finally $T_{10} \in \mathscr{T}$. Note that the successive meshes are always conforming and that REFINE_RECURSIVE bisects elements in $\mathscr{C}(T_{10}, \mathscr{T})$ twice before getting back to T_{10}

Proof. We first observe that T has maximal generation within $\mathscr{C}(T, \mathscr{T})$. So recursion is applied to elements with generation $\leq g(T)$, whence the recursion terminates. We also note that this procedure creates children of T and either children or grandchildren of triangles $T_k \in \mathscr{C}(T, \mathscr{T}) = \{T_i\}_{i=0}^m$ with $k \geq 1$. If T' is a child of T there is nothing to prove. If not, we consider first $m = 1$, in which case T' is a child of T_1 because T_0 and T_1 are compatibly divisible and so have the same generation; thus $g(T') = g(T_1) + 1 = g(T_0) + 1$. Finally, if $m > 1$, then $g(T_k) < g(T)$ and we apply Corollary 6 to deduce

$$g(T') \leq g(T_k) + 2 \leq g(T) + 1,$$

as asserted. □

The following crucial lemma links generation and distance between T and $T' \in \mathscr{C}_*(T, \mathscr{T})$, the latter being defined as

$$\operatorname{dist}(T', T) := \inf_{x' \in T', x \in T} |x' - x|.$$

Lemma 18 (Distance and generation). *Let $T \in \mathscr{M}$. Any newly created $T' \in \mathscr{C}_*(T, \mathscr{T})$ by REFINE_RECURSIVE $(\mathscr{T}, T)$ satisfies*

$$\operatorname{dist}(T', T) \leq D_2 \frac{2}{\sqrt{2} - 1} 2^{-g(T')/2}, \tag{107}$$

where $D_2 > 0$ is the constant in (5).

Proof. Suppose $T' \subset T_i \in \mathscr{C}(T, \mathscr{T})$ have been created by subdividing T_i (see Fig. 18). If $i \leq 1$ then $\operatorname{dist}(T', T) = 0$ and there is nothing to prove. If $i > 1$, then we observe that $\operatorname{dist}(T', T_{i-1}) = 0$, whence

$$\operatorname{dist}(T', T) \leq \operatorname{dist}(T_{i-1}, T) + \operatorname{diam}(T_{i-1}) \leq \sum_{k=1}^{i-1} \operatorname{diam}(T_k)$$

$$\leq D_2 \sum_{k=1}^{i-1} 2^{-g(T_k)/2} < D_2 \frac{1}{1 - 2^{-1/2}} 2^{-g(T_{i-1})/2},$$

because the generations decrease exactly by 1 along the chain $\mathscr{C}(T)$ according to Corollary 5(b). Since T' is a child or grandchild of T_i, we deduce

$$g(T') \leq g(T_i) + 2 = g(T_{i-1}) + 1,$$

whence

$$\operatorname{dist}(T', T) < D_2 \frac{2^{1/2}}{1 - 2^{-1/2}} 2^{-g(T')/2}.$$

This is the desired estimate. □

The recursive procedure REFINE_RECURSIVE is the core of the routine REFINE of Sect. 1.3: given a conforming mesh $\mathscr{T} \in \mathbb{T}$ and a subset $\mathscr{M} \subset \mathscr{T}$ of marked elements, REFINE creates a conforming refinement $\mathscr{T}_* \geq \mathscr{T}$ of $\mathscr{T}$ such that all elements of $\mathscr{M}$ are bisected at least once:

REFINE $(\mathscr{T}, \mathscr{M})$
for all $T \in \mathscr{M} \cap \mathscr{T}$ do
  $\mathscr{T} :=$ REFINE_RECURSIVE $(\mathscr{T}, T)$;
end
return $(\mathscr{T})$

It may happen that an element $T' \in \mathscr{M}$ is scheduled prior to T for refinement and $T \in \mathscr{C}(T', \mathscr{T})$. Since the call REFINE_RECURSIVE $(\mathscr{T}, T')$ bisects T, its two children replace T in $\mathscr{T}$. This implies that $T \notin \mathscr{M} \cap \mathscr{T}$, which prevents further refinement of T.

In practice, one often likes to bisect selected elements several times, for instance each marked element is scheduled for $b \geq 1$ bisections. This can be done by assigning the number $b(T) = b$ of bisections that have to be executed for each marked element T. If T is bisected then we assign $b(T) - 1$ as the number of pending bisections to its children and the set of marked elements is $\mathscr{M} := \{T \in \mathscr{T} \mid b(T) > 0\}$.

6.3 *Conforming Meshes: Proof of Theorem 1*

Figure 19 reveals that the issue of propagation of mesh refinement to keep conformity is rather delicate. In particular, an estimate of the form

$$\#\mathscr{T}_k - \#\mathscr{T}_{k-1} \leq \Lambda \#\mathscr{M}_{k-1}$$

is not valid with a constant Λ independent of k; in fact the constant can be proportional to k according to Fig. 19.

Binev et al. [7] for $d = 2$ and Stevenson [53] for $d > 2$ show that control of the propagation of refinement by bisection is possible when considering the collective effect:

$$\#\mathcal{T}_k - \#\mathcal{T}_0 \le \Lambda_0 \sum_{j=0}^{k-1} \#\mathcal{M}_j.$$

This can be heuristically motivated as follows. Consider the set $\mathcal{M} := \bigcup_{j=0}^{k-1} \mathcal{M}_j$ used to generate the sequence $\mathcal{T}_0 \le \mathcal{T}_1 \le \cdots \le \mathcal{T}_k =: \mathcal{T}$. Suppose that each element $T_* \in \mathcal{M}$ is assigned a fixed amount C_1 of money to spend on refined elements in $\mathcal{T}$, i.e., on $T \in \mathcal{T} \setminus \mathcal{T}_0$. Assume further that $\lambda(T, T_*)$ is the portion of money spent by T_* on T. Then it must hold

$$\sum_{T \in \mathcal{T} \setminus \mathcal{T}_0} \lambda(T, T_*) \le C_1 \quad \text{for all } T_* \in \mathcal{M}. \tag{108a}$$

In addition, we suppose that the investment of all elements in $\mathcal{M}$ is fair in the sense that each $T \in \mathcal{T} \setminus \mathcal{T}_0$ gets at least a fixed amount C_2, whence

$$\sum_{T_* \in \mathcal{M}} \lambda(T, T_*) \ge C_2 \quad \text{for all } T \in \mathcal{T} \setminus \mathcal{T}_0. \tag{108b}$$

Therefore, summing up (108b) and using the upper bound (108a) we readily obtain

$$C_2(\#\mathcal{T} - \#\mathcal{T}_0) \le \sum_{T \in \mathcal{T} \setminus \mathcal{T}_0} \sum_{T_* \in \mathcal{M}} \lambda(T, T_*) = \sum_{T_* \in \mathcal{M}} \sum_{T \in \mathcal{T} \setminus \mathcal{T}_0} \lambda(T, T_*) \le C_1 \, \#\mathcal{M},$$

which proves Theorem 1 for $\mathcal{T}$ and $\mathcal{M}$. In the remainder of this section we design such an allocation function $\lambda \colon \mathcal{T} \times \mathcal{M} \to \mathbb{R}^+$ in several steps and prove that recurrent refinement by bisection yields (108) provided $\mathcal{T}_0$ satisfies (6).

Construction of the allocation function: The function $\lambda(T, T_*)$ is defined with the help of two sequences $\big(a(\ell)\big)_{\ell=-1}^{\infty}, \big(b(\ell)\big)_{\ell=0}^{\infty} \subset \mathbb{R}^+$ of positive numbers satisfying

$$\sum_{\ell \ge -1} a(\ell) = A < \infty, \qquad \sum_{\ell \ge 0} 2^{-\ell/2} b(\ell) = B < \infty, \qquad \inf_{\ell \ge 1} b(\ell)\, a(\ell) = c_* > 0,$$

and $b(0) \ge 1$. Valid instances are $a(\ell) = (\ell + 2)^{-2}$ and $b(\ell) = 2^{\ell/3}$.

With these settings we are prepared to define $\lambda \colon \mathcal{T} \times \mathcal{M} \to \mathbb{R}^+$ by

$$\lambda(T, T_*) := \begin{cases} a(g(T_*) - g(T)) & \operatorname{dist}(T, T_*) < D_3\, B\, 2^{-g(T)/d} \text{ and} \\ & g(T) \le g(T_*) + 1 \\ 0 & \text{else,} \end{cases}$$

where $D_3 := D_2\big(1 + 2(\sqrt{2} - 1)^{-1}\big)$. Therefore, the investment of money by $T_* \in \mathcal{M}$ is restricted to cells T that are sufficiently close and are of generation $g(T) \leq g(T_*) + 1$. Only elements of such generation can be created during refinement of T_* according to Lemma 17. We stress that except for the definition of B, this construction is mutidimensional and we refer to [45, 53] for details.

The following lemma shows that the total amount of money spend by the allocation function $\lambda(T, T_*)$ per marked element T_* is bounded.

Lemma 19 (Upper bound). *There exists a constant $C_1 > 0$ only depending on $\mathcal{T}_0$ such that λ satisfies* (108a), *i.e.,*

$$\sum_{T \in \mathcal{T} \setminus \mathcal{T}_0} \lambda(T, T_*) \leq C_1 \quad \textit{for all } T_* \in \mathcal{M}.$$

Proof. $\boxed{1}$ Given $T_* \in \mathcal{M}$ we set $g_* = g(T_*)$ and we let $0 \leq g \leq g_* + 1$ be a generation of interest in the definition of λ. We claim that for such g the cardinality of the set

$$\mathcal{T}(T_*, g) = \{T \in \mathcal{T} \mid \operatorname{dist}(T, T_*) < D_3\, B\, 2^{-g/2} \text{ and } g(T) = g\}$$

is uniformly bounded, i.e., $\#\mathcal{T}(T_*, g) \leq C$ with C solely depending on D_1, D_2, D_3, B.

From (5) we learn that $\operatorname{diam}(T_*) \leq D_2 2^{-g_*/2} \leq 2D_2 2^{-(g_*+1)/2} \leq 2D_2 2^{-g/2}$ as well as $\operatorname{diam}(T) \leq D_2 2^{-g/2}$ for any $T \in \mathcal{T}(T_*, g)$. Hence, all elements of the set $\mathcal{T}(T_*, g)$ lie inside a ball centered at the barycenter of T_* with radius $(D_3 B + 3D_2)2^{-g/2}$. Again relying on (5) we thus conclude

$$\#\mathcal{T}(T_*, g) D_1 2^{-g} \leq \sum_{T \in \mathcal{T}(T_*, g)} |T| \leq c(D_3 B + 3D_2)^2 2^{-g},$$

whence $\#\mathcal{T}(T_*, g) \leq c\, D_1^{-1}\, (D_3 B + 3D_2)^2 =: C$.

$\boxed{2}$ Accounting only for non-zero contributions $\lambda(T, T_*)$ we deduce

$$\sum_{T \in \mathcal{T} \setminus \mathcal{T}_0} \lambda(T, T_*) = \sum_{g=0}^{g_*+1} \sum_{T \in \mathcal{T}(T_*, g)} a(g_* - g) \leq C \sum_{\ell=-1}^{\infty} a(\ell) = CA =: C_1,$$

which is the desired upper bound. □

The definition of λ also implies that each refined element receives a fixed amount of money. We show this next.

Lemma 20 (Lower bound). *There exists a constant $C_2 > 0$ only depending on $\mathcal{T}_0$ such that λ satisfies* (108b), *i.e.,*

$$\sum_{T_* \in \mathcal{M}} \lambda(T, T_*) \geq C_2 \quad \textit{for all } T \in \mathcal{T} \setminus \mathcal{T}_0.$$

Proof. [1] Fix an arbitrary $T_0 \in \mathcal{T} \setminus \mathcal{T}_0$. Then there is an iteration count $1 \leq k_0 \leq k$ such that $T_0 \in \mathcal{T}_{k_0}$ and $T_0 \notin \mathcal{T}_{k_0-1}$. Therefore there exists a $T_1 \in \mathcal{M}_{k_0-1} \subset \mathcal{M}$ such that T_0 is generated during REFINE_RECURSIVE $(\mathcal{T}_{k_0-1}, T_1)$. Iterating this process we construct a sequence $\{T_j\}_{j=1}^J \subset \mathcal{M}$ with corresponding iteration counts $\{k_j\}_{j=1}^J$ such that T_j is created by REFINE_RECURSIVE $(\mathcal{T}_{k_j-1}, T_{j+1})$. The sequence is finite since the iteration counts are strictly decreasing and thus $k_J = 0$ for some $J > 0$, or equivalently $T_J \in \mathcal{T}_0$.

Since T_j is created during refinement of T_{j+1} we infer from (106) that

$$g(T_{j+1}) \geq g(T_j) - 1.$$

Accordingly, $g(T_{j+1})$ can decrease the previous value of $g(T_j)$ at most by 1. Since $g(T_J) = 0$ there exists a smallest value s such that $g(T_s) = g(T_0) - 1$. Note that for $j = 1, \ldots, s$ we have $\lambda(T_0, T_j) > 0$ if $\operatorname{dist}(T_0, T_j) \leq D_3 B g^{-g(T_0)/d}$.

[2] We next estimate the distance $\operatorname{dist}(T_0, T_j)$. For $1 \leq j \leq s$ and $\ell \geq 0$ we define the set

$$\mathcal{T}(T_0, \ell, j) := \{T \in \{T_0, \ldots, T_{j-1}\} \mid g(T) = g(T_0) + \ell\}$$

and denote by $m(\ell, j)$ its cardinality. The triangle inequality combined with an induction argument yields

$$\begin{aligned} \operatorname{dist}(T_0, T_j) &\leq \operatorname{dist}(T_0, T_1) + \operatorname{diam}(T_1) + \operatorname{dist}(T_1, T_j) \\ &\leq \sum_{i=1}^{j} \operatorname{dist}(T_{i-1}, T_i) + \sum_{i=1}^{j-1} \operatorname{diam}(T_i). \end{aligned}$$

We apply (107) for the terms of the first sum and (5) for the terms of the second sum to obtain

$$\begin{aligned} \operatorname{dist}(T_0, T_j) &< D_2 \frac{2}{\sqrt{2}-1} \sum_{i=1}^{j} 2^{-g(T_{i-1})/2} + D_2 \sum_{i=1}^{j-1} 2^{-g(T_i)/2} \\ &\leq D_2 \left(1 + \frac{2}{\sqrt{2}-1}\right) \sum_{i=0}^{j-1} 2^{-g(T_i)/2} \\ &= D_3 \sum_{\ell=0}^{\infty} m(\ell, j)\, 2^{-(g(T_0)+\ell)/2} \\ &= D_3 2^{-g(T_0)/2} \sum_{\ell=0}^{\infty} m(\ell, j)\, 2^{-\ell/2}. \end{aligned}$$

For establishing the lower bound we distinguish two cases depending on the size of $m(\ell, s)$. This is done next.

3 *Case* 1: $m(\ell, s) \leq b(\ell)$ for all $\ell \geq 0$. From this we conclude

$$\operatorname{dist}(T_0, T_s) < D_3 2^{-g(T_0)/2} \sum_{\ell=0}^{\infty} b(\ell)\, 2^{-\ell/2} = D_3 B\, 2^{-g(T_0)/2}$$

and the definition of λ then readily implies

$$\sum_{T_* \in \mathscr{M}} \lambda(T_0, T_*) \geq \lambda(T_0, T_s) = a(g(T_s) - g(T_0)) = a(-1) > 0.$$

4 *Case* 2: There exists $\ell \geq 0$ such that $m(\ell, s) > b(\ell)$. For each of these ℓ's there exists a smallest $j = j(\ell)$ such that $m(\ell, j(\ell)) > b(\ell)$. We let ℓ^* be the index ℓ that gives rise to the smallest $j(\ell)$, and set $j^* = j(\ell^*)$. Consequently

$$m(\ell, j^* - 1) \leq b(\ell) \quad \text{for all } \ell \geq 0 \quad \text{and} \quad m(\ell^*, j^*) > b(\ell^*).$$

As in Case 1 we see $\operatorname{dist}(T_0, T_i) < D_3 B\, 2^{-g(T_0)/2}$ for all $i \leq j^* - 1$, or equivalently

$$\operatorname{dist}(T_0, T_i) < D_3 B\, 2^{-g(T_0)/2} \quad \text{for all } T_i \in \mathscr{T}(T_0, \ell, j^*).$$

We next show that the elements in $\mathscr{T}(T_0, \ell^*, j^*)$ spend enough money on T_0. We first consider $\ell^* = 0$ and note that $T_0 \in \mathscr{T}(T_0, 0, j^*)$. Since $m(0, j^*) > b(0) \geq 1$ we discover $j^* \geq 2$. Hence, there is an $T_i \in \mathscr{T}(T_0, 0, j^*) \cap \mathscr{M}$, which yields the estimate

$$\sum_{T_* \in \mathscr{M}} \lambda(T_0, T_*) \geq \lambda(T_0, T_i) = a(g(T_i) - g(T_0)) = a(0) > 0.$$

For $\ell^* > 0$ we see that $T_0 \notin \mathscr{T}(T_0, \ell^*, j^*)$, whence $\mathscr{T}(T_0, \ell^*, j^*) \subset \mathscr{M}$. In addition, $\lambda(T_0, T_i) = a(\ell^*)$ for all $T_i \in \mathscr{T}(T_0, \ell^*, j^*)$. From this we conclude

$$\begin{aligned}\sum_{T_* \in \mathscr{M}} \lambda(T_0, T_*) &\geq \sum_{T_* \in \mathscr{T}(T_0, \ell^*, j^*)} \lambda(T_0, T_*) = m(\ell^*, j^*)\, a(\ell^*) \\ &> b(\ell^*)\, a(\ell^*) \geq \inf_{\ell \geq 1} b(\ell)\, a(\ell) = c_* > 0.\end{aligned}$$

5 In summary we have proved the assertion since for any $T_0 \in \mathscr{T} \setminus \mathscr{T}_0$

$$\sum_{T_* \in \mathscr{M}} \lambda(T_0, T_*) \geq \min\{a(-1), a(0), c_*\} =: C_2 > 0. \qquad \square$$

Remark 11 (Complexity with $b > 1$ bisections). To show the complexity estimate when REFINE performs $b > 1$ bisections, the set $\mathcal{M}_k$ is to be understood as a sequence of *single* bisections recorded in sets $\{\mathcal{M}_k(j)\}_{j=1}^{b}$, which belong to intermediate triangulations between $\mathcal{T}_k$ and $\mathcal{T}_{k+1}$ with $\#\mathcal{M}_k(j) \leq 2^{j-1}\#\mathcal{M}_k$, $j = 1, \ldots, b$. Then we also obtain Theorem 1 because

$$\sum_{j=1}^{b} \#\mathcal{M}_k(j) \leq \sum_{j=1}^{b} 2^{j-1}\#\mathcal{M}_k = (2^b - 1)\#\mathcal{M}_k.$$

In practice, it is customary to take $b = d$ [50].

6.4 Nonconforming Meshes: Proof of Lemma 3

We now examine briefly the refinement process for quadrilaterals with one hanging node per edge, which gives rise to the so-called 1*-meshes*. The refinement of $T \in \mathcal{T}$ might affect four elements of $\mathcal{T}$ for $d = 2$ (or 2^d elements for any dimension $d \geq 2$), all contained in the *refinement patch* $R(T, \mathcal{T})$ of T in $\mathcal{T}$. The latter is defined as

$$R(T, \mathcal{T}) := \{T' \in \mathcal{T} \mid T' \text{ and } T \text{ share an edge and } g(T') \leq g(T)\},$$

and is called *compatible* provided $g(T') = g(T)$ for all $T' \in R(T, \mathcal{T})$. The generation gap between elements sharing an edge, in particular those in $R(T, \mathcal{T})$, is always ≤ 1 for 1-meshes, and is 0 if $R(T, \mathcal{T})$ is compatible. The element size satisfies

$$h_T = 2^{-g(T)} h_{T_0} \quad \forall T \in \mathcal{T},$$

where $T_0 \in \mathcal{T}_0$ is the ancestor of T in the initial mesh $\mathcal{T}_0$. Lemma 2 is thus valid

$$h_T < \bar{h}_T \leq D_2 2^{-g(T)} \quad \forall T \in \mathcal{T}. \tag{109}$$

Given an element $T \in \mathcal{M}$ to be refined, the routine REFINE_RECURSIVE $(\mathcal{T}, T)$ refines recursively $R(T, \mathcal{T})$ in such a way that the intermediate meshes are always 1-meshes, and reads as follows:

REFINE_RECURSIVE $(\mathcal{T}, T)$
if $g = \min\{g(T'') : T'' \in R(T, \mathcal{T}\} < g(T)$
  let $T' \in R(T, \mathcal{T})$ satisfy $g(T') = g$
  $\mathcal{T}$:= REFINE_RECURSIVE $(\mathcal{T}, T')$;
else
  subdivide T;
  update $\mathcal{T}$ upon replacing T by its children;
end if
return $(\mathcal{T})$

The conditional prevents the generation gap within $R(T, \mathscr{T})$ from getting larger than 1. If it fails, then the refinement patch $R(T, \mathscr{T})$ is compatible and refining T increases the generation gap from 0 to 1 without violating the 1-mesh structure. This implies Lemma 17: for all newly created elements $T' \in \mathscr{T}_*$

$$g(T') \leq g(T) + 1. \tag{110}$$

In addition, REFINE_RECURSIVE $(\mathscr{T}, T)$ creates a minimal 1-mesh $\mathscr{T}_* \geq \mathscr{T}$ refinement of $\mathscr{T}$ so that T is subdivided only *once*. This yields Lemma 18: there exist a geometric constant $D > 0$ such that for all newly created elements $T' \in \mathscr{T}_*$

$$\operatorname{dist}(T, T') \leq D2^{g(T')}. \tag{111}$$

The procedure REFINE_RECURSIVE is the core of REFINE, which is conceptually identical to that in Sect. 6.2. Suppose that each marked element $T \in \mathscr{M}$ is to be subdivided $\rho \geq 1$ times. We assign a flag $q(T)$ to each element T which is initialized $q(T) = \rho$ if $T \in \mathscr{M}$ and $q(T) = 0$ otherwise. The marked set $\mathscr{M}$ is then the set of elements T with $q(T) > 0$, and every time T is subdivided it is removed from $\mathscr{T}$ and replaced by its children, which inherit the flag $q(T) - 1$. This avoids the conflict of subdividing again an element that has been previously refined by REFINE_RECURSIVE. The procedure REFINE $(\mathscr{T}, \mathscr{M})$ reads

```
REFINE (𝒯, ℳ)
for all T ∈ ℳ ∩ 𝒯 do
    𝒯 := REFINE_RECURSIVE (𝒯, T);
end
return (𝒯)
```

and its output is a minimal 1-mesh $\mathscr{T}_* \geq \mathscr{T}$, refinement of $\mathscr{T}$, so that all marked elements of $\mathscr{M}$ are refined at least ρ times. Since $\mathscr{T}_*$ has one hanging node per side it is thus admissible in the sense (22). However, the refinement may spread outside $\mathscr{M}$ and the issue of complexity of REFINE again becomes non-trivial.

With the above ingredients in place, the proof of Lemma 3 follows along the lines of Sect. 6.3; see Problem 50.

6.5 Notes

The complexity theory for bisection hinges on the initial labeling (6) for $d = 2$. That such a labeling exists is due to Mitchell [39, Theorem 2.9] and Binev et al. [7, Lemma 2.1], but the proofs are not constructive. A couple of global bisections of $\mathscr{T}_0$, as depicted in Fig. 6, guarantee (6) over the ensuing mesh. For $d > 2$ the corresponding initial labeling is due to Stevenson [53, Sect. 4 – Condition (b)], who in turn improves upon Maubach [36] and Traxler [54] and shows how to impose it upon further refining each element of $\mathscr{T}_0$. We refer to the survey [45] for a discussion of this condition: a key consequence is that every uniform refinement of $\mathscr{T}_0$ gives a conforming bisection mesh.

The fundamental properties of chains, especially Lemmas 17 and 18, along with the clever ideas of Sect. 6.3 are due to Binev et al. [7] for $d = 2$, and Stevenson for $d > 2$; see [45]. Bonito and Nochetto [9] observed, in the context of dG methods, that such properties extend to admissible nonconforming meshes.

6.6 Problems

Problem 47 (Largest number of bisections). Show that REFINE_RECURSIVE $(\mathscr{T}, T)$ for $d = 2$ bisects T exactly once and all the elements in the chain $\mathscr{C}(T, \mathscr{T})$ at most twice. This property extends to $d > 2$ provided the initial labeling of Stevenson [53, Sect. 4 – Condition (b)] is enforced.

Problem 48 (Properties of generation for quad-refinement). Prove (110) and (111).

Problem 49 (Largest number of subdivisions for quads). Show that the procedure REFINE_RECURSIVE $(\mathscr{T}, T)$ subdivides T exactly once and never subdivides any other quadrilateral of $\mathscr{T}$ more than once.

Problem 50 (Lemma 3). Combine (110) and (111) to prove Lemma 3 for any $\rho \geq 1$.

7 Convergence Rates

We have already realized in Sect. 1.6 that we can a priori accommodate the degrees of freedom in such a way that the finite element approximation retains optimal energy error decay for a class of singular functions. This presumes knowledge of the exact solution u. At the same time, we have seen numerical evidence in Sect. 5.4 that the standard AFEM of Sect. 5.1, achieves such a performance without direct access to the exact solution u. Practical experience strongly suggests that this is even true for a much larger class of problems and adaptive methods. The challenge ahead is to reconcile these two distinct aspects of AFEM.

A crucial insight in such a connection for the simplest scenario, the Laplacian and piecewise constant forcing f, is due to Stevenson [52]:

Any marking strategy that reduces the energy error relative to the current value must contain a substantial portion of $\mathscr{E}_{\mathscr{T}}(U)$, and so it can be related to Dörfler Marking. (112)

This allows one to compare meshes produced by AFEM with optimal ones and to conclude a quasi-optimal error decay. We discuss this issue in Sect. 7.3. However, this is not enough to handle the model problem (87) with variable $\boldsymbol{A}$ and f.

The objective of this section is to study (87) for general data $\boldsymbol{A}$ and f. This study hinges on the total error and its relation with the quasi error, which is contracted by AFEM. This approach allows us to improve upon and extend Stevenson [52] to variable data. In doing so, we follow closely Cascón et al. [14]. The present theory, however, does not extend to noncoercive problems and marking strategies other than Dörfler's. These remain important open questions.

As in Sect. 5, u will always be the weak solution of (87) and, except when stated otherwise, any explicit constant or hidden constant in $\lesssim$ may depend on the uniform shape-regularity of $\mathbb{T}$, the dimension d, the polynomial degree n, the (global) eigenvalues of $\boldsymbol{A}$, and the oscillation $\operatorname{osc}_{\mathscr{T}_0}(\boldsymbol{A})$ of $\boldsymbol{A}$ on the initial mesh $\mathscr{T}_0$, but not on a specific grid $\mathscr{T} \in \mathbb{T}$.

7.1 The Total Error

We first introduce the concept of *total error* for the Galerkin function $U \in \mathbb{V}(\mathscr{T})$

$$|||u - U|||_{\Omega}^2 + \operatorname{osc}_{\mathscr{T}}^2(U) \tag{113}$$

(see Mekchay and Nochetto [37]), and next assert its equivalence to the quasi error (99). In fact, in view of the upper and lower a posteriori error bounds (94), and

$$\operatorname{osc}_{\mathscr{T}}^2(U) \leq \mathscr{E}_{\mathscr{T}}^2(U),$$

we have

$$\begin{aligned} C_2\, \mathscr{E}_{\mathscr{T}}^2(U) &\leq |||u - U|||_{\Omega}^2 + \operatorname{osc}_{\mathscr{T}}^2(U) \\ &\leq |||u - U|||_{\Omega}^2 + \mathscr{E}_{\mathscr{T}}^2(U) \leq (1 + C_1)\, \mathscr{E}_{\mathscr{T}}^2(U), \end{aligned}$$

whence

$$\mathscr{E}_{\mathscr{T}}^2(U) \approx |||u - U|||_{\Omega}^2 + \operatorname{osc}_{\mathscr{T}}^2(U). \tag{114}$$

Since AFEM selects elements for refinement based on information extracted exclusively from the error indicators $\{\mathscr{E}_{\mathscr{T}}(U, T)\}_{T \in \mathscr{T}}$, we realize that the decay rate of AFEM must be characterized by the total error. Moreover, on invoking the upper bound (94a) again, we also see that the total error is equivalent to the quasi error

$$|||u - U|||_{\Omega}^2 + \operatorname{osc}_{\mathscr{T}}^2(U) \approx |||u - U|||_{\Omega}^2 + \mathscr{E}_{\mathscr{T}}^2(U).$$

The latter is the quantity being strictly reduced by AFEM (Theorem 9). Finally, the total error satisfies the following Cea's type-lemma, or equivalently AFEM is quasi-optimal regarding the total error. In fact, if the oscillation vanishes, then this is Cea's Lemma stated in Theorem 4; see also Problem 12.

Lemma 21 (Quasi-optimality of total error). *There exists an explicit constant Λ_2, which depends on $\boldsymbol{A}$ $\mathscr{T}_0$, n and d, such that for any $\mathscr{T} \in \mathbb{T}$ and the corresponding Galerkin solution $U \in \mathbb{V}(\mathscr{T})$ there holds*

$$|\!|\!| u - U |\!|\!|_\Omega^2 + \operatorname{osc}_{\mathscr{T}}^2(U) \le \Lambda_2 \inf_{V \in \mathbb{V}(\mathscr{T})} \Big(|\!|\!| u - V |\!|\!|_\Omega^2 + \operatorname{osc}_{\mathscr{T}}^2(V) \Big).$$

Proof. For $\epsilon > 0$ choose $V_\epsilon \in \mathbb{V}(\mathscr{T})$, with

$$|\!|\!| u - V_\epsilon |\!|\!|_\Omega^2 + \operatorname{osc}_{\mathscr{T}}^2(V_\epsilon) \le (1+\epsilon) \inf_{V \in \mathbb{V}(\mathscr{T})} \Big(|\!|\!| u - V |\!|\!|_\Omega^2 + \operatorname{osc}_{\mathscr{T}}^2(V) \Big).$$

Applying Problem 46 with $\mathscr{T}_* = \mathscr{T}$, $V = U$, and $V_* = V_\epsilon$ yields

$$\operatorname{osc}_{\mathscr{T}}^2(U) \le 2 \operatorname{osc}_{\mathscr{T}}^2(V_\epsilon) + C_3 \, |\!|\!| U - V_\epsilon |\!|\!|_\Omega^2,$$

with

$$C_3 := \Lambda_1 \operatorname{osc}_{\mathscr{T}_0}(\boldsymbol{A})^2.$$

Since $U \in \mathbb{V}(\mathscr{T})$ is the Galerkin solution, $U - V_\epsilon \in \mathbb{V}(\mathscr{T})$ is orthogonal to $u - U$ in the energy norm, whence $|\!|\!| u - U |\!|\!|_\Omega^2 + |\!|\!| U - V_\epsilon |\!|\!|_\Omega^2 = |\!|\!| u - V_\epsilon |\!|\!|_\Omega^2$ and

$$\begin{aligned} |\!|\!| u - U |\!|\!|_\Omega^2 + \operatorname{osc}_{\mathscr{T}}^2(U) &\le \big(1 + C_3\big) \, |\!|\!| u - V_\epsilon |\!|\!|_\Omega^2 + 2 \operatorname{osc}_{\mathscr{T}}^2(V_\epsilon) \\ &\le (1+\epsilon) \, \Lambda_2 \inf_{V \in \mathbb{V}(\mathscr{T})} \Big(|\!|\!| u - U |\!|\!|_\Omega^2 + \operatorname{osc}_{\mathscr{T}}^2(V) \Big), \end{aligned}$$

with $\Lambda_2 = \max\big\{2, 1 + C_3\big\}$. The assertion follows upon taking $\epsilon \to 0$. □

7.2 Approximation Classes

In view of (114) and Lemma 21, the definition of approximation class $\mathbb{A}_s$ hinges on the concept of best total error:

$$\inf_{V \in \mathbb{V}(\mathscr{T})} \Big(|\!|\!| u - V |\!|\!|_\Omega^2 + \operatorname{osc}_{\mathscr{T}}^2(V) \Big).$$

We first let $\mathbb{T}_N \subset \mathbb{T}$ be the set of all possible conforming refinements of $\mathscr{T}_0$ with at most N elements more than $\mathscr{T}_0$, i.e.,

$$\mathbb{T}_N = \{ \mathscr{T} \in \mathbb{T} \mid \#\mathscr{T} - \#\mathscr{T}_0 \le N \}.$$

The quality of the best approximation in $\mathbb{T}_N$ with respect to the total error is characterized by

$$\sigma(N; u, f, \boldsymbol{A}) := \inf_{\mathcal{T}\in\mathbb{T}_N} \inf_{V\in\mathbb{V}(\mathcal{T})} \left(|||u - V|||_\Omega^2 + \operatorname{osc}_{\mathcal{T}}^2(V)\right)^{1/2},$$

and the approximation class $\mathbb{A}_s$ for $s > 0$ is defined by

$$\mathbb{A}_s := \left\{ (v, f, \boldsymbol{A}) \mid |v, f, \boldsymbol{A}|_s := \sup_{N>0} \left(N^s \sigma(N; v, f, \boldsymbol{A})\right) < \infty \right\}.$$

Therefore, if $(v, f, \boldsymbol{A}) \in \mathbb{A}_s$, then $\sigma(N; v, f, \boldsymbol{A}) \lesssim N^{-s}$ decays with rate N^{-s}. We point out the upper bound $s \leq n/d$ for polynomial degree $n \geq 1$; this can be seen with full regularity and uniform refinement (see (14)). Note that if $(v, f, \boldsymbol{A}) \in \mathbb{A}_s$ then for all $\varepsilon > 0$ there exist $\mathcal{T}_\varepsilon \geq \mathcal{T}_0$ conforming and $V_\varepsilon \in \mathbb{V}(\mathcal{T}_\varepsilon)$ such that (see Problem 51)

$$|||v - V_\varepsilon|||_\Omega^2 + \operatorname{osc}_{\mathcal{T}_\varepsilon}^2(V_\varepsilon) \leq \varepsilon^2 \quad \text{and} \quad \#\mathcal{T}_\varepsilon - \#\mathcal{T}_0 \leq |v, f, \boldsymbol{A}|_s^{1/s}\, \varepsilon^{-1/s}. \tag{115}$$

In addition, thanks to Lemma 21, the solution u with data $(f, \boldsymbol{A})$ satisfies

$$\sigma(N; u, f, \boldsymbol{A}) \approx \inf_{\mathcal{T}\in\mathbb{T}_N} \left\{ \mathcal{E}_{\mathcal{T}}(U, \mathcal{T}) \mid U = \mathsf{SOLVE}(\mathbb{V}(\mathcal{T})) \right\}. \tag{116}$$

This establishes a direct connection between $\mathbb{A}_s$ and AFEM.

Mesh overlay: For the subsequent discussion it will be convenient to merge two conforming meshes $\mathcal{T}_1, \mathcal{T}_2 \in \mathbb{T}$. Given the corresponding forests $\mathcal{F}_1, \mathcal{F}_2 \in \mathbb{F}$ we consider the set $\mathcal{F}_1 \cup \mathcal{F}_2 \in \mathbb{F}$, which satisfies $\mathcal{T}_0 \subset \mathcal{F}_1 \cup \mathcal{F}_2$. Then $\mathcal{F}_1 \cup \mathcal{F}_2$ is a forest and its leaves are called the *overlay* of $\mathcal{F}_1$ and $\mathcal{F}_2$:

$$\mathcal{T}_1 \oplus \mathcal{T}_2 = \mathcal{T}(\mathcal{F}_1 \cup \mathcal{F}_2).$$

We next bound the cardinality of $\mathcal{T}_1 \oplus \mathcal{T}_2$ in terms of that of $\mathcal{T}_1$ and $\mathcal{T}_2$; see [14,52].

Lemma 22 (Overlay). *The overlay $\mathcal{T} = \mathcal{T}_1 \oplus \mathcal{T}_2$ is conforming and*

$$\#\mathcal{T} \leq \#\mathcal{T}_1 + \#\mathcal{T}_2 - \#\mathcal{T}_0. \tag{117}$$

Proof. See Problem 52. □

Discussion of $\mathbb{A}_s$: We now would like to show a few examples of membership in $\mathbb{A}_s$ and highlight some important open questions. We first investigate the class $\mathbb{A}_s$ for piecewise polynomial coefficient matrix $\boldsymbol{A}$ of degree $\leq n$ over $\mathcal{T}_0$. In this simplified scenario, the oscillation $\operatorname{osc}_{\mathcal{T}}(U)$ reduces to *data oscillation* (see (58) and (93)):

$$\operatorname{osc}_{\mathcal{T}}(U) = \operatorname{osc}_{\mathcal{T}}(f) := \|h(f - P_{2n-2}\, f)\|_{L^2(\Omega)}.$$

We then have the following characterization of $\mathbb{A}_s$ in terms of the standard approximation classes [7, 8, 52]:

$$\mathscr{A}_s := \Big\{ v \in \mathbb{V} \mid |v|_{\mathscr{A}_s} := \sup_{N>0} \big(N^s \inf_{\mathscr{T} \in \mathbb{T}_N} \inf_{V \in \mathbb{V}(\mathscr{T})} \| v - V \|_\Omega \big) < \infty \Big\},$$

$$\bar{\mathscr{A}_s} := \Big\{ g \in L^2(\Omega) \mid |g|_{\bar{\mathscr{A}_s}} := \sup_{N>0} \big(N^s \inf_{\mathscr{T} \in \mathbb{T}_N} \operatorname{osc}_{\mathscr{T}}(g) \big) < \infty \Big\}.$$

Lemma 23 (Equivalence of classes). *Let A be piecewise polynomial of* degree $\le n$ *over $\mathscr{T}_0$. Then $(u, f, A) \in \mathbb{A}_s$ if and only if $(u, f) \in \mathscr{A}_s \times \bar{\mathscr{A}_s}$ and*

$$|u, f, A|_s \approx |u|_{\mathscr{A}_s} + |f|_{\bar{\mathscr{A}_s}}. \tag{118}$$

Proof. It is obvious that $(u, f, A) \in \mathbb{A}_s$ implies $(u, f) \in \mathscr{A}_s \times \bar{\mathscr{A}_s}$ as well as the bound $|u|_{\mathscr{A}_s} + |f|_{\bar{\mathscr{A}_s}} \lesssim |u, f, A|_s$.

In order to prove the reverse inequality, let $(u, f) \in \mathscr{A}_s \times \bar{\mathscr{A}_s}$. Then there exist $\mathscr{T}_1, \mathscr{T}_2 \in \mathbb{T}_N$ so that $\|u - U_{\mathscr{T}_1}\|_\Omega \le |u|_{\mathscr{A}_s} N^{-s}$ where $U_{\mathscr{T}_1} \in \mathbb{V}(\mathscr{T}_1)$ is the best approximation and $\operatorname{osc}_{\mathscr{T}_2}(f, \mathscr{T}_2) \le |f|_{\bar{\mathscr{A}_s}} N^{-s}$.

The overlay $\mathscr{T} = \mathscr{T}_1 \oplus \mathscr{T}_2 \in \mathbb{T}_{2N}$ according to (117), and

$$\|u - U_{\mathscr{T}}\|_\Omega + \operatorname{osc}_{\mathscr{T}}(f) \le \|u - U_{\mathscr{T}_1}\|_\Omega + \operatorname{osc}_{\mathscr{T}_2}(f) \le 2^s \big(|u|_{\mathscr{A}_s} + |f|_{\bar{\mathscr{A}_s}} \big)(2N)^{-s}.$$

This yields $(u, f, A) \in \mathbb{A}_s$ together with the bound $|u, f, A|_s \lesssim |u|_{\mathscr{A}_s} + |f|_{\bar{\mathscr{A}_s}}$. □

Corollary 7 (Membership in $\mathbb{A}_{1/2}$ with piecewise linear A). *Let $d = 2$, $n = 1$, and $u \in H_0^1(\Omega)$ be the solution of the model problem with piecewise linear A and $f \in L^2(\Omega)$. If $u \in W_p^2(\Omega; \mathscr{T}_0)$ is piecewise W_p^2 over the initial grid $\mathscr{T}_0$ and $p > 1$, then $(u, f, A) \in \mathbb{A}_{1/2}$ and*

$$|u, f, A|_{1/2} \lesssim \|D^2 u\|_{L^p(\Omega; \mathscr{T}_0)} + \|f\|_{L^2(\Omega)}. \tag{119}$$

Proof. Since $f \in L^2(\Omega)$, we realize that for all uniform refinements $\mathscr{T} \in \mathbb{T}$ we have

$$\operatorname{osc}_{\mathscr{T}}(f) = \|h(f - P_0 f)\|_{L^2(\Omega)} \le h_{\max}(\mathscr{T}) \|f\|_{L^2(\Omega)} \lesssim (\#\mathscr{T})^{-1/2} \|f\|_{L^2(\Omega)}.$$

This implies $f \in \bar{\mathscr{A}}_{1/2}$ with $|f|_{\bar{\mathscr{A}}_{1/2}} \lesssim \|f\|_{L^2(\Omega)}$. Moreover, for $u \in W_p^2(\Omega; \mathscr{T}_0)$ we learn from Corollary 2 and Remark 6 of Sect. 1.6 that $u \in \mathscr{A}_{1/2}$ and $|u|_{\mathscr{A}_{1/2}} \lesssim \|D^2 u\|_{L^2(\Omega; \mathscr{T}_0)}$. The assertion then follows from Lemma 23. □

Corollary 8 (Membership in $\mathbb{A}_{1/2}$ with variable A). *Let $d = 2$, $n = 1$, $p > 1$, $f \in L^2(\Omega)$. Let $A \in W_\infty^1(\Omega, \mathscr{T}_0)$ be piecewise Lipschitz and $u \in W_p^2(\Omega; \mathscr{T}_0) \cap H_0^1(\Omega)$ be piecewise W_p^2 over the initial mesh $\mathscr{T}_0$. Then $(u, f, A) \in \mathbb{A}_{1/2}$ and*

$$|u, f, A|_{1/2} \lesssim \|D^2 u\|_{L^p(\Omega; \mathscr{T}_0)} + \|f\|_{L^2(\Omega)} + \|A\|_{W_\infty^1(\Omega; \mathscr{T}_0)}. \tag{120}$$

Proof. Combine Problem 55 with Corollary 2. □

Corollary 9 (Membership in $\mathbb{A}_s$ with $s < 1/d$). *Let $d \geq 2$, $n = 1$, $1 < t < 2$, $p > 1$, and $f \in L^2(\Omega)$. Let $\boldsymbol{A} \in W^1_\infty(\Omega, \mathscr{T}_0)$ be piecewise Lipschitz and $u \in W^t_p(\Omega; \mathscr{T}_0) \cap H^1_0(\Omega)$ be piecewise W^t_p over the initial mesh $\mathscr{T}_0$ with $t - \frac{d}{p} > 1 - \frac{d}{2}$. Then $(u, f, \boldsymbol{A}) \in \mathbb{A}_{(t-1)/d}$ and*

$$|u, f, \boldsymbol{A}|_{(t-1)/d} \lesssim \|D^t u\|_{L^p(\Omega;\mathscr{T}_0)} + \|f\|_{L^2(\Omega)} + \|\boldsymbol{A}\|_{W^1_\infty(\Omega;\mathscr{T}_0)}. \tag{121}$$

Proof. Combine Problem 9 with Problem 55. □

Example 2 (Pre-asymptotics). Corollary 7 shows that oscillation decays with rate $1/2$ for $f \in L^2(\Omega)$. Since the decay rate of the total error is $s \leq 1/2$, oscillation can be ignored asymptotically; this is verified in Problems 56–58. However, oscillation may dominate the total error, or equivalently the class $\mathbb{A}_s$ may fail to describe the behavior of $|||u - U_k|||_\Omega$, in the early stages of adaptivity. In fact, we recall from Problem 32 that the discrete solution $U_k = 0$, and $|||u - U_k|||_\Omega \approx 2^{-K}$ is constant for as many steps $k \leq K$ as desired. In contrast, $\mathscr{E}_k(U_k) = \text{osc}_k(U_k) = \|h(f - \bar{f})\|_{L^2(\Omega)} = \|hf\|_{L^2(\Omega)}$ reduces strictly for $k \leq K$ but overestimates $|||u - U_k|||_\Omega$. The fact that the preasymptotic regime $k \leq K$ for the energy error could be made arbitrarily long would be problematic if we were to focus exclusively on $|||u - U_k|||_\Omega$.

In practice, this effect is typically less dramatic because f is not orthogonal to $\mathbb{V}(\mathscr{T}_k)$. Figure 20 displays the behavior of the AFEM for the smooth solution u_S

$$u_S(x, y) = 10^{-2} a_i^{-1}(x^2 + y^2) \sin^2(\kappa\pi x) \sin^2(\kappa\pi y), \quad 1 \leq i \leq 4 \tag{122}$$

of the model problem (87) with discontinuous coefficients $\{a_i\}_{i=1}^4$ in checkerboard pattern as in Sect. 5.4 and frequencies $\kappa = 5, 10,$ and 15. We can see that the error

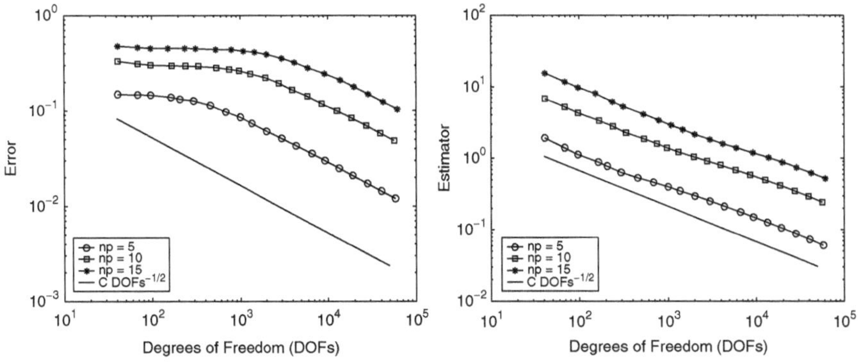

Fig. 20 Decay of the energy error (*left*) and the estimator (*right*) for the smooth solution u_S of (122) with frequencies $\kappa = 5, 10,$ and 15. The energy error exhibits a frequency-dependent plateau in the preasymptotic regime and later an optimal decay. This behavior is allowed by $\mathbb{A}_s$

exhibits a frequency-dependent plateau in the preasymptotic regime and later an optimal decay. In contrast, the estimator decays always with the optimal rate. Since all decisions of the AFEM are based on the estimator, this behavior has to be expected and is consistent with our notion of approximation class $\mathbb{A}_s$, which can be characterized just by the estimator according to (116).

7.3 Quasi-Optimal Cardinality: Vanishing Oscillation

In this section we follow the ideas of Stevenson [52] for the simplest scenario with vanishing oscillation $\mathrm{osc}_{\mathscr{T}}(U) = 0$, and thereby explore the insight (112). We recall that in this case the a posteriori error estimates (94) become

$$C_2\,\mathscr{E}^2_{\mathscr{T}}(U) \leq |||u - U|||^2_{\Omega} \leq C_1\,\mathscr{E}^2_{\mathscr{T}}(U). \tag{123}$$

It is then evident that the ratio $C_2/C_1 \leq 1$, between the *reliability* constant C_1 and the *efficiency* constant C_2, is a quality measure of the estimator $\mathscr{E}_{\mathscr{T}}(U)$: the closer to 1 the better! This ratio is usually closer to 1 for non-residual estimators, such as those discussed in Sect. 5.5, but their theory is a bit more cumbersome.

Assumptions for optimal decay rate: The following are further restrictions on AFEM to achieve optimal error decay, as predicted by the approximation class $\mathscr{A}_s$.

Assumption 1 (Marking parameter: vanishing oscillation). *The parameter ϑ of Dörfler marking satisfies $\vartheta \in (0, \vartheta_*)$ with*

$$\vartheta_* := \sqrt{\frac{C_2}{C_1}}. \tag{124}$$

Assumption 2 (Cardinality of $\mathscr{M}$). *MARK selects a set $\mathscr{M}$ with* minimal *cardinality.*

Assumption 3 (Initial labeling). *The labeling of the initial mesh $\mathscr{T}_0$ satisfies* (6) *for $d = 2$ or its multidimensional counterpart for $d > 2$ [45, 52].*

A few comments about these assumptions are now in order.

Remark 12 (Threshold $\vartheta_* < 1$). It is reasonable to be cautious in making marking decisions if the constants C_1 and C_2 are very disparate, and thus the ratio C_2/C_1 is far from 1. This justifies the upper bound $\vartheta_* \leq 1$ in Assumption 1.

Remark 13 (Minimal $\mathscr{M}$). According to the equidistribution principle (16) and the local lower bound (68) without oscillation, it is natural to mark elements with largest error indicators. This leads to a minimal set $\mathscr{M}$, as stated in Assumption 2, and turns out to be crucial to link AFEM with optimal meshes and approximation classes.

Remark 14 (Initial triangulation). Assumption 3 guarantees the complexity estimate of module REFINE stated in Theorem 1 and proved in Sect. 6.3:

$$\#\mathscr{T}_k - \#\mathscr{T}_0 \le \Lambda_0 \sum_{j=0}^{k-1} \#\mathscr{M}_j .$$

Assumption 3 is rather restrictive for dimension $d > 2$. Any other refinement giving the same complexity estimate can replace REFINE together with Assumption 3.

Even though we cannot expect local upper bounds between the continuous and discrete solution, the following crucial result shows that this is not the case between discrete solutions on nested meshes $\mathscr{T}_* \ge \mathscr{T}$: what matters is the set of elements of $\mathscr{T}$ which are no longer in $\mathscr{T}_*$.

Lemma 24 (Localized upper bound). *Let $\mathscr{T}, \mathscr{T}_* \in \mathbb{T}$ satisfy $\mathscr{T}_* \ge \mathscr{T}$ and let $\mathscr{R} := \mathscr{R}_{\mathscr{T}\to\mathscr{T}_*}$ be the refined set. If $U \in \mathbb{V}$, $U_* \in \mathbb{V}_*$ are the corresponding Galerkin solutions, then*

$$|\!|\!| U_* - U |\!|\!|_\Omega^2 \le C_1\, \mathscr{E}_\mathscr{T}^2(U, \mathscr{R}). \tag{125}$$

Proof. See Problem 53. □

We are now ready to explore Stevenson's insight (112) for the simplest scenario.

Lemma 25 (Dörfler marking: vanishing oscillation). *Let ϑ satisfy Assumption 1 and set $\mu := 1 - \vartheta^2/\vartheta_*^2 > 0$. Let $\mathscr{T}_* \ge \mathscr{T}$ and the corresponding Galerkin solution $U_* \in \mathbb{V}(\mathscr{T}_*)$ satisfy*

$$|\!|\!| u - U_* |\!|\!|_\Omega^2 \le \mu\, |\!|\!| u - U |\!|\!|_\Omega^2 . \tag{126}$$

Then the refined set $\mathscr{R} = \mathscr{R}_{\mathscr{T}\to\mathscr{T}_*}$ satisfies the Dörfler property

$$\mathscr{E}_\mathscr{T}(U, \mathscr{R}) \ge \vartheta\, \mathscr{E}_\mathscr{T}(U, \mathscr{T}). \tag{127}$$

Proof. Since $\mu < 1$ we use the lower bound in (123), in conjunction with (126) and Pythagoras equality (92), to derive

$$\begin{aligned}(1-\mu) C_2 \mathscr{E}_\mathscr{T}^2(U, \mathscr{T}) &\le (1-\mu)\, |\!|\!| u - U |\!|\!|_\Omega^2 \\ &\le |\!|\!| u - U |\!|\!|_\Omega^2 - |\!|\!| u - U_* |\!|\!|_\Omega^2 = |\!|\!| U - U_* |\!|\!|_\Omega^2 .\end{aligned}$$

In view of Lemma 24, we thus deduce

$$(1-\mu) C_2 \mathscr{E}_\mathscr{T}^2(U, \mathscr{T}) \le C_1 \mathscr{E}_\mathscr{T}^2(U, \mathscr{R}),$$

which is the assertion in disguise. □

To examine the cardinality of $\mathcal{M}_k$ in terms of $|||u - U_k|||_\Omega$ we must relate AFEM with the approximation class $\mathcal{A}_s$. Even though this might appear like an undoable task, the key to unravel this connection is given by Lemma 25. We show this now.

Lemma 26 (Cardinality of $\mathcal{M}_k$). *Let Assumptions 1 and 2 hold. If $u \in \mathcal{A}_s$ then*

$$\#\mathcal{M}_k \lesssim |u|_s^{1/s} \, |||u - U_k|||_\Omega^{-1/s} \qquad \forall k \geq 0. \tag{128}$$

Proof. We invoke that $u \in \mathcal{A}_s$, together with Problem 51 with $\varepsilon^2 = \mu \, |||u - U_k|||_\Omega^2$, to find a mesh $\mathcal{T}_\varepsilon \in \mathbb{T}$ and the Galerkin solution $U_\varepsilon \in \mathbb{V}(\mathcal{T}_\varepsilon)$ so that

$$|||u - U_\varepsilon|||_\Omega^2 \leq \varepsilon^2, \quad \#\mathcal{T}_\varepsilon - \#\mathcal{T}_0 \lesssim |u|_s^{\frac{1}{s}} \varepsilon^{-\frac{1}{s}}.$$

Since $\mathcal{T}_\varepsilon$ may be totally unrelated to $\mathcal{T}_k$, we introduce the overlay $\mathcal{T}_* = \mathcal{T}_\varepsilon \oplus \mathcal{T}_k$. We exploit the property $\mathcal{T}_* \geq \mathcal{T}_\varepsilon$ to conclude that the Galerkin solution $U_* \in \mathbb{V}(\mathcal{T}_*)$ satisfies (127):

$$|||u - U_*|||_\Omega^2 \leq |||u - U_\varepsilon|||_\Omega^2 \leq \varepsilon^2 = \mu \, |||u - U|||_\Omega^2 .$$

Therefore, Lemma 25 implies that the refined set $\mathcal{R} = \mathcal{R}_{\mathcal{T} \to \mathcal{T}_*}$ satisfies a Dörfler marking with parameter $\vartheta < \vartheta_*$. But MARK delivers a minimal set $\mathcal{M}_k$ with this property, according to Assumption 2, whence

$$\#\mathcal{M}_k \leq \#\mathcal{R} \leq \#\mathcal{T}_* - \#\mathcal{T}_k \leq \#\mathcal{T}_\varepsilon - \#\mathcal{T}_0 \lesssim |u|_s^{\frac{1}{s}} \varepsilon^{-\frac{1}{s}},$$

where we use Lemma 22 to account for the overlay. The proof is complete. □

Proposition 4 (Quasi-optimality: vanishing oscillation). *Let Assumptions 1–3 hold. If $u \in \mathcal{A}_s$, then AFEM gives rise to a sequence $(\mathcal{T}_k, \mathbb{V}_k, U_k)_{k=0}^\infty$ such that*

$$|||u - U_k|||_\Omega \lesssim |u|_s \, (\#\mathcal{T}_k - \#\mathcal{T}_0)^{-s} \qquad \forall k \geq 1.$$

Proof. We make use of Assumption 3, along with Theorem 1, to infer that

$$\#\mathcal{T}_k - \#\mathcal{T}_0 \leq \Lambda_0 \sum_{j=0}^{k-1} \#\mathcal{M}_j \lesssim |u|_s^{\frac{1}{s}} \sum_{j=0}^{k-1} |||u - U_j|||_\Omega^{-\frac{1}{s}} .$$

We now use the contraction property (97) of Lemma 15

$$|||u - U_k|||_\Omega \leq \alpha^{k-j} \, |||u - U_j|||_\Omega$$

to replace the sum above by

$$\sum_{j=0}^{k-1} |||u - U_j|||_{\Omega}^{-\frac{1}{s}} \leq |||u - U_k|||_{\Omega}^{-\frac{1}{s}} \sum_{j=0}^{k-1} \alpha^{\frac{k-j}{s}} < \frac{\alpha^{\frac{1}{s}}}{1-\alpha^{\frac{1}{s}}} |||u - U_k|||_{\Omega}^{-\frac{1}{s}},$$

because $\alpha < 1$ and the series is summable. This completes the proof. □

7.4 *Quasi-Optimal Cardinality: General Data*

In this section we remove the restriction $\operatorname{osc}_{\mathcal{T}}(U) = 0$, and thereby make use of the basic ingredients developed in Sects. 7.1 and 7.2. Therefore, we replace the energy error by the total error and the linear approximation class $\mathcal{A}_s$ for u by the nonlinear class $\mathbb{A}_s$ for the triple $(u, f, \boldsymbol{A})$. To account for the presence of general f and $\boldsymbol{A}$, we need to make an even more stringent assumption on the threshold ϑ_*.

Assumption 4 (Marking parameter: general data). *Let* $C_3 = \Lambda_1 \operatorname{osc}^2_{\mathcal{T}_0}(\boldsymbol{A})$ *be the constant in Problem 46 and Lemma 21. The marking parameter* ϑ *satisfies* $\vartheta \in (0, \vartheta_*)$ *with*

$$\vartheta_* = \sqrt{\frac{C_2}{1 + C_1(1 + C_3)}}. \tag{129}$$

We now proceed along the same lines as those of Sect. 7.3.

Lemma 27 (Dörfler marking: general data). *Let Assumption 4 hold and set* $\mu := \frac{1}{2}(1 - \frac{\vartheta^2}{\vartheta_*^2}) > 0$. *If* $\mathcal{T}_* \geq \mathcal{T}$ *and the corresponding Galerkin solution* $U_* \in \mathbb{V}(\mathcal{T}_*)$ *satisfy*

$$|||u - U_*|||_{\Omega}^2 + \operatorname{osc}^2_{\mathcal{T}_*}(U_*) \leq \mu\big(|||u - U|||_{\Omega}^2 + \operatorname{osc}^2_{\mathcal{T}}(U)\big), \tag{130}$$

then the refined set $\mathcal{R} = \mathcal{R}_{\mathcal{T} \to \mathcal{T}_*}$ *satisfies the Dörfler property*

$$\mathcal{E}_{\mathcal{T}}(U, \mathcal{R}) \geq \vartheta\, \mathcal{E}_{\mathcal{T}}(U, \mathcal{T}). \tag{131}$$

Proof. We split the proof into four steps.

[1] In view of the global lower bound (94b)

$$C_2\, \mathcal{E}^2_{\mathcal{T}}(U) \leq |||u - U|||_{\Omega}^2 + \operatorname{osc}^2_{\mathcal{T}}(U)$$

and (130), we can write

$$\begin{aligned}(1 - 2\mu)\, C_2\, \mathcal{E}^2_{\mathcal{T}}(U) &\leq (1 - 2\mu)\big(|||u - U|||_{\Omega}^2 + \operatorname{osc}^2_{\mathcal{T}}(U)\big) \\ &\leq \big(|||u - U|||_{\Omega}^2 - 2\, |||u - U_*|||_{\Omega}^2 \big) + \big(\operatorname{osc}^2_{\mathcal{T}}(U) - 2 \operatorname{osc}^2_{\mathcal{T}_*}(U_*)\big).\end{aligned}$$

[2] Combining the Pythagoras orthogonality relation (92)

$$|||u - U|||_{\Omega}^2 - |||u - U_*|||_{\Omega}^2 = |||U - U_*|||_{\Omega}^2$$

with the localized upper bound Lemma 24 yields

$$|||u-U|||_\Omega^2 - 2\,|||u-U_*|||_\Omega^2 \le |||U-U_*|||_\Omega^2 \le C_1\,\mathscr{E}_{\mathscr{T}}^2(U,\mathscr{R}).$$

$\boxed{3}$ To deal with oscillation we decompose the elements of $\mathscr{T}$ into two disjoint sets: $\mathscr{R}$ and $\mathscr{T}\setminus\mathscr{R}$. In the former case, we have

$$\operatorname{osc}_{\mathscr{T}}^2(U,\mathscr{R}) - 2\operatorname{osc}_{\mathscr{T}_*}^2(U_*,\mathscr{R}) \le \operatorname{osc}_{\mathscr{T}}^2(U,\mathscr{R}) \le \mathscr{E}_{\mathscr{T}}^2(U,\mathscr{R}),$$

because $\operatorname{osc}_{\mathscr{T}}(U,T) \le \mathscr{E}_{\mathscr{T}}(U,T)$ for all $T\in\mathscr{T}$. On the other hand, we use that $\mathscr{T}\setminus\mathscr{R} = \mathscr{T}\cap\mathscr{T}_*$ and apply Problem 46 in conjunction with Lemma 24 to arrive at

$$\operatorname{osc}_{\mathscr{T}}^2(U,\mathscr{T}\setminus\mathscr{R}) - 2\operatorname{osc}_{\mathscr{T}_*}^2(U_*,\mathscr{T}\setminus\mathscr{R}) \le C_3\,|||U-U_*|||_\Omega^2 \le C_1C_3\mathscr{E}_{\mathscr{T}}^2(U,\mathscr{R}).$$

Adding these two estimates gives

$$\operatorname{osc}_{\mathscr{T}}^2(U) - 2\operatorname{osc}_{\mathscr{T}_*}^2(U_*) \le (1+C_1C_3)\mathscr{E}_{\mathscr{T}}^2(U,\mathscr{R}).$$

$\boxed{4}$ Returning to $\boxed{1}$ we realize that

$$(1-2\mu)\,C_2\,\mathscr{E}_{\mathscr{T}}^2(U,\mathscr{T}) \le \big(1+C_1(1+C_3)\big)\,\mathscr{E}_{\mathscr{T}}^2(U,\mathscr{R}),$$

which is the asserted estimate (131) in disguise. □

Lemma 28 (Cardinality of $\mathscr{M}_k$: general data). *Let Assumptions 2 and 4 hold. If $(u,f,\boldsymbol{A})\in\mathbb{A}_s$, then*

$$\#\mathscr{M}_k \lesssim |u,f,\boldsymbol{A}|_s^{1/s}\,\big(\,|||u-U_k|||_\Omega + \operatorname{osc}_k(U_k)\big)^{-1/s} \quad \forall k\ge 0. \tag{132}$$

Proof. We split the proof into three steps.
$\boxed{1}$ We set $\varepsilon^2 := \mu\Lambda_2^{-1}\big(\,|||u-U_k|||_\Omega^2 + \operatorname{osc}_k^2(U_k)\big)$ with $\mu = \frac{1}{2}\big(1-\frac{\vartheta^2}{\vartheta_*^2}\big) > 0$ as in Lemma 27 and Λ_2 given Lemma 21. Since $(u,f,\boldsymbol{A})\in\mathbb{A}_s$, in view of Problem 51 there exists $\mathscr{T}_\varepsilon\in\mathbb{T}$ and $U_\varepsilon\in\mathbb{V}(\mathscr{T}_\varepsilon)$ such that

$$|||u-U_\varepsilon|||_\Omega^2 + \operatorname{osc}_\varepsilon^2(U_\varepsilon) \le \varepsilon^2 \quad \text{and} \quad \#\mathscr{T}_\varepsilon - \#\mathscr{T}_0 \lesssim |u,f,\boldsymbol{A}|_s^{1/2}\,\varepsilon^{-1/s}.$$

Since $\mathscr{T}_\varepsilon$ may be totally unrelated to $\mathscr{T}_k$ we introduce the overlay $\mathscr{T}_* = \mathscr{T}_k\oplus\mathscr{T}_\varepsilon$.
$\boxed{2}$ We claim that the total error over $\mathscr{T}_*$ reduces by a factor μ relative to that one over $\mathscr{T}_k$. In fact, since $\mathscr{T}_*\ge\mathscr{T}_\varepsilon$ and so $\mathbb{V}(\mathscr{T}_*)\supset\mathbb{V}(\mathscr{T}_\varepsilon)$, we use Lemma 21 to obtain

$$\begin{aligned}|||u-U_*|||_\Omega^2 + \operatorname{osc}_{\mathscr{T}_*}^2(U_*) &\le \Lambda_2\Big(\,|||u-U_\varepsilon|||_\Omega^2 + \operatorname{osc}_\varepsilon^2(U_\varepsilon)\Big)\\ &\le \Lambda_2\varepsilon^2 = \mu\,\big(\,|||u-U_k|||_\Omega^2 + \operatorname{osc}_k^2(U_k)\big).\end{aligned}$$

Upon applying Lemma 27 we conclude that the set $\mathscr{R} = \mathscr{R}_{\mathscr{T}_k \to \mathscr{T}_*}$ of refined elements satisfies a Dörfler marking (131) with parameter $\vartheta < \vartheta_*$.

[3] According to Assumption 2, MARK selects a minimal set $\mathscr{M}_k$ satisfying this property. Therefore, we deduce

$$\#\mathscr{M}_k \le \#\mathscr{R} \le \#\mathscr{T}_* - \#\mathscr{T}_k \le \#\mathscr{T}_\varepsilon - \#\mathscr{T}_0 \lesssim |u, f, \boldsymbol{A}|_s^{1/s}\, \varepsilon^{-1/s},$$

where we have employed Lemma 22 to account for the cardinality of the overlay. Finally, recalling the definition of ε we end up with the asserted estimate (132). □

Remark 15 (Blow-up of constant). The constant hidden in (132) blows up as $\vartheta \uparrow \vartheta_*$ because $\mu \downarrow 0$; see Problem 54.

We are ready to prove the main result of this section, which combines Theorem 9 and Lemma 28.

Theorem 10 (Quasi-optimality: general data). *Let Assumptions 2–4 hold. If $(u, f, \boldsymbol{A}) \in \mathbb{A}_s$, then AFEM gives rise to a sequence $(\mathscr{T}_k, \mathbb{V}_k, U_k)_{k=0}^\infty$ such that*

$$|||u - U_k|||_\Omega + \operatorname{osc}_k(U_k) \lesssim |u, f, \boldsymbol{A}|_s \,(\#\mathscr{T}_k - \#\mathscr{T}_0)^{-s} \quad \forall k \ge 1.$$

Proof. [1] Since no confusion arises, we use the notation $\operatorname{osc}_j = \operatorname{osc}_j(U_j)$ and $\mathscr{E}_j = \mathscr{E}_j(U_j)$. In light of Assumption 3, which yields Theorem 1, and (132) we have

$$\#\mathscr{T}_k - \#\mathscr{T}_0 \lesssim \sum_{j=0}^{k-1} \#\mathscr{M}_j \lesssim |u, f, \boldsymbol{A}|_s^{1/s} \sum_{j=0}^{k-1} \big(|||u - U_j|||_\Omega^2 + \operatorname{osc}_j^2 \big)^{-1/(2s)}.$$

[2] Let $\gamma > 0$ be the scaling factor in the (contraction) Theorem 9. The lower bound (94b) along with $\operatorname{osc}_j \le \mathscr{E}_j$ implies

$$|||u - U_j|||_\Omega^2 + \gamma \operatorname{osc}_j^2 \le |||u - U_j|||_\Omega^2 + \gamma\, \mathscr{E}_j^2 \le \Big(1 + \frac{\gamma}{C_2}\Big)\big(|||u - U_j|||_\Omega^2 + \operatorname{osc}_j^2 \big).$$

[3] Theorem 9 yields for $0 \le j < k$

$$|||u - U_k|||_\Omega^2 + \gamma\, \mathscr{E}_k^2 \le \alpha^{2(k-j)} \big(|||u - U_j|||_\Omega^2 + \gamma\, \mathscr{E}_j^2 \big),$$

whence

$$\#\mathscr{T}_k - \#\mathscr{T}_0 \lesssim |u, f, \boldsymbol{A}|_s^{1/s} \big(|||u - U_k|||_\Omega^2 + \gamma\, \mathscr{E}_k^2 \big)^{-1/(2s)} \sum_{j=0}^{k-1} \alpha^{(k-j)/s}.$$

Since $\sum_{j=0}^{k-1} \alpha^{(k-j)/s} = \sum_{j=1}^{k} \alpha^{j/s} < \sum_{j=1}^{\infty} \alpha^{j/s} < \infty$ because $\alpha < 1$, the assertion follows immediately. □

We conclude this section with several applications of Theorem 10.

Corollary 10 (Estimator decay). *Let Assumptions 2–4 be satisfied. If* $(u, f, \boldsymbol{A}) \in \mathbb{A}_s$ *then the estimator* $\mathscr{E}_k(U_k)$ *satisfies*

$$\mathscr{E}_k(U_k) \lesssim |u, f, \boldsymbol{A}|_s^{1/s} (\#\mathscr{T}_k - \#\mathscr{T}_0)^{-s} \quad \forall k \geq 1.$$

Proof. Use (114) and Theorem 10. □

Corollary 11 (W_p^2-regularity with piecewise linear A). *Let $d = 2$, the polynomial degree $n = 1$, $f \in L^2(\Omega)$, and let $\boldsymbol{A}$ be piecewise linear over $\mathscr{T}_0$. If $u \in W_p^2(\Omega; \mathscr{T}_0)$ for $p > 1$, then AFEM gives rise to a sequence $\{\mathscr{T}_k, \mathbb{V}_k, U_k\}_{k=0}^{\infty}$ satisfying* $\operatorname{osc}_k(U_k) = \|h_k(f - P_0 f)\|_{L^2(\Omega)}$ *and for all $k \geq 1$*

$$|\!|\!| u - U_k |\!|\!|_\Omega + \operatorname{osc}_k(U_k) \lesssim \left(\|D^2 u\|_{L^p(\Omega;\mathscr{T}_0)} + \|f\|_{L^2(\Omega)} \right) (\#\mathscr{T}_k - \#\mathscr{T}_0)^{-1/2}.$$

Proof. Combine Corollary 7 with Theorem 10. □

Corollary 12 (W_p^2-regularity with variable A). *Assume the setting of Corollary 11, but let $\boldsymbol{A}$ be piecewise Lipschitz over the initial grid $\mathscr{T}_0$. Then AFEM gives rise to a sequence $\{\mathscr{T}_k, \mathbb{V}_k, U_k\}_{k=0}^{\infty}$ satisfying for all $k \geq 1$*

$$\begin{aligned} &|\!|\!| u - U_k |\!|\!|_\Omega + \operatorname{osc}_k(U_k) \\ &\qquad \lesssim \left(\|D^2 u\|_{L^p(\Omega;\mathscr{T}_0)} + \|f\|_{L^2(\Omega)} + \|\boldsymbol{A}\|_{W_\infty^1(\Omega;\mathscr{T}_0)} \right) (\#\mathscr{T}_k - \#\mathscr{T}_0)^{-1/2}. \end{aligned}$$

Proof. Combine Corollary 8 with Theorem 10. □

Corollary 13 (W_p^s-regularity with $s < 1/d$). *Let $d \geq 2$, $n = 1$, $1 < t < 2$, $p > 1$, $f \in L^2(\Omega)$, and $\boldsymbol{A} \in W_\infty^1(\Omega, \mathscr{T}_0)$ be piecewise Lipschitz. If $u \in W_p^t(\Omega; \mathscr{T}_0) \cap H_0^1(\Omega)$ is piecewise W_p^t over the initial mesh $\mathscr{T}_0$ with $t - \frac{d}{p} > 1 - \frac{d}{2}$, then AFEM gives rise to a sequence $\{\mathscr{T}_k, \mathbb{V}_k, U_k\}_{k=0}^{\infty}$ satisfying for all $k \geq 1$*

$$\begin{aligned} &|\!|\!| u - U_k |\!|\!|_\Omega + \operatorname{osc}_k(U_k) \\ &\qquad \lesssim \left(\|D^t u\|_{L^p(\Omega;\mathscr{T}_0)} + \|f\|_{L^2(\Omega)} + \|\boldsymbol{A}\|_{W_\infty^1(\Omega;\mathscr{T}_0)} \right) (\#\mathscr{T}_k - \#\mathscr{T}_0)^{-(t-1)/d}. \end{aligned}$$

Proof. Combine Corollary 9 with Theorem 10. □

7.5 Extensions and Restrictions

We conclude with a brief discussion of extensions of the theory and some of its restrictions.

Optimal complexity: inexact solvers, quadrature, and storage: We point out that we have never mentioned the notion of *complexity* so far. This is because complexity estimates entail crucial issues that we have ignored: inexact solvers to approximate the Galerkin solution; quadrature; and optimal storage. We comment on them now.

Multilevel solvers are known to deliver an approximate solution with cost proportional to the number of degrees of freedom. Even though the theory is well developed for uniform refinement, it is much less understood for adaptive refinement. This is due to the fact that the adaptive bisection meshes do not satisfy the so-called nested refinement assumption. Recently, Xu et al. [60] have bridged the gap between graded and quasi-uniform grids exploiting the geometric structure of bisection grids and a resulting new space decomposition. They designed and analyzed optimal additive and multiplicative multilevel methods for any dimension $d \geq 2$ and polynomial degree $n \geq 1$, thereby improving upon Wu and Chen [59]. The theories of Sects. 5 and 7 can be suitably modified to account for optimal iterative solvers; we refer to Stevenson [52].

Quadrature is a very delicate issue in a purely a posteriori context, that is without a priori knowledge of the functions involved. Even if we were to replace both data f and $\boldsymbol{A}$ by piecewise polynomials so that quadrature would be simple, we would need to account for the discrepancy in adequate norms between exact and approximate data, again a rather delicate matter. This issue is to a large extend open.

Optimal storage is an essential, but often disregarded, aspect of a complexity analysis. For instance, ALBERTA is an excellent library for AFEM but does not have optimal storage capabilities [50].

Non-residual estimators: The cardinality analysis of this section extends to estimators other than the residual; we refer to Cascón and Nochetto [15] and Kreuzer and Siebert [35]. They include the hierarchical, Zienkiewicz–Zhu [2, 27, 55, 58], and Braess–Schoerbel estimators, as well as those based on the solution of local problems [12, 42]. Even though the contraction property of Theorem 9 is no longer valid between consecutive iterates, it is true after a fixed number of iterations, which is enough for the arguments in Proposition 4 and Theorem 10 to apply. The resulting error estimates possess constants proportional to this gap.

Nonconforming meshes: Since REFINE exhibits optimal complexity for admissible nonconforming meshes, according to Sect. 6.4, and this is the only ingredient where nonconformity might play a role, the theory of this section extends. We refer to Bonito and Nochetto [9].

Discontinuous Galerkin methods (dG): The study of cardinality for adaptive dG methods is rather technical. This is in part due to the fact that key Lemmas 26 and 28 hinge on mesh overlay, which in turn does not provide control of the level of refinement. This makes it difficult to compare broken energy norms

$$|||v|||_{\mathscr{T}}^2 = \|\boldsymbol{A}^{1/2}\nabla v\|_{L^2(\Omega;\mathscr{T})}^2 + \|h^{-1/2} [\![v]\!] \|_{L^2(\Sigma)}^2,$$

which contain jump terms with negative powers of the mesh-size over the scheleton Σ of $\mathscr{T}$. Consequently, the monotonicity of energy norms used in Lemmas 26 and 28 is no longer true!

To circumvent this difficulty, Bonito and Nochetto [9] resorted to continuous finite elements $\mathbb{V}^0(\mathscr{T})$ over the (admissible nonconforming) mesh $\mathscr{T}$, which have the same degree as their discontinuous counterpart $\mathbb{V}(\mathscr{T})$. This leads to a cardinality theory very much in the spirit of this section. However, it raises the question whether discontinuous elements deliver a better asymptotic rate over admissible nonconforming meshes. Since this result is of intrinsic interest, we report it now.

Lemma 29 (Equivalence of classes). *Let $\mathbb{A}_s$ be the approximation class using discontinuous elements of degree $\le n$ and $\mathbb{A}_s^0$ be the continuous counterpart. Then, for $0 < s \le n/d$, total errors are equivalent on the same mesh, whence $\mathbb{A}_s = \mathbb{A}_s^0$.*

Proof. We use the notation of Problem 11. Since $\mathbb{V}^0(\mathscr{T}) \subset \mathbb{V}(\mathscr{T})$, the inclusion $\mathbb{A}_s^0 \subset \mathbb{A}_s$ is obvious. To prove the converse, we let $(u, f, \boldsymbol{A}) \in \mathbb{A}_s$ and, for $N > \#\mathscr{T}_0$, let $\mathscr{T}_* \in \mathbb{T}_N$ be an admissible nonconforming grid and $U_* \in \mathbb{V}(\mathscr{T}_*)$ be so that

$$|\!|\!| u - U_* |\!|\!|_{\mathscr{T}_*} + \operatorname{osc}_{\mathscr{T}_*}(U_*) = \inf_{\mathscr{T}\in\mathbb{T}_N} \inf_{V\in\mathbb{V}(\mathscr{T})} \Big(|\!|\!| u - V |\!|\!|_{\mathscr{T}} + \operatorname{osc}_{\mathscr{T}}(V) \Big) \lesssim N^{-s}.$$

Let $I_{\mathscr{T}} : \mathbb{V}(\mathscr{T}) \to \mathbb{V}^0(\mathscr{T})$ be the interpolation operator of Problem 11. Since $I_{\mathscr{T}_*} U_* \in \mathbb{V}^0(\mathscr{T}_*)$, if we were able to prove

$$|\!|\!| u - I_{\mathscr{T}_*} U_* |\!|\!|_{\mathscr{T}_*} + \operatorname{osc}_{\mathscr{T}_*}(I_{\mathscr{T}_*} U_*) \lesssim N^{-s},$$

then $(u, f, \boldsymbol{A}) \in \mathbb{A}_s^0$. Using the triangle inequality, we get

$$|\!|\!| u - I_{\mathscr{T}_*} U_* |\!|\!|_{\mathscr{T}_*} \le \| \boldsymbol{A}^{1/2} \nabla (u - U_*) \|_{L^2(\Omega;\mathscr{T}_*)} + \| \boldsymbol{A}^{1/2} \nabla (U_* - I_{\mathscr{T}_*} U^*) \|_{L^2(\Omega;\mathscr{T}_*)},$$

because $[\![u - I_{\mathscr{T}_*} U_*]\!]$ vanish on Σ. Problem 11 implies the estimate

$$\| \boldsymbol{A}^{1/2} \nabla (U_* - I_{\mathscr{T}_*} U^*) \|_{L^2(\Omega;\mathscr{T}_*)} \lesssim \| h^{-1/2} [\![U_*]\!] \|_{L^2(\Sigma_*)} \le |\!|\!| u - U_* |\!|\!|_{\mathscr{T}_*},$$

whence

$$|\!|\!| u - I_{\mathscr{T}_*} U_* |\!|\!|_{\mathscr{T}_*} \lesssim |\!|\!| u - U_* |\!|\!|_{\mathscr{T}_*}.$$

Since $\| \boldsymbol{A}^{1/2} \nabla (U_* - I_{\mathscr{T}_*} U_*) \|_{L^2(\Omega;\mathscr{T}_*)} \lesssim |\!|\!| U_* - I_{\mathscr{T}_*} U_* |\!|\!|_{\mathscr{T}_*}$, the oscillation term can be treated similarly. In fact, Problem 46 adapted to discontinuous functions yields

$$\operatorname{osc}_{\mathscr{T}_*}(I_{\mathscr{T}_*} U_*) \lesssim \operatorname{osc}_{\mathscr{T}_*}(U_*) + |\!|\!| u - U_* |\!|\!|_{\mathscr{T}_*}.$$

Coupling the two estimates above, we end up with

$$|\!|\!| u - I_{\mathscr{T}_*} U_* |\!|\!|_{\mathscr{T}_*} + \operatorname{osc}_{\mathscr{T}_*}(I_{\mathscr{T}_*} U_*) \lesssim |\!|\!| u - U_* |\!|\!|_{\mathscr{T}_*} + \operatorname{osc}_{\mathscr{T}_*}(U_*) \lesssim N^{-s}.$$

Therefore, $(u, f, \boldsymbol{A}) \in \mathbb{A}_s^0$ as desired. □

7.6 Notes

The theory presented in this section is rather recent. It started with the breakthrough (112) by Stevenson [52] for vanishing oscillation. If f is variable and $\boldsymbol{A}$ is piecewise constant, then Stevenson extended this idea upon adding an inner loop to handle data oscillation to the usual AFEM. This idea does not extend to the model problem (87) with variable $\boldsymbol{A}$, because the oscillation then depends on the Galerkin solution.

The next crucial step was made by Cascón et al. [14], who dealt with the notion of total error of Sect. 7.1, as previously done by Mekchay and Nochetto [37], and introduced the nonlinear approximation class $\mathbb{A}_s$ of Sect. 7.2. They derived the convergence rates of Sect. 7.4.

The analysis for nonconforming meshes is due to Bonito and Nochetto [9], who developed this theory in the context of dG methods for which they also derived convergence rates. The study of non-residual estimators is due to Kreuzer and Siebert [35] and Cascón and Nochetto [15].

The theory is almost exclusively devoted to the energy norm, except for the L^2-analysis of Demlow and Stevenson [21], who proved an optimal convergence rate for mildly varying graded meshes. Convergence rates have been proved for Raviart–Thomas mixed FEM by Chen et al. [18].

7.7 Problems

Problem 51 (Alternative definition of $\mathbb{A}_s$). Show that $(v, f, \boldsymbol{A}) \in \mathbb{A}_s$ if and only there exists a constant $\Lambda > 0$ such that for all $\varepsilon > 0$ there exist $\mathscr{T}_\varepsilon \geq \mathscr{T}_0$ conforming and $V_\varepsilon \in \mathbb{V}(\mathscr{T}_\varepsilon)$ such that

$$|||v - V_\varepsilon|||_\Omega^2 + \operatorname{osc}^2_{\mathscr{T}_\varepsilon}(V_\varepsilon) \leq \varepsilon^2 \quad \text{and} \quad \#\mathscr{T}_\varepsilon - \#\mathscr{T}_0 \leq \Lambda^{1/s}\, \varepsilon^{-1/s};$$

in this case $|v, f, \boldsymbol{A}|_s \leq \Lambda$. Hint: Let $\mathscr{T}_\varepsilon$ be minimal for $|||v - V_\varepsilon|||_\Omega^2 + \operatorname{osc}^2_{\mathscr{T}_\varepsilon}(V_\varepsilon) \leq \varepsilon^2$. This means that for all $\mathscr{T} \in \mathbb{T}$ such that $\#\mathscr{T} = \#\mathscr{T}_\varepsilon - 1$ we have $|||v - V_\varepsilon|||_\Omega^2 + \operatorname{osc}^2_{\mathscr{T}_\varepsilon}(V_\varepsilon) > \varepsilon$.

Problem 52 (Lemma 22). Prove that the overlay $\mathscr{T} = \mathscr{T}_1 \oplus \mathscr{T}_2$ is conforming and

$$\#\mathscr{T} \leq \#\mathscr{T}_1 + \#\mathscr{T}_2 - \#\mathscr{T}_0.$$

Hint: for each $T \in \mathscr{T}_0$, consider two cases $\mathscr{T}_1(T) \cap \mathscr{T}_2(T) \neq \emptyset$ and $\mathscr{T}_1(T) \cap \mathscr{T}_2(T) = \emptyset$, where $\mathscr{T}_i(T)$ is the portion of the mesh $\mathscr{T}_i$ contained in T.

Problem 53 (Lemma 24). Prove that if $\mathscr{T}, \mathscr{T}_* \in \mathbb{T}$ satisfy $\mathscr{T}_* \geq \mathscr{T}$, $\mathscr{R} := \mathscr{R}_{\mathscr{T} \to \mathscr{T}_*}$ is the refined set to go from $\mathscr{T}$ to $\mathscr{T}_*$, and $U \in \mathbb{V}$, $U_* \in \mathbb{V}_*$ are the corresponding Galerkin solutions, then

$$|||U_* - U|||_\Omega^2 \leq C_1\, \mathscr{E}^2_{\mathscr{T}}(U, \mathscr{R}).$$

To this end, write the equation fulfilled by $U_* - U \in \mathbb{V}_*$ and use as a test function the local quasi-interpolant $I_{\mathcal{T}}(U_* - U)$ of $U_* - U$ introduced in Proposition 2.

Problem 54 (Explicit dependence on ϑ and s). Trace the dependence on ϑ and s, as $\vartheta \to \vartheta_*$ and $s \to 0$, in the hidden constants in Lemma 28 and Theorem 10.

Problem 55 (Asymptotic decay of oscillation). Let $A \in W^1_\infty(\Omega; \mathcal{T}_0)$ be piecewise Lipschitz over the initial grid $\mathcal{T}_0$ and $f \in L^2(\Omega)$. Show that

$$\inf_{\mathcal{T} \in \mathbb{T}_N} \operatorname{osc}_{\mathcal{T}}(U) \lesssim \Big(\|f\|_{L^2(\Omega)} + \|A\|_{W^1_\infty(\Omega;\mathcal{T}_0)} \Big) N^{-1/d}$$

is attained with uniform meshes.

Problem 56 (Faster decay of data oscillation). Let $d = 2$ and $n = 1$. Let f be piecewise W^1_1 over the initial mesh $\mathcal{T}_0$, namely $f \in W^1_1(\Omega; \mathcal{T}_0)$. Show that

$$\inf_{\mathcal{T} \in \mathbb{T}_N} \|h_{\mathcal{T}}(f - P_0 f)\|_{L^2(\Omega)} \lesssim \|f\|_{W^1_1(\Omega;\mathcal{T}_0)} N^{-1},$$

using the thresholding algorithm of Sect. 1.6. Therefore, data oscillation decays twice as fast as the energy error asymptotically on suitably graded meshes.

Problem 57 (Faster decay of coefficient oscillation). Consider the coefficient oscillation weighted locally by the energy of the discrete solution U:

$$\eta^2_{\mathcal{T}}(A, U) = \sum_{T \in \mathcal{T}} \operatorname{osc}^2_{\mathcal{T}}(A, T) \|\nabla U\|^2_{L^2(\omega_T)},$$

where $\operatorname{osc}_{\mathcal{T}}(A, T)$ is defined in Problem 45. Let $d = 2, n = 1, p > 2$, and $A \in W^2_p(\Omega; \mathcal{T}_0)$ be piecewise in W^2_p over the initial grid $\mathcal{T}_0$. Use the thresholding algorithm of Sect. 1.6 to show that $\eta_{\mathcal{T}}(A, U)$ decays with a rate twice as fast as the energy error:

$$\inf_{\mathcal{T} \in \mathbb{T}_N} \eta_{\mathcal{T}}(A, U) \lesssim \|A\|_{W^2_p(\Omega;\mathcal{T}_0)} \|\nabla U\|_{L^2(\Omega)} N^{-1}.$$

Problem 58 (Faster decay of oscillation). Combine Problems 29, 56 and 57 for $d = 2, n = 1$ and $p > 2$ to prove that if $f \in W^1_1(\Omega; \mathcal{T}_0)$ and $A \in W^2_p(\Omega; \mathcal{T}_0)$, then the oscillation $\operatorname{osc}_{\mathcal{T}}(U, \mathcal{T})$ decays with a rate twice as fast as the energy error:

$$\inf_{\mathcal{T} \in \mathbb{T}_N} \operatorname{osc}_{\mathcal{T}}(U) \lesssim \Big(\|f\|_{W^1_1(\Omega;\mathcal{T}_0)} + \|A\|_{W^2_p(\Omega;\mathcal{T}_0)} \Big) N^{-1}.$$

Acknowledgements Ricardo H. Nochetto was partially supported by NSF Grant DMS-0807811. Andreas Veeser was partially supported by Italian PRIN 2008 "Analisi e sviluppo di metodi numerici avanzati per EDP".

References

1. M. Ainsworth, D. W. Kelly, A posteriori error estimators and adaptivity for finite element approximation of the non-homogeneous Dirichlet problem. Adv. Comput. Math. **15**, 3–23 (2001)
2. M. Ainsworth, J.T. Oden, *A Posteriori Error Estimation in Finite Element Analysis, Pure and Applied Mathematics* (Wiley, New York, 2000)
3. I. Babuška, R.B. Kellogg, J. Pitkäranta, Direct and inverse error estimates for finite elements with mesh refinements. Numer. Math. **33**, 447–471 (1979)
4. I. Babuška, A. Miller, A feedback finite element method with a posteriori error estimation. I. The finite element method and some basic properties of the a posteriori error estimator. Comput. Methods Appl. Mech. Eng. **61**(1), 1–40 (1987)
5. I. Babuška, W. Rheinboldt, Error estimates for adaptive finite element computations. SIAM J. Numer. Anal. **15**, 736–754 (1978)
6. E. Bänsch, P. Morin, R.H. Nochetto, An adaptive Uzawa FEM for the Stokes problem: convergence without the inf-sup condition. SIAM J. Numer. Anal. **40**, 1207–1229 (electronic) (2002)
7. P. Binev, W. Dahmen, R. DeVore, Adaptive finite element methods with convergence rates. Numer. Math. **97**, 219–268 (2004)
8. P. Binev, W. Dahmen, R. DeVore, P. Petrushev, Approximation classes for adaptive methods. Serdica Math. J. **28**, 391–416 (2002). Dedicated to the memory of Vassil Popov on the occasion of his 60th birthday
9. A. Bonito, R.H. Nochetto, Quasi-optimal convergence rate for an adaptive discontinuous Galerkin method. SIAM J. Numer. Anal. **48**, 734–771 (2010)
10. D. Braess, *Finite Elements: Theory, Fast Solvers, and Applications in Solid Mechanics*, 2nd edn. (Cambridge University Press, Cambridge, 2001)
11. S. Brenner, L.R. Scott, *The Mathematical Theory of Finite Element Methods, Springer Texts in Applied Mathematics*, vol. 15 (Springer, New York, 2008)
12. C. Carstensen, S.A. Funken, Fully reliable localized error control in the FEM. SIAM J. Sci. Comput. **21**, 1465–1484 (1999)
13. C. Carstensen, R.H.W. Hoppe, Error reduction and convergence for an adaptive mixed finite element method. Math. Comp. **75**, 1033–1042 (2006)
14. J.M. Cascón, C. Kreuzer, R.H. Nochetto, K.G. Siebert, Quasi-optimal convergence rate for an adaptive finite element method. SIAM J. Numer. Anal. **46**, 2524–2550 (2008)
15. J.M. Cascón, R.H. Nochetto, Convergence and quasi-optimality for AFEM based on non-residual a posteriori error estimators. To appear in IMA J. Numer. Anal.
16. J.M. Cascón, R.H. Nochetto, K.G. Siebert, Design and convergence of AFEM in $H(\text{div})$. Math. Models Methods Appl. Sci. **17**, 1849–1881 (2007)
17. Z. Chen, J. Feng, An adaptive finite element algorithm with reliable and efficient error control for linear parabolic problems. Math. Comp. **73**, 1167–1042 (2006)
18. L. Chen, M. Holst, J. Xu, Convergence and optimality of adaptive mixed finite element methods. Math. Comp. **78**, 35–53 (2009)
19. P.G. Ciarlet, *The Finite Element Method for Elliptic Problems. Classics in Applied Mathematics*, vol. 40, SIAM, (North-Holland, Amsterdam, 2002)
20. A. Demlow, Convergence of an adaptive finite element method for controlling local energy errors. Submitted for publication
21. A.. Demlow, R.P. Stevenson, Convergence and quasi-optimality of an adaptive finite element method for controlling L_2 errors. Submitted for publication
22. R.A. DeVore, Nonlinear approximation, in *Acta Numerica*, vol. 7, ed. by A. Iserles (Cambridge University Press, Cambridge, 1998), pp. 51–150.
23. L. Diening, Ch. Kreuzer, Convergence of an adaptive finite element method for the p-Laplacian equation. SIAM J. Numer. Anal. **46**, 614–638 (2008)
24. W. Dörfler, A convergent adaptive algorithm for Poisson's equation. SIAM J. Numer. Anal. **33**, 1106–1124 (1996)

25. W. Dörfler, M. Rumpf, An adaptive strategy for elliptic problems including a posteriori controlled boundary approximation. Math. Comp. **67**, 1361–1382 (1998)
26. T. Dupont, L.R. Scott, Polynomial approximation of functions in Sobolev spaces. Math. Comp. **34**, 441–463 (1980)
27. F. Fierro, A. Veeser, A posteriori error estimators, gradient recovery by averaging, and superconvergence. Numer. Math. **103**, 267–298 (2006)
28. E.M. Garau, P. Morin, C. Zuppa, Convergence of adaptive finite element methods for eigenvalue problems. Math. Models Methods Appl. Sci. **19**, 721–747 (2009)
29. F. Gaspoz, P. Morin, Approximation classes for adaptive higher order finite element approximation. Submitted for publication
30. P. Grisvard, *Elliptic Problems in Nonsmooth Domains, Monographs and Studies in Mathematics*, vol. 24 (Pitman Advanced Publishing Program, Boston, MA, 1985)
31. M. Holst, G. Tsogtgerel, Y. Zhu, Local convergence of adaptive methods for nonlinear partial differential equations. Preprint, arXiv:math.NA/1001.1382 (2009)
32. R.H.W. Hoppe, G. Kanschat, T. Warburton, Convergence analysis of an adaptive interior penalty discontinuous Galerkin method. SIAM J. Numer. Anal. **47**, 534–550 (2009)
33. O.A. Karakashian, F. Pascal, Convergence of adaptive discontinuous Galerkin approximations of second-order elliptic problems. SIAM J. Numer. Anal. **45**, 641–665 (2007)
34. R.B. Kellogg, On the Poisson equation with intersecting interfaces. Applicable Anal. **4**, 101–129 (1974/75)
35. Ch. Kreuzer, K.G. Siebert, Decay rates of adaptive finite elements with Dörfler marking. Numer. Math. 117, 679–716 (2011)
36. J.M. Maubach, Local bisection refinement for n-simplicial grids generated by reflection. SIAM J. Sci. Comput. **16**, 210–227 (1995)
37. K. Mekchay, R.H. Nochetto, Convergence of adaptive finite element methods for general second order linear elliptic PDEs. SIAM J. Numer. Anal. **43**, 1803–1827 (electronic) (2005)
38. K. Mekchay, P. Morin, R.H. Nochetto, AFEM for the Laplace–Beltrami operator on graphs: design and conditional contraction property. Math. Comp. **80**, 625–648 (2011)
39. W.F. Mitchell, Unified multilevel adaptive finite element methods for elliptic problems. Ph.D. thesis, Department of Computer Science, University of Illinois, Urbana, 1988
40. P. Morin, R.H. Nochetto, K.G. Siebert, Data oscillation and convergence of adaptive FEM. SIAM J. Numer. Anal. **38**, 466–488 (2000)
41. P. Morin, R.H. Nochetto, K.G. Siebert, Convergence of adaptive finite element methods. SIAM Review **44**, 631–658 (2002)
42. P. Morin, R.H. Nochetto, K.G. Siebert, Local problems on stars: a posteriori error estimators, convergence, and performance. Math. Comp. **72**, 1067–1097 (electronic) (2003)
43. P. Morin, K.G. Siebert, A. Veeser, Convergence of finite elements adapted for weak norms, in *Applied and Industrial Mathematics in Italy II, Series on Advances in Mathematics for Applied Sciences*, vol. 75, ed. by V. Cutello, G. Fotia, L. Puccio, (World Scientific, Singapore, 2007), pp. 468–479
44. P. Morin, K.G. Siebert, A. Veeser, A basic convergence result for conforming adaptive finite elements. Math. Mod. Meth. Appl. Sci. **5**, 707–737 (2008)
45. R.H. Nochetto, K.G. Siebert, A. Veeser, Theory of adaptive finite element methods: an introduction, in *Multiscale, Nonlinear and Adaptive Approximation* (Springer, New York, 2009), pp. 409–542
46. R.H. Nochetto, A. Schmidt, K.G. Siebert, A. Veeser, Pointwise a posteriori error estimates for monotone semi-linear equations. Numer. Math. **104**, 515–538 (2006)
47. R. Sacchi, A. Veeser, Locally efficient and reliable a posteriori error estimators for Dirichlet problems. Math. Models Methods Appl. **16**, 319–346 (2006)
48. L.R. Scott, S. Zhang, Finite element interpolation of nonsmooth functions satisfying boundary conditions. Math. Comp. **54**, 483–493 (1990)
49. K.G. Siebert, A convergence proof for adaptive finite elements without lower bound. IMA J. Numer. Anal. (2010), first published online May 30, 2010. doi:10.1093/imanum/drq001 (in this volume)

50. K.G. Siebert, Mathematically founded design of adaptive finite element software, in *Multiscale and Adaptivity: Modeling, Numerics and Applications, CIME-EMS Summer School in Applied Mathematics*, ed. by G. Naldi, G. Russo (Springer, New York, 2011), pp. 227–309
51. K.G. Siebert, A. Veeser, A unilaterally constrained quadratic minimization with adaptive finite elements. SIAM J. Optim. **18**, 260–289 (2007)
52. R. Stevenson, Optimality of a standard adaptive finite element method. Found. Comput. Math. **7**, 245–269 (2007)
53. R. Stevenson, The completion of locally refined simplicial partitions created by bisection. Math. Comput. **77**, 227–241 (2008)
54. C.T. Traxler, An algorithm for adaptive mesh refinement in n dimensions. Computing **59**, 115–137 (1997)
55. A. Veeser, Convergent adaptive finite elements for the nonlinear Laplacian. Numer. Math. **92**, 743–770 (2002)
56. A. Veeser, R. Verfürth, Explicit upper bounds for dual norms of residuals. SIAM J. Numer. Anal. **47**, 2387–2405 (2009)
57. R. Verfürth, A posteriori error estimators for the Stokes equations. Numer. Math. **55**, 309–325 (1989)
58. R. Verfürth, *A Review of A Posteriori Error Estimation and Adaptive Mesh-Refinement Techniques*, Adv. Numer. Math. John Wiley, Chichester, UK (1996).
59. H. Wu, Z. Chen, Uniform convergence of multigrid V-cycle on adaptively refined finite element meshes for second order elliptic problems. Science in China: Series A Mathematics **49**, 1–28 (2006)
60. J. Xu, L. Chen, R.H. Nochetto, Adaptive multilevel methods on graded bisection grids, in *Multiscale, Nonlinear and Adaptive Approximation* (Springer, New York, 2009), pp. 599–659

Mathematically Founded Design of Adaptive Finite Element Software

Kunibert G. Siebert

Abstract In these lecture notes we derive from the mathematical concepts of adaptive finite element methods basic design principles of adaptive finite element software. We introduce finite element spaces, discuss local refinement of simplical grids, the assemblage and structure of the discrete linear system, the computation of the error estimator, and common adaptive strategies. The mathematical discussion naturally leads to appropriate data structures and efficient algorithms for the implementation. The theoretical part is complemented by exercises giving an introduction to the implementation of solvers for linear and nonlinear problems in the adaptive finite element toolbox ALBERTA.

1 Introduction

In these lecture notes we discuss the design of data structures and algorithms of adaptive finite element software for the efficient approximation of solutions to partial differential equations given by a variational problem. It builds on several courses devoted to the implementation of adaptive finite elements, in particular

- Graduate Course "Implementation of Adaptive Finite Elements", Dipartimento di Matematica "Federigo Enriques", Università degli Studi di Milano, Italy;
- Sussex Summer School on Scientific Computation S4C 07: "ALBERTA Finite Elements", University of Sussex at Brighton, England;
- Winter School "Implementation of Adaptive Finite Elments" at the Institut für Mathematik, Universität Augsburg, Germany;

K.G. Siebert (✉)
Institut für Angewandte Analysis und Numerische Simulation, Fachbereich Mathematik, Universität Stuttgart, Pfaffenwaldring 57, 70569 Stuttgart, Germany
e-mail: kg.siebert@ians.uni-stuttgart.de; www.ians.uni-stuttgart.de/nmh/

S. Bertoluzza et al., *Multiscale and Adaptivity: Modeling, Numerics and Applications*, Lecture Notes in Mathematics 2040, DOI 10.1007/978-3-642-24079-9_4,

- C.I.M.E Summer School Summer School in Applied Mathematics "Multiscale and Adaptivity: Modeling, Numerics and Applications" in Cetraro, Italy.

Nowadays, there exists a great variety of numerical software. On one hand there are commercial packages, on the other hand there is a huge number of public domain software that is often even open source. When looking specifically for adaptive finite element packages one realizes that most software is developed at universities or academic research centers and is usually public domain.

Experience strongly suggests that the design and implementation of numerical methods enormously profits from a precise mathematical definition and description of basic data structures and algorithms. In fact, designing data structures and algorithms without a clear mathematical structure usually turns out to result in knotty implementations with many adhoc solutions that do not reflect the real structure of the underlying problem.

To this end, the basic philosophy of this course is to first give a precise mathematical definition and description of adaptive finite elements. These principles should be met by any adaptive finite element software. In this respect, the course should be helpful for getting started to work with any available adaptive finite element package. In particular we discuss how these mathematical principles have entered the design of the adaptive finite element toolbox ALBERTA, which is open source software and can be downloaded from

`http://www.alberta-fem.de/.`

ALBERTA is a C library for the implementation of adaptive finite element solvers. Additional course material like some start-up files for using ALBERTA can be downloaded from

`http://www.ians.uni-stuttgart.de/nmh/downloads/iafem/.`

Literature. There exists a vast variety of books about finite elements. Here, we only want to mention the books by Ciarlet [16], and Brenner and Scott [11] as the most prominent ones. There are only few books about adaptive finite elements available. We refer to the books of Verfürth [42], and Ainsworth and Oden [1] for basic ingredients of adaptive methods such as error estimators and adaptive strategies. Recent results in the analysis of adaptive finite elements including convergence and optimal error decay in terms of degrees of freedom are subject of the lectures by Nochetto and Veeser in this summer school; compare with the chapter "Primer of adaptive finite element methods" in this volume [31]. More details can be found in the overview article by Nochetto, Siebert, and Veeser [30]. Finally, when it comes to aspects of the implementation one is usually restricted to documentations of existing software. In this course we follow the ideas of Schmidt and Siebert in the design of the finite element toolbox ALBERTA [33, 34]. We also refer to [22, 35] for recent developments of ALBERTA.

1.1 The Variational Problem

We introduce the foundation of finite elements, the variational formulation of partial differential equations (PDEs), where we restrict ourselves to linear problems. We state the variational problem, characterize well-posedness, and give two basic examples.

Problem 1 (Variational Problem). Let $(\mathbb{V}, \|\cdot\|_{\mathbb{V}})$ be an Hilbert space, $\mathbb{V}^*$ its dual with norm

$$\|f\|_{\mathbb{V}^*} = \sup_{\|v\|_{\mathbb{V}}=1} \langle f, v\rangle .$$

Let $\mathscr{B}\colon \mathbb{V}\times\mathbb{V}\to\mathbb{R}$ be a continuous bilinear form. For $f \in \mathbb{V}^*$ we are looking for a solution of the variational problem

$$u \in \mathbb{V}: \qquad \mathscr{B}[u, v] = \langle f, v\rangle \qquad \text{for all } v \in \mathbb{V}. \tag{1}$$

Theorem 2 (Existence and Uniqueness). *Problem 1 is well-posed, that means that for any $f \in \mathbb{V}^*$ there exists a unique solution $u \in \mathbb{V}$ if and only if the bilinear form $\mathscr{B}$ satisfies the* inf-sup condition:

$$\exists \alpha > 0: \qquad \inf_{\substack{v\in\mathbb{V}\\ \|v\|_{\mathbb{V}}=1}} \sup_{\substack{w\in\mathbb{V}\\ \|w\|_{\mathbb{V}}=1}} \mathscr{B}[v, w] \ge \alpha, \qquad \inf_{\substack{w\in\mathbb{V}\\ \|w\|_{\mathbb{V}}=1}} \sup_{\substack{v\in\mathbb{V}\\ \|v\|_{\mathbb{V}}=1}} \mathscr{B}[v, w] \ge \alpha. \tag{2}$$

In addition, the solution u satisfies the stability bound $\|u\|_{\mathbb{V}} \le \alpha^{-1}\|f\|_{\mathbb{V}^*}$.

This characterization is due to Nečas [29, Theorem 3.1], compare also the contributions of Babuška [3, Theorem 2.1] and Brezzi [12, Corollary 0.1]. We also refer to [30] for a more detailed discussion.

Example 3 (2nd Order Elliptic Equation). Given $f \in L^2(\Omega)$ solve the 2nd order elliptic PDE

$$-\operatorname{div}(\boldsymbol{A}(x)\nabla u) + \boldsymbol{b}(x)\cdot\nabla u + c(x)u = f \quad \text{in } \Omega, \qquad u = 0 \quad \text{on } \partial\Omega.$$

For the variational formulation we let $\mathbb{V} = H_0^1(\Omega)$ and set

$$\mathscr{B}[w, v] := \int_\Omega \nabla v\cdot\boldsymbol{A}(x)\nabla w + v\,\boldsymbol{b}\cdot\nabla w + c\,v\,w\,dx, \quad \langle f, v\rangle = \int_\Omega f\,w\,dx.$$

Assuming that $\boldsymbol{A} \in L^\infty(\Omega;\mathbb{R}^d)$ is strictly symmetric positive definite (spd), i.e., $\boldsymbol{A}(x) = \boldsymbol{A}(x)^\top$ and $\xi\cdot\boldsymbol{A}(x)\xi \ge c_*\,|\xi|_2^2$ for all $\xi \in \mathbb{R}^d$ and almost all $x \in \Omega$, and $\boldsymbol{b} \in L^\infty(\Omega;\mathbb{R}^d)$, $c \in L^\infty(\Omega)$ with $c - \frac{1}{2}\operatorname{div}\boldsymbol{b} \ge 0$ in Ω one can prove in combination with Friedrich's inequality that $\mathscr{B}$ is continuous and coercive on $\mathbb{V}$, i.e., $\mathscr{B}[v, v] \ge \alpha\|v\|_{\mathbb{V}}$ for all $v \in \mathbb{V}$. Coercivity implies the inf-sup condition (2).

Therefore, there exists for any right hand side $f \in L^2(\Omega) \subset H^{-1}(\Omega) = \big(H^1(\Omega)\big)^*$ a unique weak solution $u \in H_0^1(\Omega)$; compare for instance with [30, Sect. 2.5.2].

Example 4 (Stokes Problem). Given an external force $\boldsymbol{f} = \big[f_1, \dots, f_d\big]^\top \in L^2(\Omega; \mathbb{R}^d)$ and the viscosity $\nu > 0$ of a stationary, viscous, and incompressible fluid find its *velocity* field $\boldsymbol{u} = \big[u_1, \dots, u_d\big]^\top$ and *pressure* p such that

$$-\nu\Delta\boldsymbol{u} + \nabla p = \boldsymbol{f} \quad \text{in } \Omega, \qquad \operatorname{div}\boldsymbol{u} = 0 \quad \text{in } \Omega, \qquad \boldsymbol{u} = \boldsymbol{0} \quad \text{on } \partial\Omega,$$

where $\Delta\boldsymbol{u} = \big[\Delta u_1, \dots, \Delta u_d\big]^\top$. The variational problem looks for a weak solution $u = (\boldsymbol{u}, p)$ in the space $\mathbb{V} = H_0^1(\Omega; \mathbb{R}^d) \times \{q \in L^2(\Omega) \mid \int_\Omega q\,dx = 0\}$ with the bilinear form

$$\mathscr{B}[w, v] := \nu \int_\Omega \nabla\boldsymbol{v} : \nabla\boldsymbol{w}\,dx - \int_\Omega \operatorname{div}\boldsymbol{w}\,q\,dx - \int_\Omega \operatorname{div}\boldsymbol{v}\,r\,dx$$

for all $w = (\boldsymbol{w}, r), v = (\boldsymbol{v}, q) \in \mathbb{V}$, and the right hand side $\langle f, v\rangle := \int_\Omega \boldsymbol{f} \cdot \boldsymbol{v}\,dx$.

The Stokes problem has the structure of a saddle point problem and existence and uniqueness follows from an inf-sup condition of $\mathscr{B}$ on $\mathbb{V}$ derived by Nečas [14]; compare also with [19, Theorem III.3.1] and [30, Sect. 2.5.2].

1.2 The Basic Adaptive Algorithm

We next discuss the basic adaptive iteration. From the discussion we derive the basic objectives to be tackled when implementing adaptive finite elements. We close with an overview of the organization of the lecture.

The adaptive approximation of the solution u to (1) is an iteration of the form

$$\textsf{SOLVE} \;\longrightarrow\; \textsf{ESTIMATE} \;\longrightarrow\; \textsf{ENLARGE}$$

with the following modules.

- SOLVE computes an approximation to u in a finite dimensional subspace $\mathbb{V}_k \subset \mathbb{V}$, for instance the (Ritz-)Galerkin approximation

$$U_k \in \mathbb{V}_k: \qquad \mathscr{B}[U_k, V] = \langle f, V\rangle \qquad \forall\, V \in \mathbb{V}_k.$$

- ESTIMATE computes an error bound for a suitable error notion $\|U_k - u\|$, usually $\|U_k - u\|_{\mathbb{V}}$.
- ENLARGE enlarges the space $\mathbb{V}_k$ into $\mathbb{V}_{k+1}$ to achieve a better approximation in the next iteration.

The step ENLARGE obviously needs structure of the discrete spaces for including new directions in $\mathbb{V}_{k+1}$. Additionally, the module ESTIMATE has to provide information about new directions that should be included.

A suitable choice for the discrete spaces are finite element discretizations. Finite element spaces are defined over a *decomposition* of Ω into (closed) elements T of a grid $\mathcal{T}$ such that

$$\bar{\Omega} = \bigcup_{T \in \mathcal{T}} T \qquad \text{and} \qquad |\Omega| = \sum_{T \in \mathcal{T}} |T|.$$

On each single element $T \in \mathcal{T}$ a finite element function belongs to a *local function space* $\mathbb{P}(T) \subset \mathbb{V}(T)$. Each finite element function V fulfills some *global continuity condition* such that $V \in \mathbb{V}$. We are going to focus on:

- a *triangulation* of Ω for a decomposition, i. e., elements T are *simplices* (intervals in 1d, triangles in 2d, and tetrahedra in 3d);
- the space of all *polynomials* of degree $p \geq 1$ as local function space;
- *globally continuous functions* to deal with H^1 conforming approximations.

We restrict ourselves to so-called *h-adaptive finite elements*, i. e., new directions are included by local refinement of selected grid elements. Starting with an initial triangulation $\mathcal{T}_0$, the adaptive h approximation of the solution u to (1) is an iteration of the form

$$\mathsf{SOLVE} \quad \longrightarrow \quad \mathsf{ESTIMATE} \quad \longrightarrow \quad \mathsf{MARK} \quad \longrightarrow \quad \mathsf{REFINE},$$

where SOLVE and ESTIMATE are as above, and

- MARK: Selects a subset $\mathcal{M}_k \subset \mathcal{T}_k$ subject for refinement.
- REFINE: Refines at least all elements in $\mathcal{M}_k$ and outputs a new triangulation $\mathcal{T}_{k+1}$.

Aspects for the Implementation

For an implementation of the adaptive algorithm we have to address the following issues:

1. SOLVE: For a given triangulation $\mathcal{T}$ and finite element space $\mathbb{V}(\mathcal{T})$ efficiently solve for the discrete solution $U \in \mathbb{V}(\mathcal{T})$.
2. ESTIMATE: For a given triangulation $\mathcal{T}$ and discrete solution $U \in \mathbb{V}(\mathcal{T})$ compute an error estimator of the form

$$\mathcal{E}_{\mathcal{T}}^2(U, \mathcal{T}) := \sum_{T \in \mathcal{T}} \mathcal{E}_{\mathcal{T}}^2(U, T).$$

3. MARK: For a given triangulation $\mathcal{T}$ and error indicators $\{\mathcal{E}_{\mathcal{T}}(U, T)\}_{T \in \mathcal{T}}$ select elements in $\mathcal{M} \subset \mathcal{T}$ for refinement.
4. REFINE: Given a given triangulation $\mathcal{T}$ and a set of marked elements $\mathcal{M}$ refine at least all elements in $\mathcal{M}$.

SOLVE, ESTIMATE, MARK work an a given fixed grid like any finite element code. The design of the basic data structures is inspired by the mathematical definition of finite element spaces. Main work in the implementation has to be done for SOLVE and REFINE. ESTIMATE and especially MARK are then easy to realize.

This introduction puts us into the position to give an overview of the organisation of the lectures held at the C.I.M.E summer school. We start with the definition of triangulations and finite element spaces. In the second part we review algorithms for local mesh refinement and coarsening of simplicial grids with main focus on refinement by bisection. The third part treats the assemblage of the discrete linear system. We conclude the course in the last part with the adaptive algorithm and final remarks. In addition to the C.I.M.E lectures we have added a supplement dealing with a nonlinear problem and a saddlepoint problem. We demonstrate hereby that the principles introduced for a model problem in the lectures are easily applicable to more complex problems.

2 Triangulations and Finite Element Spaces

We define triangulations and finite element spaces, and state basic properties important for the implementation.

2.1 Triangulations

The use of triangulations for the decomposition of the domain has several advantages. In context of adaptive refinement the most important one is that simplicial grids easily allow for global *and* local refinement while this is not that easy for other types of grids; compare with Figs. 1 and 2 with an example of rectangular grids. Furthermore, simplicial grids allow for a good approximation of curved geometries.

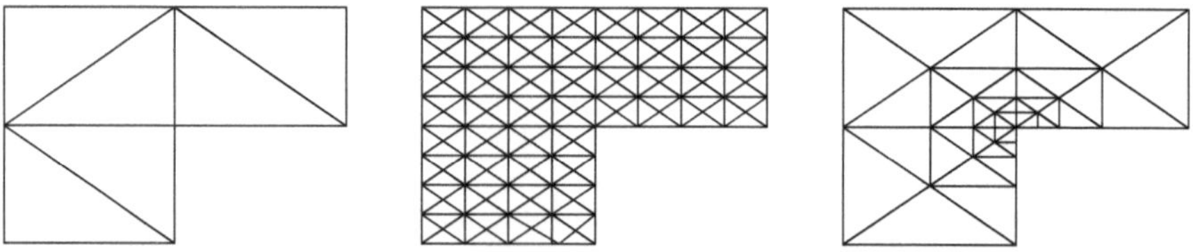

Fig. 1 Global and local refinement of a triangular grid

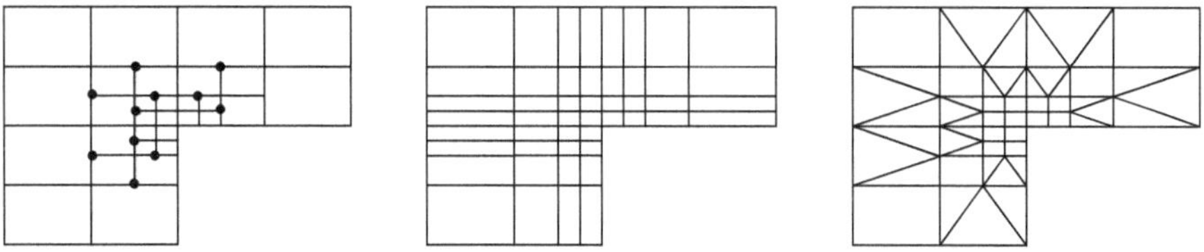

Fig. 2 Local refinement of a rectangular grid: local refinement results either in having hanging nodes (*left*), global effects (*middle*), or mixed type of elements (*right*)

We start with the definition of the basic object, the simplex.

Definition 5 (Simplex). Let $z_0, \ldots, z_d \in \mathbb{R}^d$ be given such that $z_1 - z_0, \ldots, z_d - z_0$ are linearly independent.

1. A (non-degenerate) d*-simplex* in $\mathbb{R}^d$ is the convex set

$$T = \text{conv hull}\{z_0, \ldots, z_d\}.$$

2. For $0 \le n < d$ let $z'_0, \ldots, z'_n \in \{z_0, \ldots, z_d\}$. The set

$$T' = \text{conv hull}\{z'_0, \ldots, z'_n\} \subset \partial T$$

is called n*-sub-simplex* of T.

3. The *diameter* and *inball-diameter* of T is

$$\overline{h}_T := \text{diam}(T), \quad \underline{h}_T := \sup\{2r \mid B_r \subset T \text{ is a ball of radius } r\}.$$

A measure of the element's "quality" is the *shape coefficient*

$$\sigma_T := \overline{h}_T / \underline{h}_T.$$

In what follows we call a $(d-1)$-sub-simplex of a d-simplex *side*. Other special sup-simplices are: *vertex* (0-sub-simplex), *edge* (1-sub-simplex for $d \ge 2$), and *face* (2-sub-simplex for $d \ge 3$).

Definition 6 (Triangulation, Conformity). Let $\Omega \subset \mathbb{R}^d$ be a domain with a polygonal boundary.

1. A *triangulation* of Ω is a set $\mathcal{T}$ of d-simplices, such that

$$\Omega = \text{interior} \bigcup_{T \in \mathcal{T}} T \qquad \text{and} \qquad |\Omega| = \sum_{T \in \mathcal{T}} |T|$$

2. $\mathcal{T}$ is a *conforming triangulation*, iff for two simplices $T_1, T_2 \in \mathcal{T}$ with $T_1 \neq T_2$ the intersection $T_1 \cap T_2$ is either empty or a complete n-sub-simplex of both T_1 and T_2 for some $0 \le n < d$.

Remark 7. Condition (1) ensures that the closure $\bar{\Omega}$ of Ω is covered by the elements of the triangulation. Condition (2) is in particular important for defining continuous finite elements. It excludes the existence of so-called *irregular nodes*. An irregular node is a vertex of $\mathcal{T}$ that belongs to some element T but is not a vertex of T.

Aspects for the Implementation

The mathematical definition of a simplex and triangulation already gives some indication on implementation issues:

1. We need a data structure holding information about a triangulation, this is a list/vector/... of elements. In case of adaptive methods elements have to be added or deleted (refinement/coarsening) efficiently.
2. We need a data structure for storing element information:
    - Vertex coordinates describing the geometrical shape;
    - Neighbor information;
    - ...

Neighbor information is needed to check for conformity.

Definition 8 (Shape Regularity). A sequence of triangulations $\{\mathcal{T}_k\}_{k\geq 0}$ is called *shape regular* if and only if

$$\sup_{k\geq 0}\max_{T\in\mathcal{T}_k}\sigma_T = \sup_{k\geq 0}\max_{T\in\mathcal{T}_k}\frac{\overline{h}_T}{\underline{h}_T} < \infty.$$

Shape regularity plays an essential role when analyzing finite elements. In particular, interpolation constants strongly depend on the shape coefficients of elements. Uniform estimates for a sequence of triangulations thus strongly rely on shape regularity as stated in Definition 8; compare with [16, Sect. 3]. The question of shape regularity will be a crucial aspect in deriving refinement algorithms in Sect. 3.

2.2 *Finite Element Spaces*

We next define finite element spaces, which are constructed from local function spaces and glued together by a global continuity condition. In doing this we follow the fundamental concept of finite elements:

Everything is done from local to global.

We first introduce two special simplices that are important in the subsequent discussion.

Definition 9 (Standard and Reference Simplex). Denoting by $\{e_1, \dots, e_d\}$ the unit vectors in $\mathbb{R}^d$, the *standard simplex* in $\mathbb{R}^d$ is

$$\hat{T} := \text{conv hull}\{\hat{z}_0 = 0, \hat{z}_1 = e_1, \dots, \hat{z}_d = e_d\}.$$

We next let $\{\bar{e}_0, \dots, \bar{e}_d\}$ be the unit vectors in $\mathbb{R}^{d+1}$. The *reference simplex* $\bar{T}$ is the so-called *Gibbs simplex* in $\mathbb{R}^{d+1}$, namely

$$\bar{T} := \text{conv hull}\{\bar{z}_0 = \bar{e}_0, \dots, \bar{z}_d = \bar{e}_d\}.$$

Any simplex T is *affine equivalent* to the standard element $\hat{T}$. This simplex plays a relevant role in the analysis of finite elements as well as in practice when dealing with numerical integration; compare with Sect. 4.3.

Lemma 10 (Affine Equivalence). *A simplex T with vertices $\{z_0, \dots, z_d\}$ is* affine equivalent *to $\hat{T}$, i. e., the affine linear mapping $F_T : \hat{T} \to T$ given by*

$$F_T(\hat{x}) = \boldsymbol{A}_T\,\hat{x} + \boldsymbol{b}_T := \begin{bmatrix} | & & | \\ z_1 - z_0 & \cdots & z_d - z_0 \\ | & & | \end{bmatrix} \hat{x} + z_0$$

is one to one and it holds $F_T(\hat{z}_i) = z_i$, $i = 0, \dots, d$.

The reference simplex $\bar{T}$ is the domain of the *barycentric coordinates*, which are used as the natural coordinate system on a simplex T for defining local functions (Fig. 3).

Definition 11 (Barycentric Coordinates). Let T be an arbitrary d-simplex with vertices $\{z_0, \dots, z_d\}$. For $x \in \mathbb{R}^d$ the *barycentric coordinates* $\lambda = \lambda^T(x) \in \mathbb{R}^{d+1}$ on T are the unique solution to

$$\sum_{k=0}^{d} \lambda_k\, z_k = x \qquad \text{and} \qquad \sum_{k=0}^{d} \lambda_k(x) = 1. \tag{3}$$

The definition (3) of the barycentric coordinates is equivalent to the linear system

$$\begin{bmatrix} | & & | \\ z_0 & \cdots & z_d \\ | & & | \\ 1 & \cdots & 1 \end{bmatrix} \begin{bmatrix} \lambda_0 \\ \lambda_1 \\ \vdots \\ \lambda_d \end{bmatrix} = \begin{bmatrix} x_1 \\ x_2 \\ \vdots \\ 1 \end{bmatrix}. \tag{4}$$

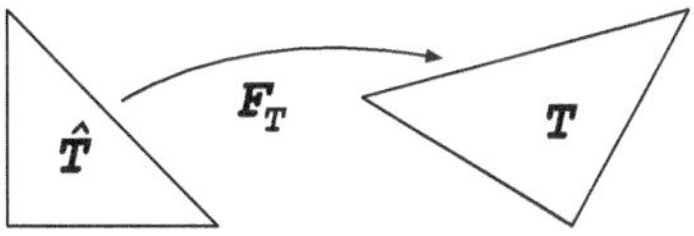

Fig. 3 Affine equivalence of simplices given by the affine mapping $F_T : \hat{T} \to T$

Linear independence of $z_1 - z_0, \dots, z_d - z_0$ implies

$$\begin{vmatrix} | & & | \\ z_0 & \cdots & z_d \\ | & & | \\ 1 & \cdots & 1 \end{vmatrix} = \begin{vmatrix} | & | & & | \\ z_0 & z_1 - z_0 & \cdots & z_d - z_0 \\ | & | & & | \\ 1 & 0 & \cdots & 0 \end{vmatrix} = \begin{vmatrix} | & & | \\ z_1 - z_0 & \cdots & z_d - z_0 \\ | & & | \end{vmatrix} \neq 0.$$

Therefore, for any $x \in \mathbb{R}^d$ its barycentric coordinates $\lambda^T(x)$ are uniquely determined.

Recalling the definition of the reference simplex $\bar{T}$ we see that $\bar{T}$ lies in the hypersurface

$$H = \Big\{\lambda \in \mathbb{R}^{d+1} \mid \sum_{k=0}^{d} \lambda_k = 1\Big\}.$$

The definition of $\bar{T}$ also implies that any component λ_k is non-negative, whence

$$\bar{T} = \left\{[\lambda_0, \dots, \lambda_d]^T \in \mathbb{R}^{d+1} \mid \lambda_k \geq 0, \sum_{k=0}^{d} \lambda_k = 1\right\}.$$

Given a simplex T, a point $x \in \mathbb{R}^d$ and its barycentric coordinates $\lambda = \lambda^T(x) \in \mathbb{R}^{d+1}$ we conclude $\lambda \in H$, and furthermore, $x \in T$ if and only if $\lambda \in \bar{T}$. In summary, the mapping $\lambda^T: T \to \bar{T}$ defined as $x \mapsto \lambda^T(x)$ is an affine bijection. Denoting by $x^T: \bar{T} \to T$ its inverse, we see

$$x^T(\lambda) = \sum_{k=0}^{d} \lambda_k \, z_k.$$

Aspects for the Implementation

We call $[\lambda_0, \dots, \lambda_d]^\top$ *local coordinates* on T and $x = [x_1, \dots, x_d]^\top$ *world coordinates*. Calculating world coordinates x from local coordinates λ only requires the computation of a convex combination of the element's vertices. The computation of the local coordinates λ from given world coordinates x requires the solution of the $(d + 1) \times (d + 1)$ linear system (4), which is computationally more expensive.

We next discuss the relation of the mappings $F_T: \hat{T} \to T$, $\lambda^T: T \to \bar{T}$, and its inverse $x^T: \bar{T} \to T$. As depicted in Fig. 4 we can compute the local coordinates $\lambda^T(x)$ by first mapping x to $\hat{x} = F_T^{-1}(x) \in \hat{T}$ and then computing the local coordinates of $\hat{x}$ with respect to $\hat{T}$, i.e., $\lambda^T = \lambda^{\hat{T}} \circ F_T^{-1}$. In the same vein, we

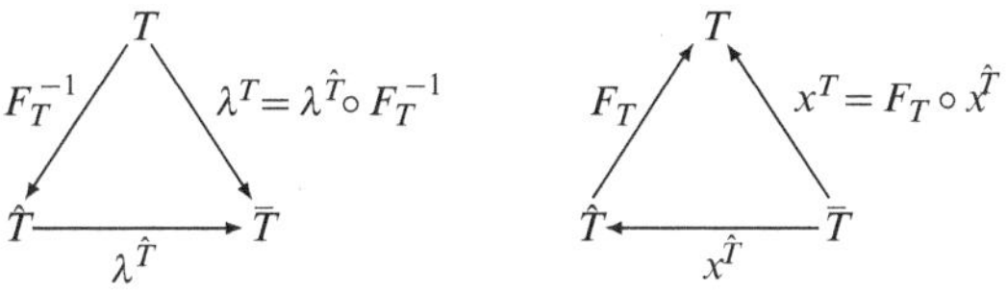

Fig. 4 Mappings on T, $\bar{T}$, and $\hat{T}$

can calculate the world coordinates $x^T(\lambda)$ by first computing the world coordinates $\hat{x} = x^{\hat{T}}(\lambda)$ on $\hat{T}$ and then mapping $\hat{x}$ to T by F_T, i. e., $x^T = F_T \circ x^{\hat{T}}$.

Therefore, every function $\phi: T \to \mathbb{R}$ *uniquely defines*

$$\begin{aligned} \bar{\phi}: \bar{T} &\to \mathbb{R} \\ \lambda &\mapsto \phi(x^T(\lambda)) \end{aligned} \qquad \text{and} \qquad \begin{aligned} \hat{\phi}: \hat{T} &\to \mathbb{R} \\ \hat{x} &\mapsto \phi(F_T(\hat{x})) \end{aligned}$$

and, accordingly,

- $\hat{\phi}: \hat{T} \to \mathbb{R}$ uniquely defines $\phi = \hat{\phi} \circ F_T^{-1}: T \to \mathbb{R}$ and $\bar{\phi} = \hat{\phi} \circ x^{\hat{T}}: \bar{T} \to \mathbb{R}$,
- $\bar{\phi}: \bar{T} \to \mathbb{R}$ uniquely defines $\phi = \bar{\phi} \circ \lambda^T: T \to \mathbb{R}$ and $\hat{\phi} = \bar{\phi} \circ \lambda^{\hat{T}}: \hat{T} \to \mathbb{R}$.

This brings us in the position to define a $\mathbb{V}$-conforming finite element over a triangulating $\mathscr{T}$ via a function space on $\bar{T}$. Consider a given function space $\bar{\mathbb{P}} \subset \mathbb{V}(\bar{T})$. This uniquely defines the function spaces

$$\mathbb{P}(T) := \{\bar{\phi} \circ \lambda^T \in \mathbb{V}(T) \mid \bar{\phi} \in \bar{\mathbb{P}}\} \qquad \text{and} \qquad \hat{\mathbb{P}} := \left\{\bar{\phi} \circ \lambda^{\hat{T}} \in \mathbb{V}(\hat{T}) \mid \bar{\phi} \in \bar{\mathbb{P}}\right\}$$

A conforming finite element space is then defined as follows.

Definition 12 (Conforming Finite Element Space). For given *function space* $\mathbb{V}$, *conforming triangulation* $\mathscr{T}$, and *local function space* $\bar{\mathbb{P}} \subset \mathbb{V}(\bar{T})$ on $\bar{T}$ we define

$$\mathbb{V}(\mathscr{T}) = \mathsf{FES}(\mathscr{T}, \bar{\mathbb{P}}, \mathbb{V}) := \left\{V \in \mathbb{V} \mid V_{|T} \circ x^T \in \bar{\mathbb{P}},\ T \in \mathscr{T}\right\}.$$

Once the local function space is fixed, we keep in $\mathbb{V}(\mathscr{T})$ the dependence of the finite element space on $\mathscr{T}$ and $\mathbb{V}$. The dependence of the discrete space on $\mathscr{T}$ is essential in the discussion of adaptive methods.

The following characterization is useful for H^1 conforming finite elements.

Lemma 13 (H^1 Conforming Finite Elements). *For* $\bar{\mathbb{P}} \subset C^1(\bar{T})$ *we have*

$$\mathsf{FES}(\mathscr{T}, \bar{\mathbb{P}}, H^1(\Omega)) = \mathsf{FES}(\mathscr{T}, \bar{\mathbb{P}}, C^0(\bar{\Omega})),$$

i. e., H^1 conforming finite elements are globally continuous.

Proof. Follows from piecewise integration by parts and the trace theorem. □

In case $\bar{\mathbb{P}} = \mathbb{P}_p$, the local functions on T are polynomials of degree p.

Lemma 14 (Polynomial Finite Elements). *For $p \geq 0$ let $\mathbb{P}_p$ be the space of all polynomial of degree $\leq p$ and set $\bar{\mathbb{P}} = \mathbb{P}_p$. Then holds $\mathbb{P}(T) = \mathbb{P}_p$ for all $T \in \mathscr{T}$, i. e.,*

$$\mathsf{FES}(\mathscr{T}, \mathbb{P}_p, \mathbb{V}) = \{V \in \mathbb{V} \mid V_{|T} \in \mathbb{P}_p \ \forall\, T \in \mathscr{T}\}.$$

Proof. Follows from the fact, that $\lambda^T : T \to \bar{T}$ is affine, and thus $\Phi = \bar{\phi} \circ \lambda^T$ is again a polynomial of degree $\leq p$. □

Polynomials and Barycentric Coordinates. Any polynomial P of degree $\leq p$ in $\mathbb{R}^d$ can be written as

$$P(x) = \sum_{|\alpha|=0}^{p} c_\alpha\, x^\alpha = \sum_{|\alpha|=0}^{p} c_\alpha \prod_{i=1}^{d} x_i^{\alpha_i}$$

with $c_\alpha \in \mathbb{R}$ using multi-index notation.

Utilizing the barycentric coordinates λ of $x(\lambda) = \sum_{j=0}^{d} z_j \lambda_j$ on a simplex T yields

$$\bar{P}(\lambda) = P(x(\lambda)) = \sum_{|\alpha|=0}^{p} c_\alpha \prod_{i=1}^{d} \left[\sum_{j=0}^{d} z_{ji} \lambda_j \right]^{\alpha_i} = \sum_{|\beta|=0}^{p} \tilde{c}_\beta \lambda^\beta.$$

The barycentric coordinates sum up to 1, i. e., $\sum_{j=0}^{d} \lambda_j = 1$, and thus

$$\bar{P}(\lambda) = c_0 + \sum_{|\beta|=1}^{p} \tilde{c}_\beta \lambda^\beta = \sum_{|\beta|=1}^{p} \bar{c}_\beta \lambda^\beta,$$

i. e., we can write $\bar{P}$ without constant term.

We next state two important examples of H^1-conforming finite element spaces.

Theorem 15 (Linear Finite Elements). (1) *On a simplex T with vertices $z_0, \dots, z_d$ any $P \in \mathbb{P}_1(T)$ is uniquely determined by the values at the vertices of T:*

$$P(x) = \bar{P}(\lambda) = \sum_{j=0}^{d} P(z_j)\, \lambda_j.$$

The dimension of $\mathbb{P}_1(T)$ is $d + 1$.

(2) *Let $\mathscr{T}$ be a conforming triangulation of Ω with vertices $z_1, \dots, z_N$. Then, any function $V \in \mathsf{FES}(\mathscr{T}, \mathbb{P}_1, C^0(\bar{\Omega}))$ is uniquely determined by the values at the vertices of $\mathscr{T}$, i. e., by $V(z_j)$, $j = 0, \dots, N$.*

(3) *A basis of $\mathsf{FES}(\mathscr{T}, \mathbb{P}_1, C^0(\bar{\Omega}))$ is given by the so-called* hat functions

$$\Phi_i \in \mathsf{FES}(\mathscr{T}, \mathbb{P}_1, C^0(\bar{\Omega})), \qquad \Phi_i(z_j) = \delta_{ij} \quad \textit{for } i, j = 1, \dots, N.$$

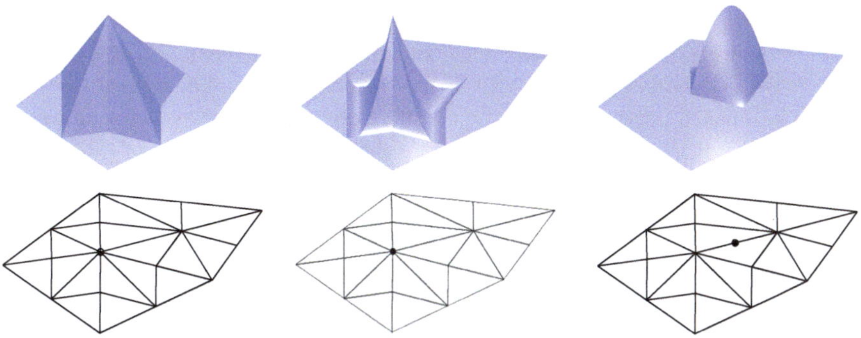

Fig. 5 A typical piecewise linear basis function in 2d, the so-called *hat-function* (*left*) and typical piecewise quadratic basis functions in 2d (*middle* and *right*)

This basis is called Lagrange basis *or* nodal basis. *A typical basis function in 2d is shown in Fig. 5 (left).*

Theorem 16 (Quadratic Finite Elements). (1) *Let T be a simplex with vertices $z_0, \ldots, z_d$ and edge midpoints $z_{ij} = \frac{1}{2}(z_i + z_j)$ $(0 \le i < j \le d)$. Any $P \in \mathbb{P}_2(T)$ is uniquely determined by the values of P at the vertices and edge midpoints of T:*

$$P(x) = \bar{P}(\lambda) = \sum_{j=0}^{d} P(z_j)\,(2\,\lambda_j^2 - \lambda_j) + \sum_{0 \le i < j \le d} P(z_{ij})\,4\,\lambda_i\,\lambda_j .$$

The dimension of $\mathbb{P}_2(T)$ is $\frac{1}{2}(d+1)(d+2)$.

(2) *Let $\mathcal{T}$ be a conforming triangulation of Ω with vertices $z_1, \ldots, z_N$ and edge midpoints $z_{N+1}, \ldots, z_{\bar{N}}$. Then, any function $V \in \mathsf{FES}(\mathcal{T}, \mathbb{P}_2, C^0(\bar{\Omega}))$ is uniquely determined by the values at the vertices and edge midpoints of $\mathcal{T}$, i. e., by $V(z_j)$, $j = 1, \ldots, \bar{N}$.*

(3) *The* nodal *or* Lagrange basis *of $\mathsf{FES}(\mathcal{T}, \mathbb{P}_2, C^0(\bar{\Omega}))$ is given by the functions*

$$\Phi_i \in \mathsf{FES}(\mathcal{T}, \mathbb{P}_2, C^0(\bar{\Omega})), \qquad \Phi_i(z_j) = \delta_{ij} \quad \text{for } i, j = 1, \ldots, \bar{N}.$$

Typical basis functions in 2d are depicted in Fig. 5 (middle and right).

Aspects for the Implementation

(1) On simplices, polynomials can be easily expressed in terms of barycentric coordinates, i. e., on the reference simplex $\bar{T} \subset \mathbb{R}^{d+1}$. They are the

"natural coordinate system" on T and make an implementation of $\bar{\mathbb{P}}$ easy. The standard element $\hat{T} \subset \mathbb{R}^d$ is the "natural element" for *integration*.

(2) A linear finite element function is uniquely determined by the values at the vertices of $\mathscr{T}$. These values are the associated *degrees of freedom (DOFs)*. Global continuity can be realized as follows: Using the local Lagrange basis, DOFs at a vertex are shared by all elements meeting at that vertex, i. e., they all access the same value.

(3) The DOFs for quadratic finite elements are the values at the vertices and edge midpoints. These DOFs are shared with all adjacent elements, which in turn implies global continuity by using the local Lagrange basis.

2.3 Basis Functions and Evaluation of Finite Element Functions

We next generalize the ideas that we have seen for linear and quadratic finite elements to higher order, i. e., we use $\bar{\mathbb{P}} = \mathbb{P}_p$ for $p \in \mathbb{N}$.

Lemma 17 (Lagrange Basis on $\bar{T}$). *Define the* Lagrange grid *on* $\bar{T}$ *as*

$$\begin{aligned}\mathscr{L}(\bar{T}) &= \left\{ [\lambda_0, \dots, \lambda_d]^\top \in \bar{T} \mid \lambda_i \in \left\{0, \tfrac{1}{p}, \dots, \tfrac{p-1}{p}, 1\right\},\ i = 0, \dots, d \right\} \\ &=: \{\lambda_1, \dots, \lambda_{\bar{n}}\}\end{aligned}$$

and set $\bar{\mathbb{P}} = \mathbb{P}_p$. *Then it holds* $\bar{n} = \#\mathscr{L}(\bar{T}) = \dim \bar{\mathbb{P}} = \binom{d+p}{p}$. *Furthermore,* $\mathscr{L}(\bar{T})$ *is uni-solvent on* $\bar{\mathbb{P}}$, *this means, each* $\bar{\phi} \in \bar{\mathbb{P}}$ *is uniquely determined by the values* $\bar{\phi}(\lambda_n)$, $n = 1, \dots, \bar{n}$. *The function set* $\{\bar{\phi}_1, \dots, \bar{\phi}_{\bar{n}}\}$ *given by*

$$\bar{\phi}_n(\lambda_m) = \delta_{nm} \qquad n, m = 1, \dots, \bar{n}$$

is the Lagrange basis *of* $\bar{\mathbb{P}}$.

Proof. Compare with [16, Theorem 2.2.1]. □

For a given simplex T the local basis is given by the mapping $\lambda^T : T \to \bar{T}$.

Lemma 18 (Lagrange Basis on T). *Let T be an affine simplex,* $\lambda^T : T \to \bar{T}$ *the barycentric coordinates, and* $x^T : \bar{T} \to T$ *its inverse. The* Lagrange grid on T *given by*

$$\mathscr{L}(T) = x^T\left(\mathscr{L}(\bar{T})\right) = \{x_1^T, \dots, x_{\bar{n}}^T\} \subset T$$

is uni-solvent on $\mathbb{P}_p(T)$. *Consequently, the function set* $\{\Phi_1^T, \dots, \Phi_{\bar{n}}^T\}$ *defined as*

$$\Phi_n^T : T \to \mathbb{R} \qquad \Phi_n^T := \bar{\phi}_n \circ \lambda^T \qquad n = 1, \dots, \bar{n}$$

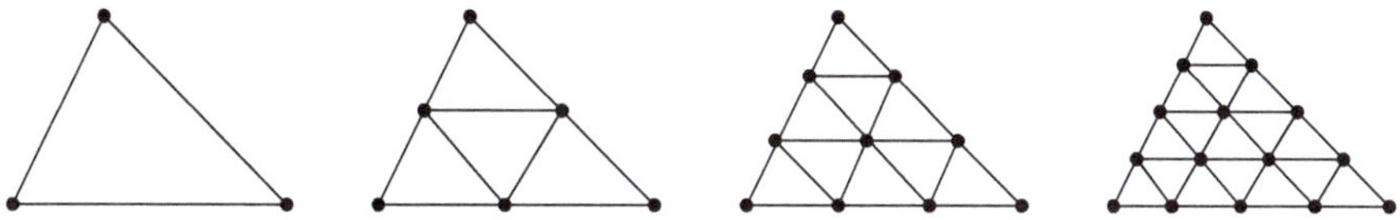

Fig. 6 Lagrange grids for polynomial degree $p = 1$ to $p = 4$ in 2d

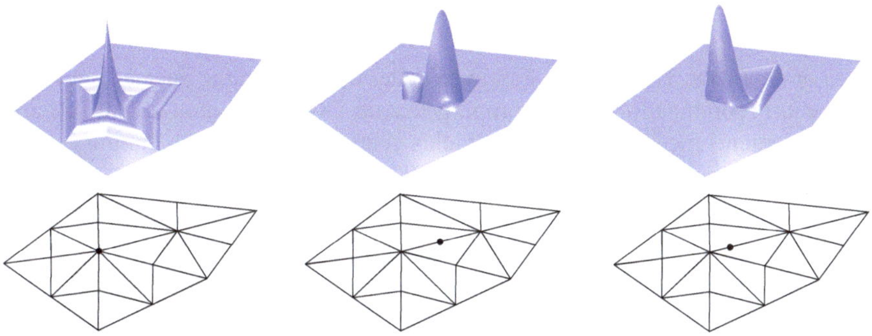

Fig. 7 Typical piecewise quartic basis functions in 2d

is the Lagrange basis *of* $\mathbb{P}(T) = \mathbb{P}_p$.

Sketch of the Proof. By construction, the Lagrange basis functions satisfy

$$\Phi_n^T(x_m^T) = \delta_{nm} \qquad n, m = 1, \ldots, \bar{n}.$$

In addition, Φ_n^T belongs to $\mathbb{P}_p$ since $\bar{\phi}_n \in \mathbb{P}_p$ and λ^T is affine. □

Examples of Lagrange grids in 2d are shown in Fig. 6 and examples of piecewise quartic basis functions in 2d are depicted in Fig. 7.

Next we construct the global basis from the local ones. Here it becomes important that conformity of $\mathscr{T}$ implies

$$\mathscr{L}(T_1) \cap T_2 = \mathscr{L}(T_2) \cap T_1 \qquad \forall T_1, T_2 \in \mathscr{T},$$

i. e., the Lagrange nodes of two elements coincide on the intersection.

Theorem 19 (Lagrange Basis of $\mathbb{V}(\mathscr{T})$). *For a conforming triangulation $\mathscr{T}$ define the Lagrange grid*

$$\mathscr{L}(\mathscr{T}) := \bigcup_{T \in \mathscr{T}} \mathscr{L}(T) =: \{x_1, \ldots, x_N\}.$$

The Lagrange grid is uni-solvent in $\mathbb{V}(\mathscr{T}) := \mathsf{FES}(\mathscr{T}, \mathbb{P}_p, H^1(\Omega))$, and therefore, the function set $\{\Phi_1, \ldots, \Phi_N\}$ given by

$$\Phi_i(x_j) = \delta_{ij} \qquad i, j = 1, \ldots, N$$

is the Lagrange basis of $\mathbb{V}(\mathscr{T})$*. Furthermore if* $x_i \in T$ *then there is* $x_n^T \in \mathscr{L}(T)$ *such that* $x_i = x_n^T$ *and*

$$\Phi_{i\,|T} = \Phi_n^T = \bar{\phi}_n \circ \lambda^T.$$

A basis of $\mathring{\mathbb{V}}(\mathscr{T}) = \mathbb{V}(\mathscr{T}) \cap H_0^1(\Omega)$ *is given by*

$$\{\Phi_i \mid x_i \in \Omega\} \subset \{\Phi_1, \ldots, \Phi_N\},$$

i. e., the basis functions related to $\mathscr{L}(\mathscr{T}) \cap \Omega$.

Idea of the Proof. The fact that Lagrange nodes coincide on the intersection of two elements in combination with the property that the DOF at a Lagrange node is shared by all elements meeting at that node yields continuity of any global finite element function. For the details compare with [16, Theorem 2.2.3]. □

Aspects for the Implementation

Generalizing the construction of Lagrange finite elements to other finite element spaces we have the following rule to define global basis functions from suitable local ones:

(1) DOFs on the boundary of an element *are shared* with other elements, i. e., the value(s) at such a node is/are the same for all adjacent elements.
(2) DOFs in the interior of an element *are not shared* with other elements and are related to locally supported finite element functions (including discontinuous finite element functions).

Evaluation of Finite Element Functions. Let $\{\Phi_1, \ldots, \Phi_N\}$ be a basis of a finite element space $\mathbb{V}(\mathscr{T})$ as constructed above. By Theorem 19 any finite element function $V \in \mathbb{V}(\mathscr{T})$ is uniquely determined by its *global coefficient vector* $\boldsymbol{v} = [v_1, \ldots, v_N]^\top$, i. e.,

$$V = \sum_{i=1}^{N} v_i \Phi_i.$$

Basis functions, and consequently V, are defined element-wise. This means, we only have a local rather than a global representation. Therefore, any access to a finite element function is only possible *element-wise*. The access of V on $T \in \mathscr{T}$ needs the representation in the local basis

$$V_{|T} = \sum_{n=1}^{\bar{n}} v_n^T \Phi_n^T = \sum_{n=1}^{\bar{n}} v_n^T \bar{\phi}_n \circ \lambda^T,$$

with the *local coefficient vector* $\boldsymbol{v}_{|T} = \left[v_1^T, \ldots, v_{\bar{n}}^T\right]^\top$. For extracting the local coefficient vector from the global one, we need access of local DOFs from global

DOFs. By construction of the global basis, there exists a unique mapping

$$I: \{1, \ldots, \bar{n}\} \times \mathcal{T} \to \{1, \ldots, N\}$$

such that

$$x_{I(n,T)} = x_n^T \qquad \forall n = 1, \ldots, \bar{n} \text{ and } T \in \mathcal{T}.$$

Therefore, the local coefficient vector $\boldsymbol{v}_{|T}$ can be extracted form the global one $\boldsymbol{v}$ by

$$v_n^T = v_{I(n,T)} \qquad n = 1, \ldots, \bar{n}.$$

Example 20. For piecewise linear elements $\mathcal{L}(\mathcal{T})$ is the set of all vertices of $\mathcal{T}$ and $\mathcal{L}(T)$ the set of vertices of T. For a vertex z_n of T with $z_n = x_i$ there holds

$$\Phi_i(x) = \lambda_n(x) \qquad x \in T.$$

Aspects for the Implementation

For the implementation we need for any finite element space a function that realizes the index mapping

$$I: \{1, \ldots, \bar{n}\} \times \mathcal{T} \to \{1, \ldots, N\},$$

i. e., we have to store for any local basis function Φ_n^T on $T \in \mathcal{T}$ the index of the global basis function Φ_i. Hereby, we strictly follow the fundamental concept: *everything is done from local to global.*

Any evaluation of a finite element function $V \in \mathbb{V}(\mathcal{T})$ in $x \in T$ is a two-step procedure.

1. Access the local coefficient vector $\boldsymbol{v}_{|T}$ on T from the global vector $\boldsymbol{v}$.
2. Evaluate $V(x)$ or $D^\alpha V(x)$ via the local basis representation

$$V(x) = \sum_{n=1}^{\bar{n}} v_n^T \Phi_n^T(x) = \sum_{n=1}^{\bar{n}} v_n^T \bar{\phi}_n(\lambda(x)),$$

Using the chain rule we obtain for $x \in T$

$$\nabla V(x) = \sum_{n=1}^{\bar{n}} v_n^T \nabla \bar{\phi}_n(\lambda(x)) = \Lambda^\top \sum_{n=1}^{\bar{n}} v_n^T \nabla_\lambda \bar{\phi}_n(\lambda(x)),$$

where

$$\nabla_\lambda \bar{\phi}(\lambda) = \left[\bar{\phi}_{,\lambda_0}(\lambda), \ldots \bar{\phi}_{,\lambda_d}(\lambda)\right] \in \mathbb{R}^{d+1}$$

and

$$\Lambda = \begin{bmatrix} \lambda_{0,x_1} & \cdots & \lambda_{0,x_d} \\ \vdots & & \vdots \\ \lambda_{d,x_1} & \cdots & \lambda_{d,x_d} \end{bmatrix} = \begin{bmatrix} - & \nabla\lambda_0 & - \\ & \vdots & \\ - & \nabla\lambda_d & - \end{bmatrix} \in \mathbb{R}^{(d+1)\times d}$$

is the *Jacobian* of the barycentric coordinates. For affine elements Λ is constant on T. Consequently, we can express derivatives of any finite element function in terms of Λ and derivatives of the basis $\{\bar{\phi}_1, \ldots \bar{\phi}_{\bar{n}}\}$. Obviously, this procedure can be generalized to higher order derivatives.

Finite element functions are usually evaluated in *barycentric coordinates* on T rather than in *world coordinates* $x \in T$:

$$V(x(\lambda)) = \sum_{n=1}^{\bar{n}} v_n^T \bar{\phi}_n(\lambda), \qquad \lambda \in \bar{T},$$

and

$$\nabla V(x(\lambda)) = \Lambda^{\mathrm{T}} \left(\sum_{n=1}^{\bar{n}} v_n^T \nabla_\lambda \bar{\phi}_n(\lambda) \right), \qquad \lambda \in \bar{T}.$$

The computation of $\lambda \mapsto x(\lambda)$ is cheap:

$$x(\lambda) = \sum_{j=0}^{d} \lambda_j z_j,$$

whereas the inverse mapping $x \mapsto \lambda(x)$ requires the solution of a small linear system. For the *computation of the Jacobian* Λ a small $d \times d$ linear system has to be inverted.

2.4 **ALBERTA** *Realization of Finite Element Spaces*

In ALBERTA all information of a finite element space is collected in the following data structure FE_SPACE reflecting the discussion above.

```
struct fe_space
{
  const DOF_ADMIN  *admin;
  const BAS_FCTS   *bas_fcts;
  MESH             *mesh;
};
```

Description:

- MESH realizes the triangulation $\mathscr{T}$:
  - geometrical information;
  - refinement and coarsening routines;
  - adding or removing DOFs together with DOF_ADMIN.

- BAS_FCTS realizes the local basis functions of $\bar{\mathbb{P}}$:
  - definition of basis functions in barycentric coordinates;
  - derivatives of basis functions with respect to barycentric coordinates;
  - access of DOFs on elements together with DOF_ADMIN.
- DOF_ADMIN realizes the continuity constraint $\mathbb{V}$:
  - gives connection between local and global DOFs;
  - administrates all vectors and matrices: enlargement or compression.

Initialization of DOFs. All DOFs used by finite element spaces have to be defined in an ALBERTA program directly in the beginning via a call to GET_MESH(). Initialization is done in an application dependent function, like for instance the following function for initializing Lagrange elements:

```
static FE_SPACE *fe_space;

void init_dof_admin(MESH *mesh)
{
  FUNCNAME("init_dof_admin");
  int             degree = 1;
  const BAS_FCTS  *lagrange;

  GET_PARAMETER(1, "polynomial degree", "%d", &degree);
  lagrange = get_lagrange(degree);
  TEST_EXIT(lagrange)("no lagrange BAS_FCTS\n");
  fe_space = get_fe_space(mesh, lagrange->name, nil, lagrange);
  return;
}
```

and in the main program

```
  ...
  mesh = GET_MESH("\ALBERTA mesh", init_dof_admin, nil);
  ...
```

In init_dof_admin() the used finite element spaces are accessed via the function get_fe_space(). Information about number and position of DOFs is then passed to MESH and DOF_ADMIN in GET_MESH(). Several finite element spaces can be used on the same grid. Each single finite element space has to be accessed via get_fe_space(). For a detailed description compare with [34, Sect. 3.6].

Structure of ALBERTA Projects. ALBERTA aims at dimension independent programming, i. e., there is one source code for dimension 1, 2, and 3. The definition of the main data structures does depend on the dimension but only via parameters defined in the ALBERTA header as macros, for instance

```
DIM
DIM_OF_WORLD
N_VERTICES
N_EDGES
...
```

Many modules strongly depend on the dimension but are hidden in the ALBERTA library. Applications can be implemented independent of the dimension by using these macro definitions. However, object files and executables strongly depend on the dimension!

For getting started, you can download a start-up archive `alberta.tgz` which creates after unpacking the directory `alberta` with the standard structure of ALBERTA projects:

```
1d/        2d/        3d/        Common/        Makefile
```

All source files are located in the sub-directory `Common`:

```
iafem.c        graphics.c           stokes-estimator.c
ellipt.c       pmc-estimator.c
```

The `?d` are used for producing the dimension dependent object files and executables. They contain the dimension dependent `Makefile` and data for the initialization (parameters and macro triangulation files):

```
Macro/        Makefile        iafem.dat        ellipt.dat
```

Basic Rule: Edit or add source files only in the `Common` sub-directory. Compilation and linking to the ALBERTA library in the `?d` sub-directory using the corresponding `Makefile`.

3 Refinement By Bisection

In this chapter we address the question about local refinement of a given triangulation with the following properties.

Problem 21 (Local Refinement). Given a conforming triangulation $\mathscr{T}$ and a subset $\mathscr{M} \subset \mathscr{T}$ of marked elements construct a *refinement* $\mathscr{T}_*$ of $\mathscr{T}$ such that

(1) all elements in $\mathscr{M}$ are refined, i. e., decomposed into sub-simplices;
(2) $\mathscr{T}_*$ is again conforming and as small as possible;
(3) recurrent refinement of some given initial grid $\mathscr{T}_0$ produces a shape regular sequence $\{\mathscr{T}_k\}_{k\geq 0}$ of conforming triangulations.

3.1 Basic Thoughts About Local Refinement

A natural decomposition of a simplex is the *regular refinement* or *red refinement*, this means a simplex T is decomposed into 2^d smaller simplices; compare with

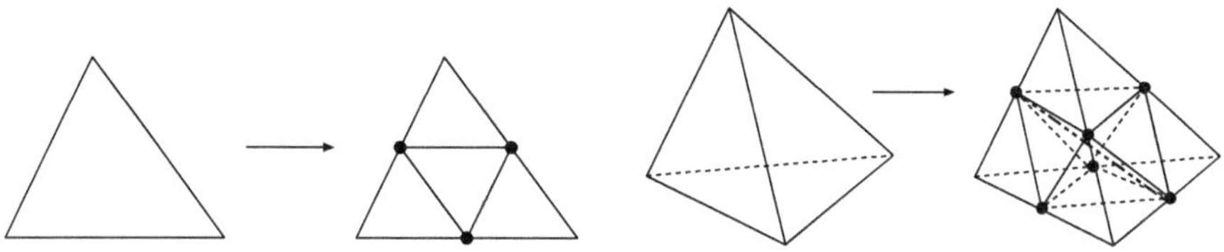

Fig. 8 Regular refinement: Decomposition of a triangle into four congruent triangles in 2d (*left*). Decomposition of a tetrahedron into four congruent tetrahedra and four additional ones in 3d (*right*). With a specific choice of the interior diagonal, the number of similarity classes can be minimized

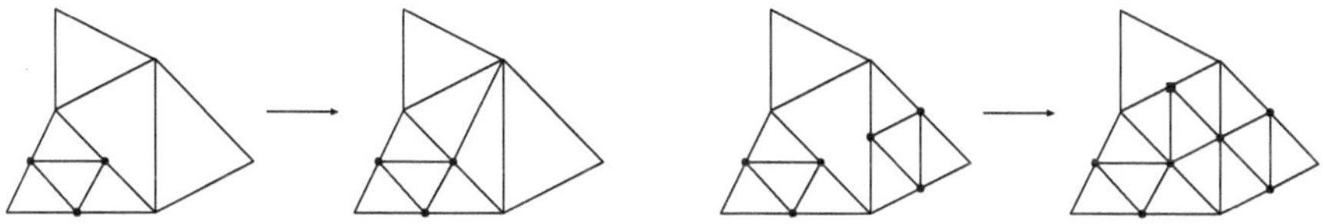

Fig. 9 Removing hanging nodes: green closure for removing one irregular node (*left*) and regular refinement for removing two irregular nodes and creating a new one (*right*)

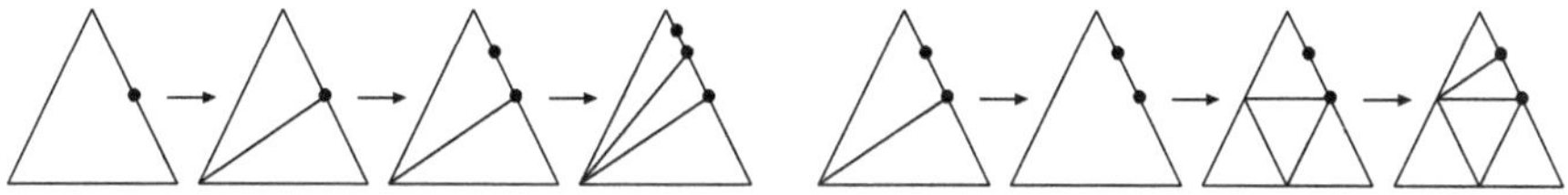

Fig. 10 Recurrently applying the green closure may lead to flat elements (*left*). Replacing the green closure by regular refinement prior to an additional green closure (*right*)

Fig. 8. All descendants of some initial element produced by recurrent regular refinement belong to one similarity class in 2d. In 3d recurrent refinement produces several similarity classes but the number is bounded; compare with [6,7]. In respect thereof, regular refinement complies very well with objective (3) of Problem 21. Then again, only allowing for regular refinement and asking for conformity always results in global refinement, i. e., all elements of $\mathscr{T}$ are refined irrespective of the number of elements in $\mathscr{M}$. Therefore, only using regular refinement does not comply with (2) of Problem 21.

Alternatively one first only regularly refines the marked elements and thereby allowing for irregular nodes. These node are then removed using additional refinement rules to create a conforming triangulation; compare with Fig. 9. In 2d some triangles have to be bisected and in 3d several types of additional refinement rules are needed. Such rules are called *green closure*. This creates more similarity classes, even in two dimensions. Additionally, these green closure elements have to be removed before a further refinement of the mesh, in order to keep the triangulations shape regular as depicted in Fig. 10. This is a very elaborate procedure, in particular in higher space dimensions.

On the one hand, local refinement without *bisecting* elements is impossible. On the other hand, a naive bisection of elements easily leads to flat elements. Therefore the following question arises: Is it possible to design a bisection rule that allows for local refinement and such that recurrent refinement produces shape regular grids?

In the remainder we describe such a bisection rule. Its roots in 2d go back to Sewell in 1972 [36]. Rivara introduced 1984 the *longest edge bisection* [32] and Mitchell came up in 1988 with the *newest vertex bisection* and a recursive refinement algorithm [26]. The newest vertex bisection was generalized by Bänsch to 3d in 1991 in an iterative fashion [5]. A similar approach was published by Liu and Joe [24] and later on by Arnold et al. [2]. A recursive variant of the algorithm by Bänsch was derived by Kossaczký [21]. This algorithm was generalized to any space dimension independently by Maubach [25] and Traxler [39]. A complexity estimate for refinement by bisection was given by Binev, Dahmen, and DeVore for 2d in 2004 [8] and by Stevenson for any dimension in 2008 [38].

3.2 Bisection Rule: Bisection of One Single Simplex

The objective of this section is to design a module $\{T_1, T_2\} = \mathsf{BISECT}(T)$ based on *one fixed rule* such that

(1) T is divided into two children T_1 and T_2 of same size;
(2) recurrent bisection creates shape regular descendants.

Hereafter, recurrent bisection means: BISECT can be applied to some initial element T_0, and any simplex created by a prior application of BISECT.

For stating the refinement rule we use the idea of Kossaczký that relies on vertex ordering and element type, which we define next.

Definition 22 (Vertex Ordering and Element Type). We identify a simplex with its set of *ordered* vertices and type $t \in \{0, \dots, d-1\}$:

$$T = [z_0, \dots, z_d]_t$$

A few remarks about the vertex ordering are in order.

(1) Let $\pi \neq I$ be a permutation of the indices $\{0, \dots, d\}$. Then $T = [z_0, \dots, z_d]_t$ and $T' = [z_{\pi(0)}, \dots, z_{\pi(d)}]_t$ have the same shape but are different elements. The same applies to $T = [z_0, \dots, z_d]_t$ and $T' = [z_0, \dots, z_d]_{t'}$ if $t \neq t'$.
(2) The ordering is very natural for the implementation: vertices are stored as vectors $z_0, \dots, z_d$ in an ordered fashion. For instance in the implementation $[z_0, \dots, z_d]_t$ could be a $(d+1) \times d$ matrix and an integer.
(3) The element type is only used for $d \geq 3$. It is needed to formulate the bisection rule as *one fixed rule*. The dependence of this rule on the type is essential for shape regularity.

We next turn to the precise definition of the bisection rule.

Definition 23 (Refinement Edge and New Vertex). The *refinement edge* of an element $T = [z_0, \ldots, z_d]_t$ is $\overline{z_0 z_d}$ and in the bisection step the midpoint $\bar{z} = \frac{1}{2}(z_0 + z_d)$ is the new vertex.

This rule already uniquely defines the shape of the two children

$$T_1 = \text{conv hull}\{z_0, \bar{z}, z_1, \ldots, z_{d-1}\}, \quad T_2 = \text{conv hull}\{z_d, \bar{z}, z_1, \ldots, z_{d-1}\}.$$

It remains to prescribe an ordering of the children's vertices.

Definition 24 (Bisection Rule). $\mathsf{BISECT}([z_0, \ldots, z_d]_t)$ outputs the two children

$$T_1 := [z_0, \bar{z}, \underbrace{z_1, \ldots, z_t}_{\rightarrow}, \underbrace{z_{t+1}, \ldots, z_{d-1}}_{\rightarrow}]_{(t+1) \bmod d},$$
$$T_2 := [z_d, \bar{z}, \underbrace{z_1, \ldots, z_t}_{\rightarrow}, \underbrace{z_{d-1}, \ldots, z_{t+1}}_{\leftarrow}]_{(t+1) \bmod d},$$

where arrows point in the direction of increasing indices.

Note that BISECT determines the children's refinement edges by the local ordering of their vertices. This bisection rule thereby determines the refinement edge of any descendant produced by recurrent bisection of a given initial element T_0 from the local ordering of vertices and the type of T_0. We also observe that only the labeling of the second child's vertices depends on the element type.

Example 25 (Bisection Rule in 2d and 3d). In two space dimensions the bisection rule can be implemented as follows:

function BISECT$([z_0, z_1, z_2]_t)$
  compute new vertex $\bar{z} = \frac{1}{2}(z_0 + z_2)$;
  $T_1 = [z_0, \bar{z}, z_1]_{(t+1) \bmod 2}$;
  $T_2 = [z_2, \bar{z}, z_1]_{(t+1) \bmod 2}$;
  return$(\{T_1, T_2\})$;

We see that the refinement edge of both children is opposite the new vertex $\bar{z}$, whence this is the *newest vertex bisection.* The 3d version reads

function BISECT$([z_0, z_1, z_2, z_3]_t)$
  compute new vertex $\bar{z} = \frac{1}{2}(z_0 + z_3)$;
  $T_1 = [z_0, \bar{z}, z_1, z_3]_{(t+1) \bmod 3}$;
  if $t = 0$ then
    $T_2 = [z_3, \bar{z}, z_2, z_1]_1$;
  else
    $T_2 = [z_3, \bar{z}, z_1, z_2]_{(t+1) \bmod 3}$;
  end if
  return$(\{T_1, T_2\})$;

Compare with Fig. 11.

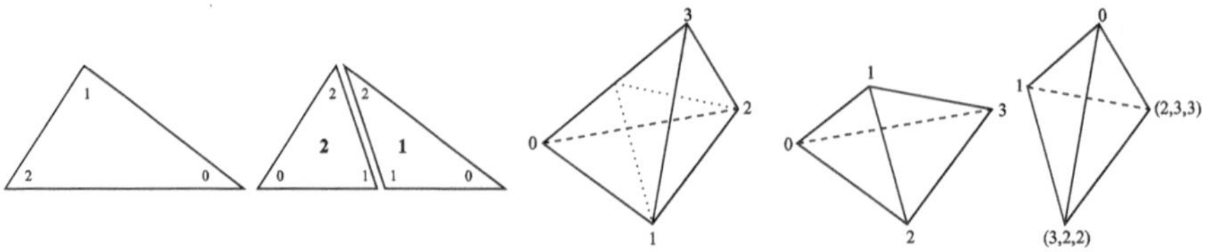

Fig. 11 Vertex ordering on parent and children in 2d (*left*) and 3d (*right*). In 3d the vertex ordering of the second child depends on the parent's type

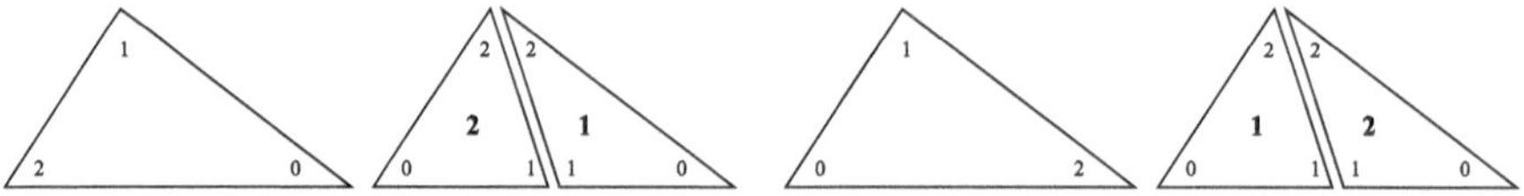

Fig. 12 Refinement of a triangle T and its reflected element T_R

Exchanging the vertices z_0 and z_d of a simplex T results in the same geometric shape of the two children. Adjusting the other vertices of T properly results in the same descendants.

Lemma 26 (Reflected Element). *Given an element* $T = [z_0, \cdots, z_d]_t$ *we denote by*

$$T_R := [z_d, \underbrace{z_1, \ldots, z_t}_{\rightarrow}, \underbrace{z_{d-1}, \ldots, z_{t+1}}_{\leftarrow}, z_0]_t$$

the reflected element. *If* $\{T_1, T_2\} = \mathsf{BISECT}(T)$ *then*

$$\{T_2, T_1\} = \mathsf{BISECT}(T_R),$$

i. e., the refinement of T *and* T_R *creates the same set of children; see Fig. 12 for 2d.*

Binary Tree. Given an initial simplex $T_0 = [z_0, \ldots, z_d]_t$ recurrent bisection induces the structure of an *infinite binary tree* $\mathbb{F}(T_0)$ (compare with Fig. 13):

(1) any node T inside the tree is an element generated by recurrent application of BISECT;
(2) the two successors of a node T are the children T_1, T_2 created by BISECT(T).

Remark 27 (The Initial Element and the Binary Tree). The ordering of the vertices on T_0 and its type in combination with the bisection rule completely determine $\mathbb{F}(T_0)$, especially the shape of any descendant of T_0. For any T_0 there exists $d(d+1)!$ different binary trees that can be associated with T_0 and BISECT. Recalling the property of the reflected element there are only $\frac{d(d+1)!}{2}$ binary trees that are essentially different.

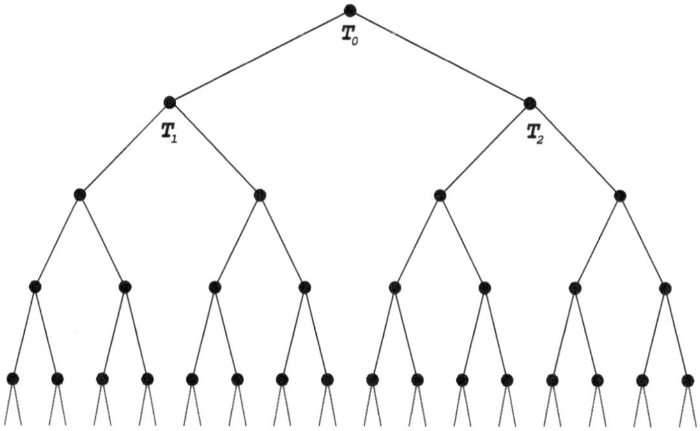

Fig. 13 Infinite binary tree associated with recurrent bisection of some initial element

Definition 28 (Generation). The *generation* $g(T) \in \mathbb{N}_0$ of an element $T \in \mathbb{F}(T_0)$ is the distance of T to T_0 within $\mathbb{F}(T_0)$, i. e., the minimal number of calls to BISECT needed to create T from T_0.

Some information about T can uniquely be deduced from $g(T)$ in $\mathbb{F}(T_0)$, for instance

(1) the type of T is $(g(T) + t_0) \bmod d$;
(2) the local mesh size $h_T := |T|^{1/d} = 2^{-g(T)/d} h_{T_0}$.

Shape Regularity. We turn to the question of shape regularity. In doing this we first consider a special simplex, the so-called *Kuhn-simplex* and then treat the general case.

Definition 29 (Kuhn-Simplex). Let π be any permutation of $\{1, \dots, d\}$. The simplex with ordered vertices

$$z_i^\pi := \sum_{j=1}^{i} e_{\pi(j)} \qquad \forall i = 0, \dots, d,$$

is a *Kuhn-Simplex.*

It holds $z_0^\pi = 0$ and $z_d^\pi = [1, \dots, 1]^T$, whence the refinement edge of a Kuhn-simplex is the main diagonal of the unit cube in $\mathbb{R}^d$. There exist $d!$ different Kuhn-simplices.

Theorem 30 (Shape Regularity for a Kuhn-Simplex). *All 2^g descendants of generation g of a Kuhn-simplex $T_\pi = \{z_0^\pi, \dots, z_d^\pi\}_0$ are mutually congruent with at most d different shapes. Moreover, the descendants of generation d are congruent to T_0 up to a scaling with factor $\frac{1}{2}$.*

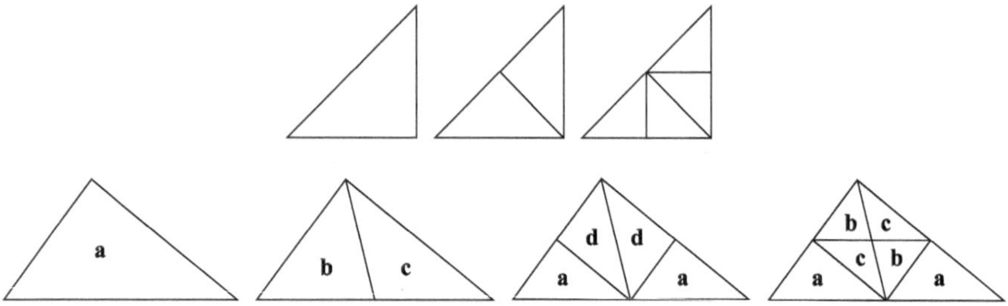

Fig. 14 Similarity classes for a Kuhn-triangle (*top*) and a generic triangle (*bottom*). Note that there is one class for a Kuhn-triangle and there are four classes for a generic triangle

Main Idea of the Proof. For any descendant of Kuhn-simplex of type 0 the refinement edge is the longest edge; compare with [25, 39]. □

Corollary 31 (Shape Regularity). *Let $T_0 = [z_0, \ldots, z_d]_t$ be an arbitrary d-simplex of type $t \in \{0, \ldots, d-1\}$. Then all descendants of T generated by bisection are shape regular, i. e.,*

$$\sup_{T \in \mathbb{F}(T_0)} \sigma_T = \sup_{T \in \mathbb{F}(T_0)} \frac{\overline{h}_T}{\underline{h}_T} \le C(T_0) < \infty.$$

Idea of the Proof. Let F_0 be an affine mapping from a Kuhn-simplex T_π to T_0. Then any descendant of T_0 is the image of a corresponding descendant of T_π under F_0. □

It is worth noticing that the number of similarity classes for a Kuhn-simplex and a generic simplex differ in general; compare with Fig. 14.

3.3 Triangulations and Refinements

We next turn to the refinement of a given conforming triangulation by bisection. In what follows $\mathscr{T}_0$ is a conforming triangulation of Ω.

Definition 32 (Master Forest and Forest). The union

$$\mathbb{F} = \mathbb{F}(\mathscr{T}_0) := \bigcup_{T_0 \in \mathscr{T}_0} \mathbb{F}(T_0).$$

is the associated *master forest* of binary trees. Any subset $\mathscr{F} \subset \mathbb{F}$ is called *forest* iff

(1) $\mathscr{T}_0 \subset \mathscr{F}$;
(2) all nodes of $\mathscr{F} \setminus \mathscr{T}_0$ have a predecessor;
(3) all nodes of $\mathscr{F}$ have either two successors or none.

A forest $\mathscr{F}$ is called *finite*, if $\max_{T\in\mathscr{F}} g(T) < \infty$. The nodes with no successors are called *leaves* of $\mathscr{F}$.

Given an initial grid $\mathscr{T}_0$ as a set of elements with ordered vertices and types, the master forest $\mathbb{F}$ holds *full information of any refinement* that can be produced by bisection. The concept of master forest and forest is very useful for theoretical results.

Lemma 33 (Forest and Triangulation). *Let $\mathscr{F} \subset \mathbb{F}$ be any finite forest. Then the set of leaves of $\mathscr{F}$ is a refinement of $\mathscr{T}_0$ giving a triangulation $\mathscr{T}(\mathscr{F})$ of Ω. In general, the triangulation $\mathscr{T}(\mathscr{F})$ is non-conforming.*

In view of this lemma, a triangulation $\mathscr{T}_*$ is a refinement of $\mathscr{T}$ iff the corresponding forests satisfy $\mathscr{F}(\mathscr{T}) \subset \mathscr{F}(\mathscr{T}_*)$ and we denote this by $\mathscr{T} \le \mathscr{T}_*$.

Aspects for the Implementation

The forest $\mathscr{F}$ of a triangulation $\mathscr{T}$ is useful for the implementation. The forest $\mathscr{F}$ can be used for a compact storage of $\mathscr{T}$ including its refinement history:

- Hierarchical information can be generated from $\mathscr{F}$: coordinates, generation, type, neighbors, …
- Only few information has to be stored explicitly on elements: information for the index mapping I, marker for refinement, …

Hierarchical information from $\mathscr{F}$ is very beneficial for implementing multigrid solvers.

Shape regularity of recurrent bisection of a single element proven in Corollary 31 directly implies the following result.

Corollary 34 (Shape Regularity). *The shape coefficients of all elements in $\mathbb{F}$ are uniformly bounded, i. e.,*

$$\sup_{T\in\mathbb{F}} \sigma_T = \sup_{T\in\mathbb{F}} \frac{\overline{h}_T}{\underline{h}_T} = \max_{T_0\in\mathscr{T}_0} \sup_{T\in\mathbb{F}(T_0)} \frac{\overline{h}_T}{\underline{h}_T} \le \max_{T_0\in\mathscr{T}_0} C(T_0) =: C(\mathscr{T}_0) < \infty.$$

On Conforming Refinements. The master forest $\mathbb{F}$ provides an infinite number of finite forests $\mathscr{F}$ such that $\mathscr{T}(\mathscr{F})$ is a triangulation of Ω. In general these triangulations are not conforming and it is a priori not clear that conforming refinements of a given conforming grid $\mathscr{T}$ exist.

So let $\mathscr{T}$ be a conforming triangulation. Considering the case $d = 2$ we realize that any refinement of $\mathscr{T}$ where all elements in $\mathscr{T}$ are bisected exactly twice is conforming. The situation changes completely when looking at $d > 2$. Here, a necessary condition for the existence of a conforming refinement is:

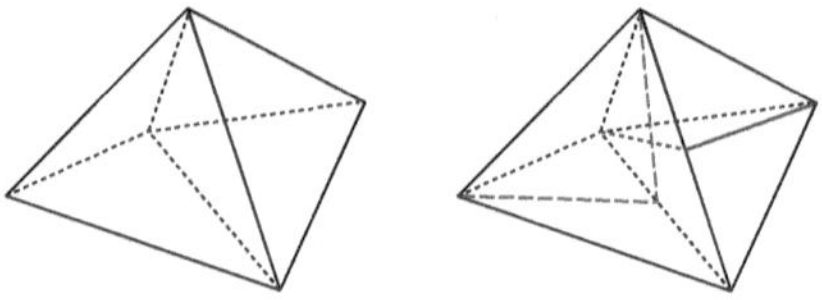

Fig. 15 Non-compatible distribution of refinement edges on neighboring elements

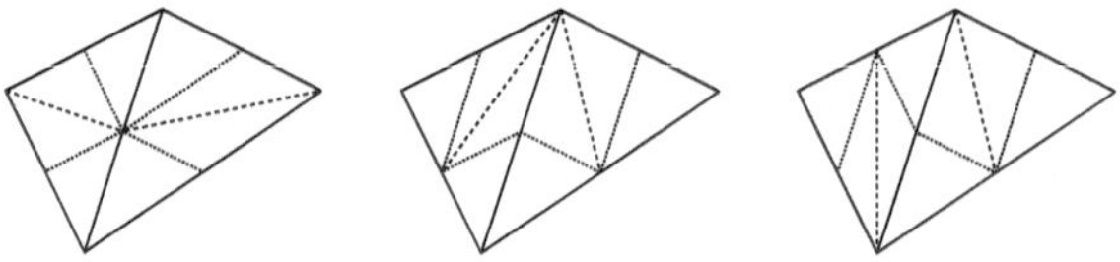

Fig. 16 Matching neighbors in 2d and their grandchildren. The elements in the *left* and *middle* picture are reflected neighbors. The elements in the rightmost picture are not reflected neighbors, but the pair of their neighboring children are

Whenever the refinement edges of two neighboring elements are both on the common side they have to coincide.

Compare with Fig. 15 for an illustration in 3d. We next have to state a condition on $\mathscr{T}_0$ that allows us to (locally) refine any conforming refinement of $\mathscr{T}_0$ into a conforming triangulation. This condition relies on the notion of *reflected neighbors*.

Definition 35 (Reflected Neighbors). Two neighboring elements $T = [z_0, \dots, z_d]_t$ and $T' = [z'_0, \dots, z'_d]_t$ are called *reflected neighbors* iff the ordered vertices of T or T_R coincide exactly with those of T' at all but one position.

If $T, T' \in \mathscr{T}$ are reflected neighbors then either the refinement edge of both elements coincides or both refinement edges are not on the common side.

Assumption 36 (Admissibility of the Initial Triangulation). *Let $\mathscr{T}_0$ satisfy*

(1) all elements are of the same type $t \in \{0, \dots, d-1\}$;
(2) all neighboring elements T and T' with common side S are matching neighbors *in the following sense:*
if $\overline{z_0 z_d} \subset S$ or $\overline{z'_0 z'_d} \subset S$ then T and T' are reflected neighbors; otherwise the pair of neighboring children of T and T' are reflected neighbors.

For instance, the set of the $d!$ Kuhn-simplices of type 0 is a conforming triangulation of the unit cube in $\mathbb{R}^d$ satisfying Assumption 36. In Fig. 16 we have given an example of matching neighbors in 2d.

Theorem 37 (Uniform Refinement). *For $g \in \mathbb{N}_0$ denote by*

$$\mathscr{T}_g := \{T \in \mathbb{F} \mid g(T) = g\}$$

the uniform refinement *of $\mathscr{T}_0$ with all elements in $\mathbb{F}$ of generation exactly g.*

(1) If $\mathscr{T}_0$ satisfies Assumption 36 then $\mathscr{T}_g$ is conforming for any $g \in \mathbb{N}_0$.

(2) *If all elements in $\mathcal{T}_0$ are of the same type, then condition (2) of Assumption 36 is necessary for $\mathcal{T}_g$ to be conforming for all g.*

The proof is a combination of [39, Sect. 4] and [38, Theorem 4.3]. Some remarks are in order.

(1) Theorem 37, and thus Assumption 36, plays a key role in the subsequent discussion. It states that two elements $T, T' \in \mathbb{F}$ of the same generation sharing a common edge are either *compatibly divisible*, i. e., $\overline{z_0 z_d} = \overline{z'_0 z'_d}$, or the refinement of T does not affect T' and vice versa. In the latter case any common edge is neither the refinement edge $\overline{z_0 z_d}$ of T nor $\overline{z'_0 z'_d}$, of T'.
(2) In 2d it is known that Assumption 36 can be satisfied for arbitrary $\mathcal{T}_0$ [26, Theorem 2.9]. Finding the right labeling of the elements' vertices in $\mathcal{T}_0$ is the so-called *perfect matching problem*, which is NP-complete. For $d \geq 3$ it is only known that the elements of any given coarse grid can be decomposed, such that the resulting triangulation satisfies Assumption 36 [38, Appendix A].
(3) There exist conforming refinements under weaker assumptions on $\mathcal{T}_0$ that can be satisfied for any conforming triangulation; compare with Bänsch [5], Liu and Jo [24], and Arnold et al. [2]. In this case only uniform refinements $\mathcal{T}_g$ with $g \bmod d = 0$ can be shown to be conforming.

3.4 Refinement Algorithms

Denoting by $\mathbb{T}$ the set of all conforming refinements of $\mathcal{T}_0$, Assumption 36 guarantees that

(1) $\#\mathbb{T} = \infty$, i. e., there are infinitely many conforming refinements of $\mathcal{T}_0$;
(2) given a conforming triangulation $\mathcal{T} \in \mathbb{T}$ and a subset $\mathcal{M} \subset \mathcal{T}$ of marked elements there exists the smallest conforming refinement $\mathcal{T}_* \in \mathbb{T}$ of $\mathcal{T}$ such that all elements in $\mathcal{M}$ are refined, i. e., $\mathcal{T}_* \cap \mathcal{M} = \emptyset$.

To show (2) let $\mathcal{T}'$ be the (non-conforming) triangulation after bisecting all elements in $\mathcal{M}$ and set $g = \max\{g(T) \mid T \in \mathcal{T}'\}$. Then $\mathcal{T}_g \in \mathbb{T}$ is a conforming refinement of $\mathcal{T}$ with $\mathcal{T}_g \cap \mathcal{M} = \emptyset$. Therefore $\mathcal{T}_* \leq \mathcal{T}_g$ and we *hope* that $\mathcal{T}_*$ is much smaller that $\mathcal{T}_g$. We want to remark, that the assumptions of Bänsch, Liu and Jo, Arnold et al. on $\mathcal{T}_0$ have the same consequences. This can be seen by using $\mathcal{T}' \leq \mathcal{T}_g$ with g sufficiently large and $g \bmod d = 0$.

We next derive refinement algorithms $\mathsf{REFINE}(\mathcal{T}, \mathcal{M})$ that output the smallest conforming refinement $\mathcal{T}_*$ of $\mathcal{T}$ with $\mathcal{T}_* \cap \mathcal{M} = \emptyset$.

Iterative Refinement. This variant first only bisects all marked elements and thereby producing irregular nodes. In the so-called *completion step* these irregular nodes are removed by bisecting additional elements. This step has to be iterated.

Algorithm 38 (Iterative Refinement). Let $\mathscr{T} \in \mathbb{T}$ and let $\mathscr{M} \subset \mathscr{T}$ be a subset of elements subject to refinement.

subroutine REFINE($\mathscr{T}, \mathscr{M}$)
  while $\mathscr{M} \neq \emptyset$ do
    for all $T \in \mathscr{M}$ do
      $\mathscr{T} := \big(\mathscr{T} \cup \mathsf{BISECT}(T)\big)\backslash\{T\}$;
    end for
    $\mathscr{M} := \emptyset$; // Completion Step
    for all $T \in \mathscr{T}$ do
      if T contains an irregular node
        $\mathscr{M} := \mathscr{M} \cup \{T\}$
      end if
    end for
  end while

Lemma 39. *The iterative refinement algorithm terminates and outputs the smallest conforming refinement $\mathscr{T}_*$ of $\mathscr{T}$ such that $\mathscr{T}_* \cap \mathscr{M} = \emptyset$.*

Idea of the Proof. After the first step the above algorithm only resolves non-conforming situations and therefore holds $\mathscr{T}_* \leq \mathscr{T}_g$ for a suitable uniform refinement $\mathscr{T}_g$. The proof does not need Assumption 36. □

The refinement procedure produces irregular nodes which leads to *not one-to-one neighbor relations*. One-to-one neighbor relation has to be re-established later on when removing an irregular node; compare with Fig. 17 for a 2d example. This is a very knotty procedure, especially in 3d. In addition, the algorithm as stated above is not efficient since there are too many iterations in the completion step. It can be tuned by directly marking all elements at the refinement edge when creating an irregular node.

Recursive Refinement. Irregular nodes can completely be avoided if the shared edge of neighboring elements is a *common refinement edge*; compare with Fig. 18. In this situation all elements at the common refinement edge can be bisected simultaneously. Such an *atomic refinement* is very convenient for the implementation.

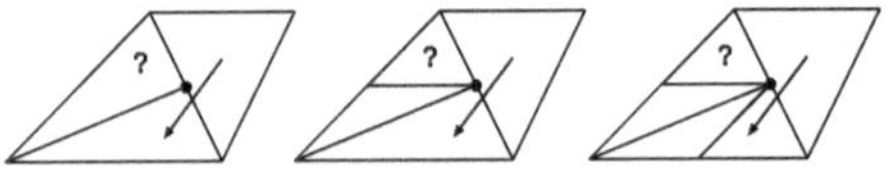

Fig. 17 The not one-to-one neighbor relation causes problems when refining the neighbor

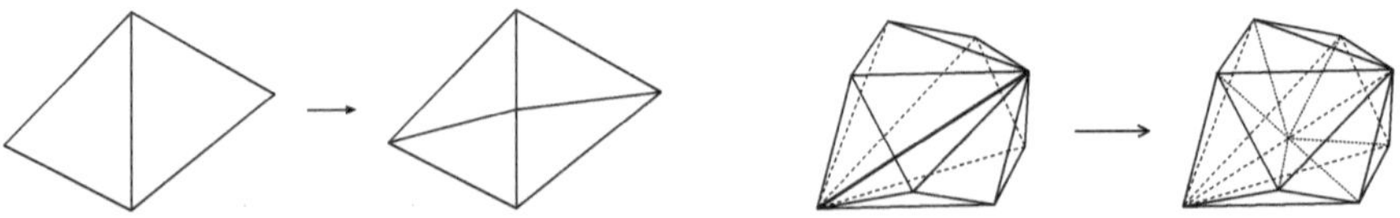

Fig. 18 Atomic refinement operation in 2d (*left*) and 3d (*right*). The common edge is the refinement edge of all elements

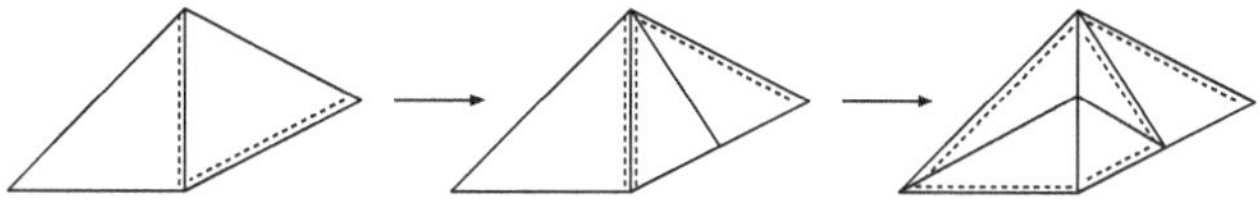

Fig. 19 Recursive refinement of a neighbor. After that the common edge is the refinement edge for both elements

In general, neighboring elements will not share a common refinement edge. However we observe the following in 2d. If the neighbor at an element's refinement edge does not share the same refinement edge, refine *recursively* first the neighbor. After that the refinement edge is shared with the neighboring child as illustrated in Fig. 19. This idea also works for $d > 2$ but involves *all elements* at the common edge and may need *several recursive refinements* of neighbors. Avoiding irregular nodes in higher dimensions is even more convenient than in 2d.

The following lemma is useful for the formulation and the analysis of the recursive refinement algorithm. It uses the notion of *refinement patch* of an element $T = [z_0, \ldots, z_d]_t \in \mathscr{T}$ defined as $R(T, \mathscr{T}) := \{T' \in \mathscr{T} \mid \overline{z_0 z_d} \subset T'\}$.

Lemma 40 (Generation in the Refinement Patch). *Any* $T \in \mathscr{T}$ *is of* locally highest generation *in* $R(\mathscr{T}; T)$*, this means*

$$g(T) = \max\{g(T') \mid T' \in R(T, \mathscr{T})\},$$

and $T' \in R(T, \mathscr{T})$ *is compatible divisible with* T *iff* $g(T') = g(T)$*. Moreover,* $\min\{g(T') \mid T' \in R(T), \mathscr{T}\} \geq g(T) - d + 1$.

The proof can for instance be found in [30, Sect. 4]. It utilizes that *any* uniform refinement of $\mathscr{T}_0$ is conforming. In respect thereof Assumption 36 is essential. We learn that only elements $T' \in R(T, \mathscr{T})$ with $g(T') < g(T)$ have to be refined recursively. Consequently, the maximal depth of recursion is $g(T)$ and recursion terminates. In addition, any element $T' \in R(T, \mathscr{T})$ is compatibly divisible after at most $(d - 1)$ recurrent bisections.

Algorithm 41 (Recursive Refinement of a Single Element). Let $\mathscr{T} \in \mathbb{T}$ and $T \in \mathscr{T}$ to be bisected.

```
function REFINE_RECURSIVE(T, 𝒯)
    do forever
        get refinement patch R(T, 𝒯);
        access T' ∈ R(T, 𝒯) with g(T') = min{g(T'') | T'' ∈ R(T, 𝒯)};
        if g(T') < g(T) then
            𝒯 := REFINE_RECURSIVE(𝒯, T');
        else
            break;
        end if
    end do
```

// Atomic Refinement Operation
get refinement patch $R(T, \mathcal{T})$;
for all $T' \in R(T, \mathcal{T})$ do
    $\mathcal{T} := \big(\mathcal{T} \cup \mathsf{BISECT}(T')\big) \setminus \{T'\}$;
end for
return($\mathcal{T}$);

The recursive refinement of a single element terminates since recursion depth is bounded by $g(T)$ and $\#R(T, \mathcal{T}) \leq C$ with a constant solely depending on $\mathcal{T}_0$. Elements are only bisected to avoid non-conforming situations. Therefore, a call of REFINE_RECURSIVE$(T, \mathcal{T})$ outputs the smallest conforming refinement of $\mathcal{T}$ such that T is bisected.

Algorithm 42 (Recursive Refinement of a Triangulation). Let $\mathcal{T} \in \mathbb{T}$ be a conforming refinement of $\mathcal{T}_0$ and let $\mathcal{M} \subset \mathcal{T}$ be a subset of elements subject to refinement.

function REFINE$(\mathcal{T}, \mathcal{M})$
    for all $T \in \mathcal{M}$ do
        $\mathcal{T} =$ REFINE_RECURSIVE$(T, \mathcal{T})$;
    end for
    return($\mathcal{T}$);

Properties of REFINE_RECURSIVE directly imply the following lemma.

Lemma 43. *The recursive refinement algorithm terminates and outputs the smallest conforming refinement $\mathcal{T}_*$ of $\mathcal{T}$ such that $\mathcal{T}_* \cap \mathcal{M} = \emptyset$.*

Aspects for the Implementation

The implementation of REFINE as stated above seems to be easy. It gets knotty because we have to collect the refinement patch from neighbor information. Furthermore, we have to take care of *shared objects*. During refinement, shared objects have to be identified and new shared objects have to be created (but only once!). Recall that elements of $\mathcal{T}$ *share objects* like

- Coordinates of vertices
- Nodes for storing DOFs of finite element functions located at
  - Vertices
  - Edges
  - Faces (3d)

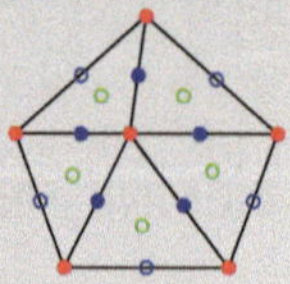

We want to close with the following remarks:

(1) Iterative and recursive refinement output the same grid $\mathcal{T}_*$ whenever both terminate.
(2) Usually, a marked element is bisected more than once, where the natural choice are d bisections.

Coarsening. The standard time-discretization of instationary problems leads to a sequence of single time-steps, where one has to solve a stationary problem. The most common strategy for adapting the grid in the new time-step is to start with the final grid from the last time-step. Since local phenomena may move in time, besides refinement also *coarsening of elements* is needed; compare with Fig. 20 for an example that is taken from [9, 10].

Coarsening of a grid $\mathcal{T}$ is *mainly the inverse operation to refinement* with the following *restriction*: Collect all children in $\mathcal{F}(\mathcal{T})$ that were created in *one atomic refinement operation*. If all these children are *leaves* of $\mathcal{F}(\mathcal{T})$ and if *all* children are marked for coarsening, undo the atomic refinement operation (Fig. 21).

Aspects for the Implementation

When implementing the coarsening operation, hierarchical information given by the associated forest $\mathcal{F}(\mathcal{T})$ dramatically simplifies the collection of all children created by a prior atomic refinement operation.

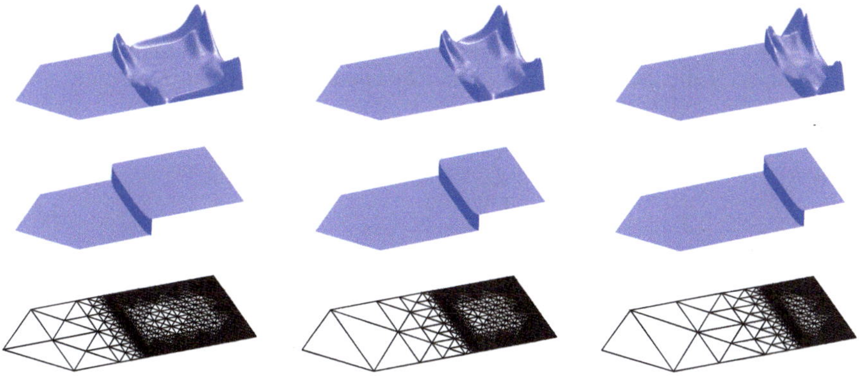

Fig. 20 Graphs of the enthalpy, modulus of the velocity above adaptive grids from a simulation of industrial crystal growth by the vertical Bridgman method for three different time instances

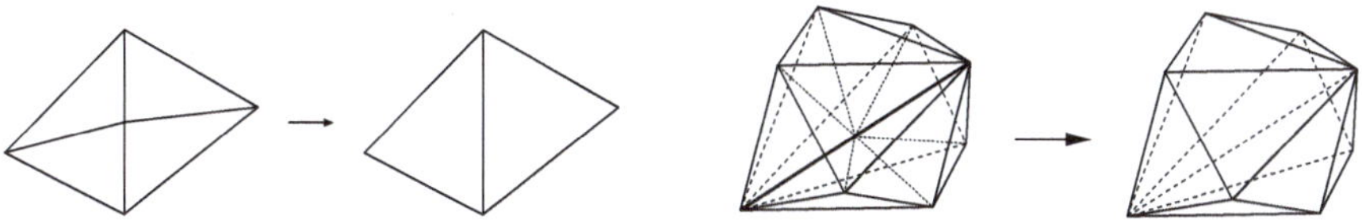

Fig. 21 Atomic coarsening operation in 2d and 3d

3.5 Complexity of Refinement By Bisection

Both variants of REFINE$(\mathscr{T}, \mathscr{M})$ output the smallest conforming refinement $\mathscr{T}_*$ of $\mathscr{T}$ such that $\mathscr{T}_* \cap \mathscr{M} = \emptyset$. Besides elements in $\mathscr{M}$ other elements are bisected in order to ensure conformity of $\mathscr{T}_*$. This raises the important question:

How large is $\mathscr{T}_*$ compared to $\mathscr{T}$ and $\mathscr{M}$?

The only thing we know is $\mathscr{T}_* \le \mathscr{T}_g$, for suitable g. This does not even imply that local refinement stays local! A first guess would be an estimate of the form

$$\#\mathscr{T}_* - \#\mathscr{T} \le C\, \#\mathscr{M}$$

with a constant C only dependent on $\mathscr{T}_0$.

Unfortunately, such an estimate is not true for refinement by bisection. To see this, let for even K

$$\mathscr{M}_k := \{T \in \mathscr{T}_k \mid 0 \in T\} \quad \text{for } k = 0, \dots, K-1$$
$$\mathscr{M}_K := \{T \in \mathscr{T}_K \mid g(T) = K \text{ and } 0 \notin T\}.$$

Examples for $K = 2, 4, 6$ are displayed in Fig. 22. There holds $\#\mathscr{M}_K = 2$ and

$$\#\mathscr{T}_{K+1} - \#\mathscr{T}_K = 4K + 2.$$

This means that in a single step the number of additionally refined elements may be proportional to the maximal level of $\mathscr{T}_K$. Then again for any even K it holds

$$\#\mathscr{T}_{K+1} - \#\mathscr{T}_0 = \sum_{k=0}^{K} (\#\mathscr{T}_{k+1} - \#\mathscr{T}_k) \le 3 \sum_{k=0}^{K} \#\mathscr{M}_k,$$

i. e., there is a chance to estimate the total number of all created elements by the total number of all marked elements. The following result due to Binev, Dahmen, and DeVore in 2d [8] and Stevenson in any dimension [38] confirms this.

Theorem 44 (Complexity of Refinement by Bisection). *Let $\mathscr{T}_0$ satisfy Assumption 36 and consider the set*

$$\mathscr{M} := \bigcup_{k=0}^{K} \mathscr{M}_k$$

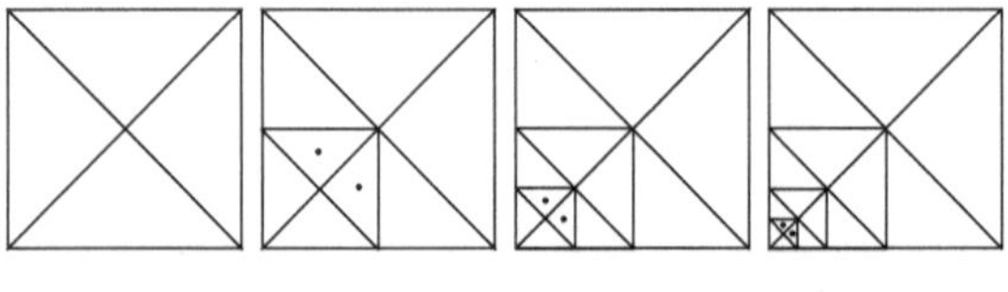

Fig. 22 Macro triangulations and triangulations $\mathscr{T}_K$ for $K = 2, 4, 6$. Elements of $\mathscr{M}_K$ are indicated by a bullet

used to generate the sequence

$$\mathscr{T}_0 \leq \mathscr{T}_1 \leq \cdots \leq \mathscr{T}_{K+1} =: \mathscr{T}.$$

There exists a constant C solely depending on $\mathscr{T}_0$ and d, such that for any $K \geq 0$ holds

$$\#\mathscr{T} - \#\mathscr{T}_0 \leq C \sum_{k=0}^{K} \#\mathscr{M}_k = C\,\#\mathscr{M}.$$

The proof of the theorem is based on the following heuristics:

(1) Assign to each element $T_* \in \mathscr{M}$ fixed amount C_1 of Euros to spend on refined elements, where $\lambda(T, T_*)$ is the portion spent by T_* on T:

$$\sum_{T \in \mathscr{T} \setminus \mathscr{T}_0} \lambda(T, T_*) \leq C_1 \qquad \forall T_* \in \mathscr{M}.$$

(2) The investment of elements in $\mathscr{M}$ is fair in the sense that each refined element gets at least a fixed amount C_2 of Euros:

$$\sum_{T_* \in \mathscr{M}} \lambda(T, T_*) \geq C_2 \qquad \forall T \in \mathscr{T} \setminus \mathscr{T}_0.$$

These assumptions obviously imply

$$C_2(\#\mathscr{T} - \#\mathscr{T}_0) \leq \sum_{T \in \mathscr{T} \setminus \mathscr{T}_0} \sum_{T_* \in \mathscr{M}} \lambda(T, T_*) = \sum_{T_* \in \mathscr{M}} \sum_{T \in \mathscr{T} \setminus \mathscr{T}_0} \lambda(T, T_*) \leq C_1\,\#\mathscr{M},$$

which directly yields $\#\mathscr{T} - \#\mathscr{T}_0 \leq C_1/C_2\,\#\mathscr{M}$.

The actual construction of the allocation function $\lambda\colon \mathscr{T} \times \mathscr{M} \to \mathbb{R}^+$ for conforming refinement by bisection is based on the following properties of the recursive algorithm relying on Assumption 36.

Lemma 45 (Basic Properties of Recursive Bisection). *Let $T \in \mathscr{T}$ and let T' be generated by REFINE_RECURSIVE(T,$\mathscr{T}$). Then there holds*

$$g(T') \leq g(T) + 1$$

and

$$\operatorname{dist}(T, T') \leq D\,2^{1/d} \sum_{g=g(T')}^{g(T)} 2^{-g/d} < D \frac{2^{1/d}}{1 - 2^{-1/d}}\, 2^{-g(T')/d}.$$

Idea of the Proof. The first claim follows from the fact that T is of locally highest generation inside the refinement patch $R(T, \mathscr{T})$. The second claim follows by an induction argument. □

Although the proof of the complexity results utilizes properties of the recursive refinement algorithm, the result holds true for any refinement algorithm outputting the smallest conforming refinement such that marked elements are refined.

3.6 ALBERTA Refinement

ALBERTA utilizes the recursive bisectioning algorithm discussed in Sect. 3.4. There is one important difference to the routine BISECT described in Sect. 3.2:

The ALBERTA refinement edge is the edge between local vertices z_0 and z_1.

Nevertheless, the ALBERTA bisection creates the same descendants by an appropriate labeling of the children's vertices that is consistent with the labeling of the routine BISECT. In fact, ALBERTA bisectioning is equivalent to exchanging the vertices z_1 and z_d of the input element and output elements of BISECT.

The notion of ALBERTA refinement edge gets important when implementing interpolation and restriction routines for coefficient vectors of finite element functions; compare with Sect. 4.4. It is also vital when describing data of the macro triangulation $\mathcal{T}_0$ since the local numbering of the elements' vertices on $\mathcal{T}_0$ determines the shape of any descendant.

Example 46 (ALBERTA Data for a Macro Triangulation). Consider the unit square $\Omega = (0,1)^2 \subset \mathbb{R}^d$ and the *macro triangulation* $\mathcal{T}_0$ built from the *two Kuhn-triangles*, where the main diagonal is the refinement edge for both triangles. Recalling that the ALBERTA refinement edge is the edge between local vertices z_0 and z_1, data of $\mathcal{T}_0$ is given as follows.

```
DIM: 2
DIM_OF_WORLD: 2

number of elements: 2
number of vertices: 4

element vertices:
2 0 1
0 2 3

vertex coordinates:
 0.0 0.0
 1.0 0.0
 1.0 1.0
 0.0 1.0
```

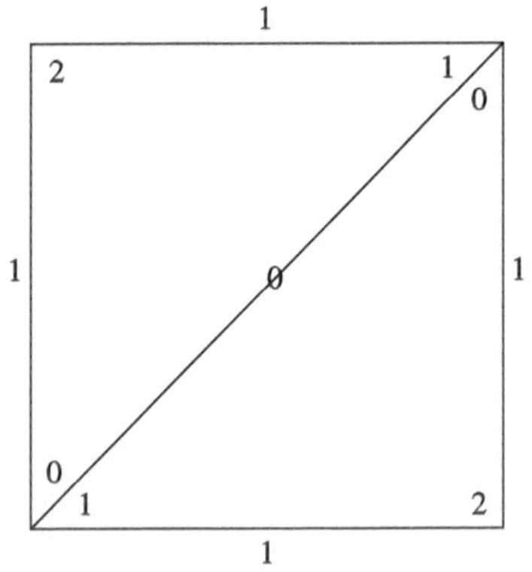

See [34, Sect. 3.2.16] for a detailed documentation of macro triangulations.

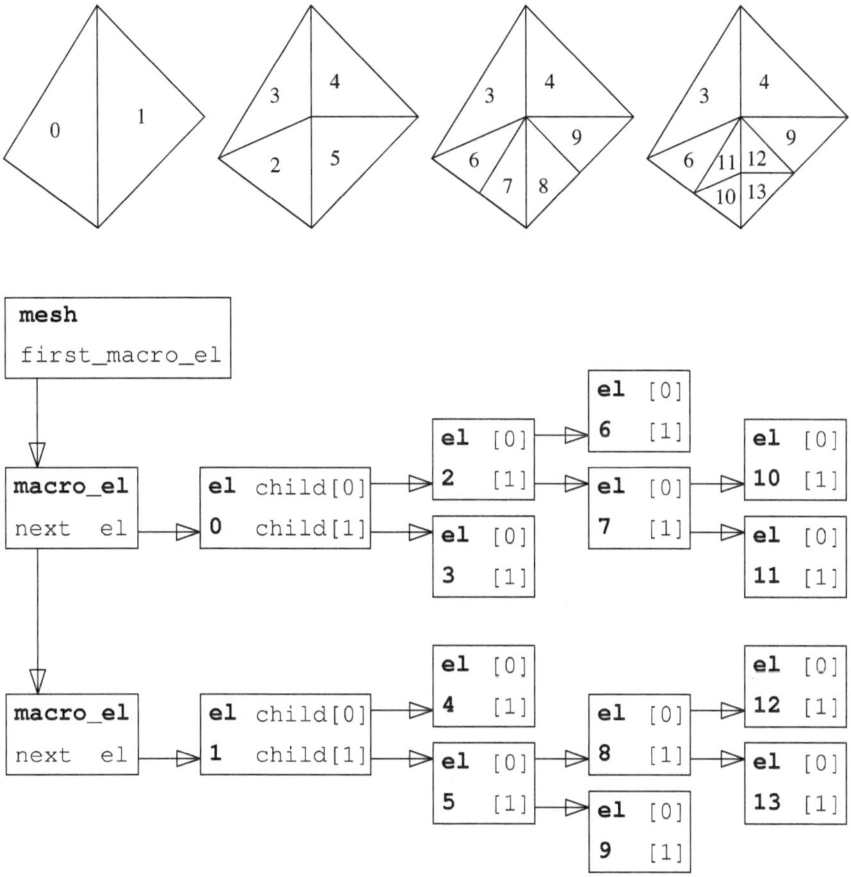

Fig. 23 Local refinement and associated binary trees as used in ALBERTA

3.7 *Mesh Traversal Routines*

As already discussed, refinement by bisection naturally induces the structure of a *binary forest*; compare Fig. 23 for an example. This binary forest can be used in the implementation for a compact storage of a refinement $\mathscr{T}$ of some initial grid $\mathscr{T}_0$ including its refinement history. Element information is split into hierarchical information, i. e., information that can be produced from $\mathscr{F}(\mathscr{T})$, and element specific information, i. e., information that can not be produced from the hierarchy. Only the latter one has to be stored explicitly for each single element of $\mathscr{T}$.

Aspects for the Implementation

Storing a triangulation in data structures reflecting the tree structure has the *following advantages*:

(1) Most part of element information must *not be stored explicitly* but can be produced from the hierarchy:

- Coordinate information
- Neighbor information
- ...

(2) The full tree provides information of the *hierarchical structure* of the sequence of triangulations needed for *multigrid methods*, e. g.,
(3) Coarsening is "easy" to implement. Just collect all leaf elements in the coarsening patch and remove them from the trees.

This way of storing triangulations has the *disadvantage* that there is no *direct access* to elements. Access to elements is only possible through the hierarchy by *mesh traversal routines*.

Mesh traversal routines *loop over elements* of the binary forest and *perform a specified operation*. Mesh traversal routines need following information:

(1) *elements to be visited*: All, leaf elements, ordering.
(2) *operation to be executed* on the selected elements.
(3) *information from the hierarchy* required for the operation on the elements.

A natural implementation of such traversal routines uses *recursion*, but some operations, like refinement, need a *non-recursive* implementation.

We look at an example in ALBERTA, where all leaf elements are marked for n refinements in a recursive and non-recursive implementation. The marker for refinement/coarsening is stored explicitly for all elements. No hierarchical information is needed. For a detailed description of the mesh and element data structures, and the traversal routines we refer to [34, Sects. 3.2.1–3.2.14 and 3.2.19].

Example 47 (Recursive Traversal Routine).

```
static int refine_global_mark;

static void refine_global_fct(const EL_INFO *el_info)
{
  el_info->el->mark = refine_global_mark;
}

static void refine_global(MESH *mesh, int mark)
{
  refine_global_mark = mark;
```

```
    mesh_traverse(mesh, -1, CALL_LEAF_EL, refine_global_fct);

  /*---  now display mesh with element markers ---------------*/
    graphics(mesh, 1);
    refine(mesh);
  /*---  now display refined mesh ----------------------------*/
    graphics(mesh, 1);
    return;
  }
```

Example 48 (Non-Recursive Traversal Routine).

```
  static void refine_global(MESH *mesh, int mark)
  {
    TRAVERSE_STACK *stack = get_traverse_stack();
    const EL_INFO  *el_info;

    el_info = traverse_first(stack, mesh, -1, CALL_LEAF_EL);
    while (el_info)
    {
      el_info->el->mark = mark;
      el_info = traverse_next(stack, el_info);
    }
    free_traverse_stack(stack);

  /*---  now display mesh with element markers ---------------*/
    graphics(mesh, 1);
    refine(mesh);
  /*---  now display refined mesh ----------------------------*/
    graphics(mesh, 1);
    return;
  }
```

In order to use these routines we have to initialize data of some macro triangulation and parameters like the number of global refinements. This is done in the main program.

Example 49 (The Main Program).

```
  int main(int argc, char **argv)
  {
    FUNCNAME("main");
    MESH     *mesh;
    int      n_refine = 2;
    char     line[256];

  /*--- first of all, init parameters of the init file  ------*/
    init_parameters(0, "alberta.dat");
  /*---------------------------------------------------------*/
  /*  get a mesh, and read the macro triangulation from file  */
  /*  name of the macro triangulation defined in alberta.dat  */
  /*---------------------------------------------------------*/
    mesh = GET_MESH("my first mesh", nil, nil);
    GET_PARAMETER(1, "macro trianulation", "%s", line);
    read_macro(mesh, line, nil);
```

```
  GET_PARAMETER(1, "global refinements", "%d", &n_refine);
  refine_global(mesh, n_refine*DIM);
  WAIT;
  return(0);
}
```

The main program initializes some parameters from a parameter file. See [34, Sect. 3.1.4] for a detailed description of parameter files, reading parameters and initializing parameters.

Example 50 (The Parameter File).

```
% Format is key: value
%
macro trianulation:  Macro/macro.amc
global refinements:  3
```

The aim of the next exercises is to get familiar with

(1) *Data structures related to mesh and elements*, for instance MESH, EL_INFO, EL, and MACRO_EL;
(2) *The mesh traversal routines*, in particular the access of elements, information needed on elements, operations performed on elements;
(3) *Data of macro triangulations.*

Exercise 51 (Meshes in 2d and 3d). Implement the following problems as an ALBERTA program. For a description of the MESH, EL and EL_INFO data structures see [34, Sects. 3.2.1–3.2.14] and for the mesh traversal routines [34, Sect. 3.2.19].

(1) Write a function `random_refine(MESH *mesh, int k)` which marks "randomly" chosen elements of `mesh` for refinement. After marking the mesh is refined by a call of `refine(mesh)`. Perform this k times.

Implement a function `random_coarsen(MESH *mesh, int k)` which marks "randomly" chosen elements of `mesh` for coarsening. After marking the mesh is coarsened by a call of `coarsen(mesh)`. Do this also k times.

Finally, write a function `coarse_to_macro(MESH *mesh)` which coarsen a mesh back to the macro triangulation. A triangulation is a macro triangulation, iff for all macro elements `macro_el`

```
macro_el->el->child[0] == nil
```

holds. Macro elements are stored as a linked list with anchor

```
mesh->first_macro_el.
```

Perform repeatedly a random refinement, followed by random coarsening and a final coarsening back to the macro triangulation. Print in each step the numbers of elements, edges, and vertices.

(2) Implement `refine_at_origin(MESH *mesh, REAL dist)` that refines all elements where the distance between the element's barycenter and the origin is at most `dist`.

Hint: Coordinate information must be available on the elements for the calculation of the barycenter. Hence, the traversal routine must be called with `CALL_LEAF_EL|FILL_COORDS` as `FILL_FLAG`. Call this function several times with a decreasing distance `dist` (bisect `dist` in each step, e.g.).

(3) Write a function `measure_omega(MESH *mesh)` that computes the measure of the triangulated domain. This is done by calculating the measure of each element and adding this value to some global variable.

Exercise 52 (Macro Triangulations). For a description of ALBERTA macro triangulations and the `BOUNDARY` data structure compare with [34, Sects. 3.2.16 and 3.2.5].

(1) Produce an ALBERTA macro triangulation file for the L-shaped domain depicted in Fig. 24 (left).
(2) Produce an ALBERTA macro triangulation for the disc depicted in Fig. 24 (right). In order to treat the curved boundaries, the functions

```
void ball_1_proj(REAL_D p);
void ball_2_proj(REAL_D p);
const BOUNDARY *ibdry(MESH *mesh, int bound);
```

have to be implemented. Description:

- `ball_1_proj(p)`: Projects point `p` onto the curved boundary of type `1` (Γ_1) by modifying the coordinate stored in `p`.

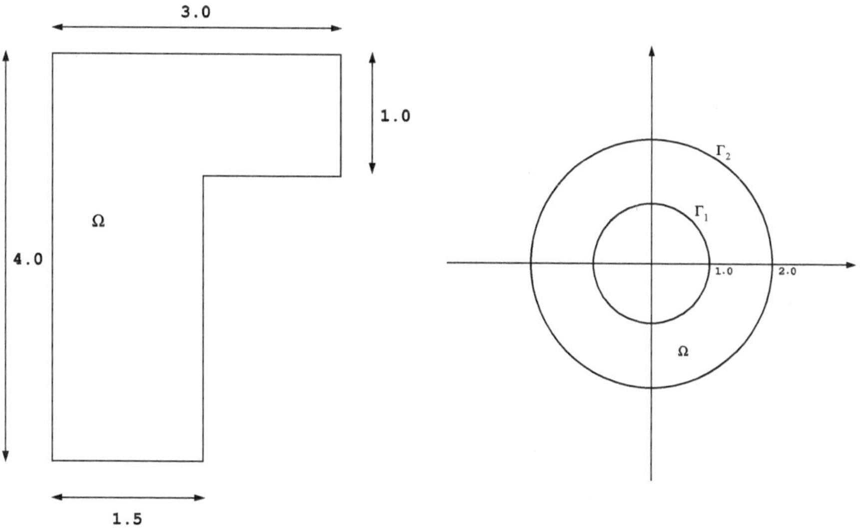

Fig. 24 Shapes of the domains for the macro triangulations: (1) L-shaped domain and (2) disc

- `ball_2_proj(p)`: Projects point `p` onto the curved boundary of type 2 (Γ_2) by modifying the coordinate stored in `p`.
- `ibdry(mesh, bound)`: Initializes the corresponding `BOUNDARY` data structures. A pointer to this function is an argument to the `read_macro()` function.

(3) Produce an ALBERTA macro triangulation file for the unit cube $(0, 1)^3$ built from the 6 Kuhn-simplices in $\mathbb{R}^3$.

Read the newly created macro triangulations with the `read_macro()` function and perform all the refinements and coarsenings implemented in Exercise 51.

How to Get Started. Prior to the installation of ALBERTA the `gltools` (version 2–4) should be installed. The `gltools` can be downloaded from

`http://www.wias-berlin.de/software/gltools/.`

Follow the installation instruction. Then download the ALBERTA library from

`http://www.alberta-fem.de/`

and create the ALBERTA library with the `configure` tools. Examples of typical `configure` commands for Linux and Mac OS X (with Snow Leopard) can be found at

`http://www.ians.uni-stuttgart.de/nmh/downloads/iafem/.`

This site also provides start-up files packed in `alberta.tgz`. Unpacking this archive creates a directory `alberta` with several sub-directories in the actual directory. The sub-directories `src/?d` (with ? = `1`, `2`, or `3`) contain

```
Macro/    Makefile    iafem.dat    ellipt.dat
```

The files `iafem.dat` and `ellipt.dat` are used for parameter declaration of the programs `iafem` and `ellipt`. `ellipt` will be subject of the next exercise. The sub-directories `Macro` contain data for several macro triangulations.

The source files `iafem.c` and `ellipt.c` (and some files for error estimators of the problems in Sect. 6) are stored in the directory `Common`. These source files are used for `1d`, `2d`, and `3d`. Compilation and linking for `?d` executables has to be done in the corresponding `?d` directory using the respective `Makefile` in the `?d` directory. Inside `Makefile` adjust the variable `ALBERTA_LIB_PATH` to the path where the ALBERTA library is installed on your system. The command `make` will then produce the executable `iafem`. Modify the file `iafem.c` in the `Common` directory for these exercises.

4 Assemblage of the Linear System

In this section we discuss the basic principles how to assemble and solve the linear system for computing the Ritz-Galerkin solution.

4.1 The Variational Problem and the Linear System

In general, one has to deal with non-homogeneous boundary data, for instance a given temperature on the boundary.

Example 51 (Elliptic PDE with Non-Homogeneous Boundary Data). The variational formulation of the elliptic PDE

$$-\Delta u = f \quad \text{in } \Omega, \qquad u = g \quad \text{on } \partial\Omega$$

reads: Find $u \in H^1(\Omega)$ such that

$$u_{|\partial\Omega} = g \qquad \text{and} \qquad \langle \nabla v, \nabla u \rangle = \langle f, v \rangle \quad \forall v \in H_0^1(\Omega).$$

A solution belongs to the set $\{w \in H^1(\Omega) \mid w = g \text{ on } \partial\Omega\}$ and this set is non-empty, iff there exists a $\bar{g} \in H^1(\Omega)$ such that $\bar{g}_{|\partial\Omega} = g$. If so, we have the identity

$$\{w \in H^1(\Omega) \mid w = g \text{ on } \partial\Omega\} = \bar{g} + H_0^1(\Omega) := \{w = \bar{g} + v \mid v \in H_0^1(\Omega)\}$$

and the *affine space* $\bar{g} + H_0^1(\Omega)$ does not depend on the particular extension $\bar{g}$ of g. Without loss of generality we denote the extension $\bar{g}$ of boundary data g by g.

We generalize the variational problem including non-homogeneous boundary data.

Problem 54 (Variational Problem for Non-Homogeneous Boundary Data). Let $(\mathbb{V}, \|\cdot\|_{\mathbb{V}})$ be a Hilbert space, $\mathring{\mathbb{V}} \subset \mathbb{V}$ a closed and *non-empty* subspace with dual $\mathring{\mathbb{V}}^*$, $\mathscr{B}\colon \mathbb{V}\times\mathbb{V} \to \mathbb{R}$ a continuous bilinear form, satisfying the inf-sup condition on $\mathring{\mathbb{V}}$

$$\exists \alpha > 0: \qquad \inf_{\substack{v\in\mathring{\mathbb{V}} \\ \|v\|_{\mathbb{V}}=1}} \sup_{\substack{w\in\mathring{\mathbb{V}} \\ \|w\|_{\mathbb{V}}=1}} \mathscr{B}[v, w] \geq \alpha, \qquad \inf_{\substack{w\in\mathring{\mathbb{V}} \\ \|w\|_{\mathbb{V}}=1}} \sup_{\substack{v\in\mathring{\mathbb{V}} \\ \|v\|_{\mathbb{V}}=1}} \mathscr{B}[v, w] \geq \alpha.$$

For $g \in \mathbb{V}$ and $f \in \mathring{\mathbb{V}}^*$ we look for a solution

$$u \in g + \mathring{\mathbb{V}}: \qquad \mathscr{B}[u, v] = \langle f, v \rangle \qquad \forall\, v \in \mathring{\mathbb{V}}, \tag{5}$$

where $g + \mathring{\mathbb{V}} := \{w = g + v \in \mathbb{V} \mid v \in \mathring{\mathbb{V}}\}$.

Lemma 55 (Existence and Uniqueness). *For any pair $(f, g) \in \mathring{\mathbb{V}}^* \times \mathbb{V}$ the variational problem* (5) *admits a unique solution $u \in \mathbb{V}$.*

Proof. We claim that for given $(f, g) \in \mathring{\mathbb{V}}^* \times \mathbb{V}$ there exists a unique solution $u_I \in \mathring{\mathbb{V}}$ of the variational problem

$$u_I \in \mathring{\mathbb{V}}: \qquad \mathscr{B}[u_I, v] = \langle f, v \rangle - \mathscr{B}[g, v] =: \langle F, v \rangle \qquad \forall\, v \in \mathring{\mathbb{V}}.$$

To see this, we observe that $\mathscr{B}$ satisfies an inf-sup condition on $\mathring{\mathbb{V}}$. Furthermore, the mapping F is linear and since $\mathscr{B}$ is continuous on $\mathbb{V}$ we can estimate for any $v \in \mathring{\mathbb{V}}$

$$|\langle F, v\rangle| \leq \|f\|_{\mathbb{V}^*}\|v\|_V + \|\mathscr{B}\|\|g\|_{\mathbb{V}}\|v\|_{\mathbb{V}} \leq (\|f\|_{\mathbb{V}^*} + \|\mathscr{B}\|\|g\|_{\mathbb{V}})\,\|v\|_{\mathbb{V}},$$

which implies $F \in \mathring{\mathbb{V}}^*$. Hence, Theorem 2 implies that u_I uniquely exists. Setting $u := u_I + g$ we see that u is the unique solution of (5). □

Problem 56 (Discrete Problem for Non-Homogeneous Boundary Data). Let $\mathbb{V}_N \subset \mathbb{V}$ be a subspace of dimension $N < \infty$ such that $\mathring{\mathbb{V}}_N := \mathbb{V}_N \cap \mathring{\mathbb{V}}$ is *non-empty* and such that $\mathscr{B}$ satisfies the *discrete inf-sup condition* on $\mathring{\mathbb{V}}_N$

$$\exists \alpha_N > 0 : \qquad \inf_{\substack{V \in \mathring{\mathbb{V}}_N \\ \|V\|_{\mathbb{V}}=1}} \sup_{\substack{W \in \mathring{\mathbb{V}}_N \\ \|W\|_{\mathbb{V}}=1}} \mathscr{B}[V, W] \geq \alpha_N.$$

Let $G \in \mathbb{V}_N$ be an approximation to $g \in \mathbb{V}$ and $f \in \mathring{\mathbb{V}}^*$. Then we look for the discrete solution

$$U \in G + \mathring{\mathbb{V}}_N : \qquad \mathscr{B}[U, V] = \langle f, V\rangle \qquad \forall\, V \in \mathring{\mathbb{V}}_N. \tag{6}$$

Some remarks about the discrete problem are in order.

(1) There is the typical mismatch that boundary data is dicretized whereas we assume that we can evaluate the right hand side for discrete functions exactly. In practice the latter one is not possible and one has to rely on numerical quadrature; compare with Sect. 4.3.
(2) The single inf-sup condition in Problem 56 in combination with dim $\mathring{\mathbb{V}}_N < \infty$ implies the second discrete inf-sup condition

$$\inf_{\substack{W \in \mathring{\mathbb{V}}_N \\ \|W\|_{\mathbb{V}}=1}} \sup_{\substack{V \in \mathring{\mathbb{V}}_N \\ \|V\|_{\mathbb{V}}=1}} \mathscr{B}[V, W] \geq \alpha_N.$$

(3) The constant α_N^{-1} enters in estimates for the condition number of the discrete linear system as well as in a priori error estimates. Consequently, *stable discretizations* with $\alpha_N \geq \underline{\alpha} > 0$ are important. In the course of this the constant $\underline{\alpha}$ is independent of the dimension N.
(4) Coercivity of the bilinear form $\mathscr{B}$ on $\mathring{\mathbb{V}}$ is inherited to any subspace $\mathring{\mathbb{V}}_N \subset \mathring{\mathbb{V}}$. Coercivity therefore implies the continuous and discrete inf-sup condition (with a uniform constant). In general, the continuous inf-sup condition on $\mathring{\mathbb{V}}$ for non-coercive $\mathscr{B}$ does not imply the discrete inf-sup condition on $\mathring{\mathbb{V}}_N$.

Lemma 57 (Existence and Uniqueness). *For any* $(f, G) \in \mathring{\mathbb{V}}^* \times \mathbb{V}_N$ *the discrete problem* (6) *admits a unique solution* $U \in \mathbb{V}_N$.

Proof. Follows exactly the lines of the proof to Lemma 55. □

Structure of the Discrete Linear System. Let $\{\Phi_1, \ldots, \Phi_N\}$ be a *basis* of $\mathbb{V}_N$ such that $\{\Phi_1, \ldots, \Phi_{\mathring{N}}\}$ is a basis of $\mathring{\mathbb{V}}_N$. We use this ordering of basis function related to $\mathring{\mathbb{V}}_N$ just for notational convenience. For unstructured grids it is advantageous that such an ordering is not mandatory. Write

$$U = \sum_{j=1}^{N} u_j \Phi_j \qquad \text{and} \qquad G = \sum_{j=1}^{N} g_j \Phi_j$$

with the *global coefficient vectors*

$$\boldsymbol{u} := \left[u_1, \ldots, u_N\right]^\top \quad \text{and} \quad \boldsymbol{g} := \left[g_1, \ldots, g_N\right]^\top.$$

Then (6) is equivalent to the N *linear equations*

$$\sum_{j=1}^{N} \mathscr{B}[\Phi_j, \Phi_i]\, u_j = \langle f, \Phi_i \rangle \qquad i = 1, \ldots, \mathring{N}$$

$$u_i = g_i \qquad i = \mathring{N} + 1, \ldots, N.$$

Defining the $N \times N$ *system matrix*

$$\boldsymbol{S} := \begin{bmatrix} \mathscr{B}[\Phi_1, \Phi_1] & \ldots & \mathscr{B}[\Phi_{\mathring{N}}, \Phi_1] & \mathscr{B}[\Phi_{\mathring{N}+1}, \Phi_1] & \ldots & \mathscr{B}[\Phi_N, \Phi_1] \\ \vdots & \ddots & \vdots & \vdots & \ddots & \vdots \\ \mathscr{B}[\Phi_1, \Phi_{\mathring{N}}], & \ldots & \mathscr{B}[\Phi_{\mathring{N}}, \Phi_{\mathring{N}}] & \mathscr{B}[\Phi_{\mathring{N}+1}, \Phi_{\mathring{N}}] & \ldots & \mathscr{B}[\Phi_N, \Phi_{\mathring{N}}], \\ 0 & \ldots & 0 & 1 \quad 0 & \ldots & 0 \\ 0 & \ldots & 0 & 0 \quad 1 & \ldots & 0 \\ \vdots & \ddots & 0 & 0 \quad 0 & \ddots & \vdots \\ 0 & \ldots & 0 & 0 \quad 0 & \ldots & 1 \end{bmatrix}$$

and the *right hand side vector*

$$\boldsymbol{f} := \left[\langle f, \Phi_1 \rangle, \ldots, \langle f, \Phi_{\mathring{N}} \rangle, g_{\mathring{N}+1}, \ldots, g_N\right]^\top \in \mathbb{R}^N$$

the discrete variational problem (6) is equivalent to *solving the linear system*

$$\boldsymbol{S}\boldsymbol{u} = \boldsymbol{f} \qquad \text{in } \mathbb{R}^{N \times N}, \tag{7}$$

where $U = \sum_{j=1}^{N} u_j \Phi_j \in \mathbb{V}_N$ is the solution to (6).

Lemma 58. *The system matrix* $\boldsymbol{S}$ *is invertible, thus* (7) *admits a unique solution* $\boldsymbol{u}$.

Proof. Equation (7) is equivalent to (6) which has for any (f, G) a unique solution. □

Aspects for the Implementation

Let $\mathbb{V}_N$ be a piecewise polynomial FE space, $\{\Phi_i\}$ the Lagrange basis, and let $g \in C^0(\partial\Omega)$. The most convenient choice for boundary data g_i is $g_i = g(z_i)$ with $z_i \in \partial\Omega$ for $i = \mathring{N} + 1, \dots, N$. Therefore, we only use boundary data g on $\partial\Omega$ and do not rely on any extension $\bar{g} \in H^1(\Omega)$. This is a highly important aspect for applications.

4.2 Assemblage: The Outer Loop

We next discuss basic principles of assembling the system matrix $\boldsymbol{S}$ and the load vector $\boldsymbol{f}$. We derive these principle from the 2nd order elliptic PDE

$$-\operatorname{div}(\boldsymbol{A}(x)\nabla u) + \boldsymbol{b}(x) \cdot \nabla u + c(x)\, u = f \quad \text{in } \Omega, \qquad u = g \quad \text{on } \partial\Omega.$$

We pose standard assumption on data:

- a bounded domain $\Omega \subset \mathbb{R}^d$ triangulated by some conforming triangulation $\mathscr{T}$;
- bounded coefficient functions $\boldsymbol{A} \in L^\infty(\Omega; \mathbb{R}^d)$, $\boldsymbol{b} \in L^\infty(\Omega; \mathbb{R}^d)$, and $c \in L^\infty(\Omega)$;
- a load function $f \in L^2(\Omega)$;
- boundary values $g \in C^0(\partial\Omega) \cap H^1(\Omega)$.

For the variational formulation we set $\mathbb{V} = H^1(\Omega)$ and $\mathring{\mathbb{V}} = H_0^1(\Omega)$ and we assume structural assumption on the coefficients that imply coercivity of the bilinear form $\mathscr{B}$ on $\mathring{\mathbb{V}}$, where $\mathscr{B}\colon \mathbb{V} \times \mathbb{V} \to \mathbb{R}$ is defined as

$$\mathscr{B}[w, v] := \int_\Omega \nabla v \cdot \boldsymbol{A}\nabla w + v\,\boldsymbol{b} \cdot \nabla w + v\,c\,w\,dx \qquad \forall v, w \in \mathbb{V};$$

compare with Example 3. Finally, we set

$$\langle f, v\rangle = \int_\Omega f\,v\,dx \qquad \forall v \in \mathbb{V}.$$

Aspects for the Implementation

Note, that the first argument w of the bilinear form always appears as the last factor of the addends inside the integral in the definition of $\mathscr{B}$. For non-symmetric $\mathscr{B}$ a typical mistake is to assemble $\boldsymbol{S}^\top$ instead of $\boldsymbol{S}$!

Assemblage of the System Matrix. Assume that v or w has local support in $\omega \subset \Omega$. Then

$$\mathscr{B}[w, v] = \int_{\omega} \nabla v \cdot \boldsymbol{A} \nabla w + v\, \boldsymbol{b} \cdot \nabla w + v\, c\, w\, dx =: \mathscr{B}_{\omega}[w, v].$$

Therefore, for $i = 1, \dots, \mathring{N}$ the jth entry in $\boldsymbol{S}$ is

$$S_{ij} = \mathscr{B}[\Phi_j, \Phi_i] = \mathscr{B}_{\omega_{ij}}[\Phi_j, \Phi_i] \qquad j = 1, \dots, N,$$

where $\omega_{ij} = \operatorname{supp}(\Phi_i) \cap \operatorname{supp}(\Phi_j)$. In addition,

$$\omega_{ij} = \emptyset \qquad \Longrightarrow \qquad S_{ij} = 0.$$

By construction, finite element basis functions are locally supported and the support of finite element basis functions is restricted to only few elements of $\mathscr{T}$. Hence,

$$C_i := \#\{\Phi_j \mid \operatorname{supp}(\Phi_i) \cap \operatorname{supp}(\Phi_j) \neq \emptyset\} \leq C,$$

where C only depends on shape regularity of $\mathscr{T}$. This implies

$$\max_{i=1,\dots,\mathring{N}} \#\{S_{ij} \neq 0 \mid j = 1, \dots, N\} \leq C,$$

i.e., the system matrix $\boldsymbol{S}$ is sparse.

Aspects for the Implementation

(1) The system matrix $\boldsymbol{S}$ is *sparse* and has to be assembled and stored with a complexity proportional to N:

$$\#\{S_{ij} \neq 0 \mid i = 1, \dots, \mathring{N},\, j = 1, \dots, N\} \leq C\, N.$$

(2) The number of entries per row strongly depends on $\bar{\mathbb{P}}$, $\mathscr{T}$ and the dimension. The values C_i may vary strongly!

For a refinement of the standard initial triangulation in 2d typical values are

- $\bar{\mathbb{P}} = \mathbb{P}_1$: $C_i = 9$
- $\bar{\mathbb{P}} = \mathbb{P}_2$: $C_i = 25$ for a basis function Φ_i at a vertex, $C_i = 9$ for a basis function Φ_i at an edge.

These variations get more pronounced for higher polynomial degree and higher space dimensions.

Consequently, sparse matrices need special data structures for efficient storage and access.

We next turn to the element-wise assemblage of the system matrix. Additivity of integrals allows us to write

$$S_{ij} = \mathscr{B}[\Phi_j, \Phi_i] = \sum_{\substack{T\in\mathscr{T} \\ T\subset \operatorname{supp}(\Phi_i)\cap\operatorname{supp}(\Phi_j)}} \mathscr{B}_T[\Phi_j, \Phi_i]$$

For the computation of S_{ij} we initialize $S_{ij} = 0$, loop over $T \in \mathscr{T}$ with $T \subset \operatorname{supp}(\Phi_i) \cap \operatorname{supp}(\Phi_j)$, compute

$$S_{ij\,|T} = \mathscr{B}_T[\Phi_j, \Phi_i],$$

and add $S_{ij\,|T}$ to S_{ij}. The *drawback* of this approach is that a loop over $\mathscr{T}$ for all S_{ij} destroys the linear complexity since this is a "from global to local approach" and for given (i, j) there is no information about $T \subset \operatorname{supp}(\Phi_i) \cap \operatorname{supp}(\Phi_j)$ available.

The "from local to global alternative" is the following approach: Set $\boldsymbol{S} := 0$, loop over $T \in \mathscr{T}$,

$$\forall i, j \text{ with } T \subset \operatorname{supp}(\Phi_i) \cap \operatorname{supp}(\Phi_j) \quad \text{add} \quad S_{ij} = S_{ij} + \mathscr{B}_T[\Phi_j, \Phi_i].$$

Linear complexity can be preserved since global basis functions are built from local basis functions. This means, on T we have precise information about (i, j) with $T \subset \operatorname{supp}(\Phi_i) \cap \operatorname{supp}(\Phi_j)$ given by the index mapping

$$I\colon \{1, \dots, \bar{n}\} \times \mathscr{T} \to \{1, \dots, N\}.$$

To be more precise: Let $\{\bar{\phi}_n\}_{n=1,\dots,\bar{n}}$ be a basis of $\bar{\mathbb{P}}$ and $\{\Phi_n^T = \bar{\phi}_i \circ \lambda^T\}_{n=1,\dots,\bar{n}}$ the transformed basis of $\mathbb{P}(T)$. For any global basis function Φ_i with $T \subset \operatorname{supp}(\Phi_i)$ there exists a unique local index n such that $\Phi_{i\,|T} = \Phi_n^T$, where Φ_n^T is a local basis function on T. The relation between i and n is given by $i = I(n, T)$.

Summarizing these ideas, we first compute on $T \in \mathscr{T}$ the *element system matrix*

$$\boldsymbol{S}^{(T)} := \left[S_{nm}^{(T)}\right]_{n,m=1,\dots,\bar{n}} := \left[\mathscr{B}_T[\bar{\phi}_m \circ \lambda^T, \bar{\phi}_n \circ \lambda^T]\right]_{n,m=1,\dots,\bar{n}}.$$

Then we access for each *local index pair* $(n, m) \in \{1, \dots, \bar{n}\}^2$ the *global index pair* $(I(n,T), I(m,T)) \in \{1, \dots, N\}^2$ and add $S_{nm}^{(T)}$ to $S_{I(n,T),I(m,T)}$.

In addition, we have to include entries related to boundary nodes, i. e., basis functions Φ_i with $i \in \mathring{N} + 1, \dots, N$. This gives the algorithm for assembling the system matrix.

Algorithm 59 (Assemblage of the System Matrix).

set $\boldsymbol{S} := \boldsymbol{0}$;
for all $T \in \mathcal{T}$ do
  for all $n \in \{1, \ldots, \bar{n}\}$ do
    get global index $i = I(n, T)$
    if $\Phi_i \in \mathring{\mathbb{V}}(\mathcal{T})$ then
      for all $m \in \{1, \ldots, \bar{n}\}$ do
        compute $S_{nm}^{(T)}$ and get global index $j = I(m, T)$;
        $S_{ij} := S_{ij} + S_{nm}^{(T)}$;
      end for
    else
      $S_{ii} := 1$; //Boundary node
    end if
  end for
end for

Aspects for the Implementation

Recall that the number of entries per row in $\boldsymbol{S}$ is small, not constant, and strongly depends on the used finite element space $\mathbb{V}(\mathcal{T})$. Therefore, $\boldsymbol{S}$ has to be stored as a sparse matrix. There are basically two ways to dynamically handle sparse matrices.

(1) Before assembling $\boldsymbol{S}$ compute the length of all rows and allocate corresponding memory.
   - *Pro:* Allows a realization as vector yielding efficient memory access.
   - *Con:* Not easy to implement for general finite element spaces.

(2) The operation $\boldsymbol{S} := \boldsymbol{0}$ removes all entries from an existing matrix. Elements are dynamically allocated when needed, i. e., when adding $S_{nm}^{(T)}$ to a non-existing entry S_{ij}.
   - *Pro:* Easy to implement for general finite element spaces.
   - *Con:* Realization does not yield the most efficient memory access.

Assemblage of the Load Vector. Recall the load vector $\boldsymbol{f}$

$$\boldsymbol{f} := \left[\langle f, \Phi_1\rangle, \ldots, \langle f, \Phi_{\mathring{N}}\rangle, g_{\mathring{N}+1}, \ldots, g_N\right]^{\top} \in \mathbb{R}^N.$$

For $i = 1, \ldots, \mathring{N}$, the values

$$f_i = \langle f, \Phi_i \rangle = \int_\Omega f\,\Phi_i\,dx = \sum_{\substack{T\in\mathcal{T}\\ T\subset \operatorname{supp}(\Phi_i)}} \int_T f\,\Phi_i\,dx$$

are also computed *element-wise* with the *local basis functions* by

$$f_n^{(T)} := \int_T f\,\Phi_{I(n,T)}\,dx = \int_T f\,(\bar{\phi}_n \circ \lambda^T)\,dx$$

As values g_i, $i = \mathring{N} + 1, \dots, N$, we take the coefficients of the *Lagrange interpolant* $I_{\mathcal{T}} g$. These coefficients are easy to compute and just need the evaluation of boundary data g at the corresponding Lagrange node.

Algorithm 60 (Assemblage of the Load Vector).

set $\boldsymbol{f} := \mathbf{0}$;
for all $T \in \mathcal{T}$ do
    for all $n \in \{1, \dots, \bar{n}\}$ do
        get global index $i = I(n, T)$
        if $\Phi_i \in \mathring{\mathbb{V}}(\mathcal{T})$ then
            compute $f_n^{(T)}$;
            $f_i := f_i + f_n^{(T)}$;
        else
            compute coefficient g_n of Lagrange interpolant;
            $f_i := g_n$;
        end if
    end for
end for

4.3 *Assemblage: Element Integrals*

In Algorithms 59 and 60 we have assumed that we can compute the element contributions $S_{nm}^{(T)}$ and $f_N^{(T)}$. We next elaborate on the issue how to actually compute element integrals. We restrict ourselves to the model problem and recall its bilinear form

$$\mathcal{B}[w, v] = \int_\Omega \nabla v \cdot \boldsymbol{A} \nabla w + v\,\boldsymbol{b} \cdot \nabla w + v\,c\,w\,dx \qquad \forall v, w \in \mathbb{V}.$$

Assuming that data $\boldsymbol{A}$, $\boldsymbol{b}$ and c is piecewise constant over $\mathcal{T}$ the computation of the element system matrix

$$S_{nm}^{(T)} = \mathcal{B}_T[\Phi_n^T, \Phi_m^T] = \int_T \nabla \Phi_m^T \cdot \boldsymbol{A} \nabla \Phi_n^T + \Phi_m^T\,\boldsymbol{b} \cdot \nabla \Phi_n^T + \Phi_m^T\,c\,\Phi_n^T\,dx$$

involves only a polynomial of degree $\leq 2p$. The following lemma states that any element integral involving only a polynomial can be computed exactly.

Lemma 61 (Integrals of Polynomials). *Let T be a simplex and $\lambda = \lambda^T$ the barycentric coordinates on T. Then for any multi-index $\alpha \in \mathbb{N}_0^{d+1}$ there holds*

$$\int_T \lambda^\alpha(x)\,dx = \frac{\alpha!}{(|\alpha|+d)!}\,|\det DF_T| = \frac{\alpha!\,d!}{(|\alpha|+d)!}\,|T|,$$

where $\lambda^\alpha = \lambda_0^{\alpha_0}\cdot\dots\cdot\lambda_d^{\alpha_d}$, $|\alpha| = \alpha_0+\dots+\alpha_d$, $\alpha! = \alpha_0!\cdot\dots\cdot\alpha_d!$.

Unfortunately, the result is more useful for symbolic computations rather then for numerical ones. Nevertheless we learn that, up to scaling by $|T|$, integrals of basis functions do not depend on T. This easily follows from the construction of basis functions in terms of barycentric coordinates. In general, element integrals involve general functions, like $\boldsymbol{A}$, $\boldsymbol{b}$, c, and f. Such integrals cannot be computed exactly without additional knowledge. For such functions we have to use numerical integration.

Definition 62 (Quadrature Formula). A *numerical quadrature formula* $\hat{Q}$ on $\hat{T}$ is a set

$$\{(w_\ell,\lambda_\ell)\in\mathbb{R}\times\mathbb{R}^{d+1}\mid \ell=1,\dots,L\}$$

of *weights* w_ℓ and *quadrature points* $\lambda_\ell \in \bar{T}$ such that

$$\int_{\hat{T}} g(\hat{x})\,d\hat{x} \approx \hat{Q}(g) := \sum_{\ell=1}^{L} w_\ell f(\hat{x}(\lambda_\ell)).$$

It is called *exact of degree p* for some $p \in \mathbb{N}$ if

$$\int_{\hat{T}} P(\hat{x})\,d\hat{x} = \hat{Q}(P) \qquad \text{for all } P\in\mathbb{P}_p.$$

It is called *stable* if $w_\ell > 0$ for all $\ell = 1,\dots,L$.

In view of the notion of exactness, Lemma 61 may be useful to derive numerical quadrature formulas of any order. In general, such formulas are not stable. Definition 62 defines a quadrature formula on $\hat{T}$. Using the affine mapping $F_T\colon\hat{T}\to T$ we can derive a quadrature rule for an arbitrary simplex T. Assume a given simplex T and a given function $g\colon T\to\mathbb{R}$, for which we want to compute an approximation to

$$\int_T g(x)\,dx.$$

Utilizing the *transformation rule* we deduce from a given quadrature rule $\hat{Q}$ on $\hat{T}$

$$\int_T g(x)\,dx = |\det DF_T| \int_{\hat{T}} (g \circ F_T)(\hat{x})\,d\hat{x} \approx |\det DF_T|\, \hat{Q}(g \circ F_T)$$
$$= |\det DF_T| \sum_{\ell=1}^{L} w_\ell g(F_T(\hat{x}(\lambda_\ell))) = |\det DF_T| \sum_{\ell=1}^{L} w_\ell g(x^T(\lambda_\ell)) =: Q_T(g).$$

Therefore, given a quadrature rule $\hat{Q}$ on $\hat{T}$ we can construct a quadrature rule Q_T on T. If $\hat{Q}$ is exact on $\mathbb{P}_p$ then Q_T is also exact on $\mathbb{P}_p$. Furthermore, if $\lambda_\ell \in \bar{T}$ for all $1 \le \ell \le L$ only values of g on T are involved. Finally, continuity of g on T is required in order to use numerical integration. Consequently, we assume from now on that data is piecewise continuous over $\mathscr{T}$.

Approximating the Load Vector. Given a fixed quadrature formula $\hat{Q}$ we compute on $T \in \mathscr{T}$:

$$f_n^{(T)} := Q_T\big(f\,(\bar{\phi}_n \circ \lambda^T)\big) = |\det DF_T| \sum_{\ell=1}^{L} w_\ell f(x^T(\lambda_\ell))\,\bar{\phi}_n(\lambda_\ell) \approx \int_T f\,(\bar{\phi}_n \circ \lambda^T)\,dx.$$

Note, that assuming that we can evaluate f at any given point $x \in \mathbb{R}^d$ the quantity $Q_T\big(f\,(\bar{\phi}_n \circ \lambda^T)\big)$ is fully computable and leads to a practical algorithm for the approximation of the load vector.

Aspects for the Implementation

(1) The values $\bar{\phi}(\lambda_\ell)$ *do not depend on* T. They only have to be computed once for each pair of quadrature formula $\hat{Q}$ and local basis $\{\bar{\phi}_n\}_{n=1,\dots\bar{n}}$.
(2) The evaluation $f(x^T(\lambda_\ell))$ requires to convert *local coordinates* λ_ℓ into *world coordinates* $x^T(\lambda_\ell)$ on T (this is fast!).
(3) The values $f(x^T(\lambda_\ell))$ should be computed first for all quadrature nodes. These computed values can then be used in combination with the precomputed values of the basis functions $\bar{\phi}_n$.

We next formulate the algorithm for assembling the load vector, where element integrals are computed using numerical quadrature.

Algorithm 63 (Assemblage of the Load Vector With Quadrature). Let $\hat{Q}$ be a numerical quadrature on $\hat{T}$ and $\{\bar{\phi}_n\}_{n=1,\dots\bar{n}}$ a basis of $\bar{\mathbb{P}}$.

compute $\bar{\phi}_n(\lambda_\ell)$ for $n = 1,\dots,\bar{n}$, $\ell = 1,\dots,L$;
set $\boldsymbol{f} := \mathbf{0}$;
for all $T \in \mathscr{T}$ do
  compute $f(x^T(\lambda_\ell))$ for $\ell = 1,\dots,L$;
  for all $n \in \{1,\dots,\bar{n}\}$ do

```
        get global index i = I(n, T)
        if Φ_i ∈ V̊(𝒯) then
            compute f_n^(T) = Q_T(f (φ̄_n ∘ λ^T));
            f_i := f_i + f_n^(T);
        else
            compute coefficient g_n of Lagrange interpolant;
            f_i := g_n;
        end if
    end for
end for
```

Approximating the System Matrix. Following the ideas developed for approximating the load vector we use numerical quadrature for the approximation of the element system matrix

$$
\begin{aligned}
S_{nm}^{(T)} &= \int_T \nabla\Phi_m^T \cdot \boldsymbol{A}\nabla\Phi_n^T \, dx + \int_T \Phi_m^T \, \boldsymbol{b}\cdot\nabla\Phi_n^T \, dx + \int_T \Phi_m^T \, c \, \Phi_n^T \, dx \\
&\approx Q_T^2\big(\nabla\Phi_m^T \cdot \boldsymbol{A}\nabla\Phi_n^T\big) + Q_T^1\big(\Phi_m^T \, \boldsymbol{b}\cdot\nabla\Phi_n^T\big) + Q_T^0\big(\Phi_m^T \, c \, \Phi_n^T\big),
\end{aligned}
$$

where $\hat{Q}^2$, $\hat{Q}^1$, $\hat{Q}^0$ are given quadrature formulas on $\hat{T}$ that may differ. We want to remark that the last line is fully computable provided that we can evaluate the coefficient functions at given world coordinates $x \in \mathbb{R}^d$.

Aspects for the Implementation

(1) The choice $\hat{Q}^2 = \hat{Q}^1 = \hat{Q}^0$ in general minimizes computational work. Nevertheless, it is beneficial to have the potential to use different formulas.
(2) Let $\boldsymbol{A}$, $\boldsymbol{b}$, and c be constant on T, $\bar{\mathbb{P}} = \mathbb{P}_p$, and assume that $\hat{Q}^\ell$ is exact on $\mathbb{P}_{2p-\ell}$. Then the element integrals are computed exactly.

For an efficient use of numerical quadrature in the approximation of the system matrix we have to identify those values that depend on the actual simplex T and those values that are independent of T. Concerning this matter we recall the computation of derivatives

$$\nabla\Phi_n^T(x) = \Lambda^\top \nabla_\lambda \bar{\phi}_n(\lambda(x))$$

with the Jacobian Λ of the barycentric coordinates. Defining the element coefficient functions

$$\begin{aligned}
\boldsymbol{A}_T &:= |\det DF_T|\, \Lambda \boldsymbol{A}_{|T} \Lambda^{\top}: T \to \mathbb{R}^{(d+1)\times(d+1)},\\
\boldsymbol{b}_T &:= |\det DF_T|\, \Lambda \boldsymbol{b}_{|T}: T \to \mathbb{R}^{d+1},\\
c_T &:= |\det DF_T|\, c_{|T}: T \to \mathbb{R}
\end{aligned}$$

we deduce for a given be a quadrature formula $\hat{Q}$ on $\hat{T}$

$$Q_T\big(\nabla\Phi_m^T \cdot \boldsymbol{A}\nabla\Phi_n^T\big) = \sum_{\ell=1}^{L} w_\ell \left(\nabla_\lambda\bar{\phi}_m(\lambda_\ell)\cdot \boldsymbol{A}_T(x^T(\lambda_\ell)\,\nabla_\lambda\bar{\phi}_n(\lambda_\ell)\right)$$

$$Q_T\big(\Phi_m^T\,\boldsymbol{b}\cdot\nabla\Phi_n^T\big) = \sum_{\ell=1}^{L} w_\ell \left(\bar{\phi}_m(\lambda_\ell)\,\boldsymbol{b}_T\big(x^T(\lambda_\ell)\big)\cdot\nabla_\lambda\bar{\phi}_n(\lambda_\ell)\right)$$

$$Q_T\big(\Phi_m^T\,c\,\Phi_n^T\big) = \sum_{\ell=1}^{L} w_\ell \left(\bar{\phi}_m(\lambda_\ell)\,c_T\big(x^T(\lambda_\ell)\big)\,\bar{\phi}_n(\lambda_\ell)\right).$$

Thereby all dependence on data and T is shifted into the element coefficient functions $\boldsymbol{A}_T$, $\boldsymbol{b}_T$, and c_T. These functions have to be evaluated at all quadrature nodes on T. All other values such as $\bar{\phi}_m(\lambda_\ell)$, $\nabla_\lambda\bar{\phi}_m(\lambda_\ell)$, etc. do not depend on T and they only have to be computed once.

If the coefficient functions $\boldsymbol{A}$, $\boldsymbol{b}$ and c are piecewise constant over $\mathscr{T}$ we can make the computation of the element system matrix even more efficient. In this case we obtain

$$Q_T\big(\Phi_m^T\,c\,\Phi_n^T\big) = c_T\big(x^T(\lambda_1)\big)\sum_{\ell=1}^{L} w_\ell \left(\bar{\phi}_m(\lambda_\ell)\,\bar{\phi}_n(\lambda_\ell)\right) = c_T\big(x^T(\lambda_1)\big)\,\hat{Q}\big(\bar{\phi}_m\,\bar{\phi}_n\big),$$

$$\begin{aligned}
Q_T\big(\Phi_m^T\,\boldsymbol{b}\cdot\nabla\Phi_n^T\big) &= \boldsymbol{b}_T\big(x^T(\lambda_1)\big)\cdot\sum_{\ell=1}^{L} w_\ell \left(\bar{\phi}_m(\lambda_\ell)\,\nabla_\lambda\bar{\phi}_n(\lambda_\ell)\right)\\
&= \boldsymbol{b}_T\big(x^T(\lambda_1)\big)\cdot\hat{Q}\big(\bar{\phi}_m\,\nabla_\lambda\bar{\phi}_n\big),
\end{aligned}$$

and

$$Q_T\big(\nabla\Phi_m^T\cdot\boldsymbol{A}\nabla\Phi_n^T\big) = \boldsymbol{A}_T\big(x^T(\lambda_1)\big)\otimes\left[\hat{Q}\big(\bar{\phi}_{m,\lambda_i}\,\bar{\phi}_{n,\lambda_j}\big)\right]_{i,j=0,\dots,d}$$

with an appropriate notion of the symbol $\otimes$. The values $\hat{Q}\big(\bar{\phi}_m\,\bar{\phi}_n\big)$, $\hat{Q}\big(\bar{\phi}_m\,\bar{\phi}_{n,\lambda_j}\big)$, and $\hat{Q}\big(\bar{\phi}_{m,\lambda_i}\,\bar{\phi}_{n,\lambda_j}\big)$ do not depend on T. They only have to be computed once for each pair of quadrature formula $\hat{Q}$ and local basis $\{\bar{\phi}_n\}_{n=1,\dots,\bar{n}}$. Using suitable quadrature formulas the corresponding integrals can be computed exactly.

Collecting the above ideas for computing in Algorithm 59 the element contributions $S_{nm}^{(T)}$ with numerical quadrature we obtain a fully practical algorithm as Algorithm 63 for the load vector.

Aspects for the Implementation

The system matrix can be computed fully automatically for a general type of PDE and for different kind of finite element spaces by collecting all element dependence in the element functions like

$$\boldsymbol{A}_T := |\det DF_T| \, \Lambda \boldsymbol{A}_{|T} \Lambda^\top, \quad \boldsymbol{b}_T := \Lambda \, |\det DF_T| \, \boldsymbol{b}_{|T}, \quad c_T := |\det DF_T| \, c_{|T}.$$

The element system matrix can be computed efficiently using the following rules:

(1) The values $\bar{\phi}_n(\lambda_\ell)$ and $\bar{\phi}_{m,\lambda_j}(\lambda_\ell)$ do not depend on T. They only have to be computed once for each pair of quadrature formula $\hat{Q}$ and local basis $\{\bar{\phi}_n\}_{n=1,\ldots,\bar{n}}$.
(2) Values of variable coefficients should be stored for all quadrature nodes. The stored values can then be used in combination with the precomputed values of all local basis functions.
(3) If $\boldsymbol{A}$, $\boldsymbol{b}$, and $\boldsymbol{c}$ are piecewise constant the values $\hat{Q}\left(\bar{\phi}_m\,\bar{\phi}_n\right)$, $\hat{Q}\left(\bar{\phi}_n\,\bar{\phi}_{m,\lambda_j}\right)$, and $\hat{Q}\left(\bar{\phi}_{n,\lambda_i}\,\bar{\phi}_{m,\lambda_j}\right)$, do not depend on T. They can be computed once and the computation of the element system matrix becomes very efficient. In addition, one can choose a quadrature formula such that these integrals are evaluated exactly.

On the Choice of the Quadrature. The starting point for the a priori error analysis including *variational crimes* like numerical integration is the first Strang Lemma. A detailed analysis can be found for instance in [16, Sect. 4].

We shortly summarize the most important results for the model problem. From the first Strang Lemma we obtain optimal order a priori error estimates provided

(1) The consistency error induced by numerical quadrature is of the same order as the approximation error;
(2) The discrete bilinear form is coercive with a constant independent of $\mathscr{T}$.

Using a quadrature formula $\hat{Q}$ that is exact on $\mathbb{P}_{2p-2}$ for the element system matrix and element load vector results in optimal order a priori error estimates of the consistency error for both the system matrix and the load vector.

For the model problem with $\boldsymbol{\xi} \cdot \boldsymbol{A}(x)\boldsymbol{\xi} \geq \alpha \, |\boldsymbol{\xi}|_2^2$ for all $\boldsymbol{\xi} \in \mathbb{R}^d$ and $x \in \Omega$, $\boldsymbol{b} = 0$, and $c \geq 0$ in Ω coercivity of $\mathscr{B}$ on $\mathring{\mathbb{V}}$ is derived by

$$\mathscr{B}[v,v] = \int_\Omega \nabla v \cdot \boldsymbol{A} \nabla v + c\, v^2 \, dx \geq \int_\Omega \alpha \, |\nabla v|_2^2 + c\, v^2 \, dx \geq \alpha \|\nabla v\|_{L^2(\Omega)}^2$$

in combination with Friedrich's inequality. The same procedure works for the discrete bilinear form provided $\hat{Q}$ is exact on $\mathbb{P}_{2p-2}$ and all weights are non-negative, i. e., $\hat{Q}$ is stable. The key point is that we use point-wise properties of $\boldsymbol{A}$ and c in combination with the fact that for $V \in \mathbb{V}(\mathscr{T})$ we have $\nabla V_{|T} \in \mathbb{P}_{2p-2}$ and therefore $Q_T(|\nabla V|^2) = \|\nabla V\|^2_{L^2(T)}$.

The proof of coercivity of $\mathscr{B}$ for $\boldsymbol{b} \not\equiv 0$ with $\operatorname{div} \boldsymbol{b} \equiv 0$ utilizes in a first step integration by parts to show

$$\int_\Omega v\boldsymbol{b} \cdot \nabla w \, dx = \frac{1}{2} \int_\Omega v\boldsymbol{b} \cdot \nabla w \, dx - \frac{1}{2} \int_\Omega w\boldsymbol{b} \cdot \nabla v \, dx$$

Therefore, we see that the first order term is skew symmetric. In particular this implies $\mathscr{B}[v, v] = \int_\Omega \nabla v \cdot \boldsymbol{A} \nabla v + c\, v^2 \, dx$ and we can proceed as above.

Including effects of numerical integration we realize that skew symmetry of the first order term in $\mathscr{B}$ is a consequence of a global argument in combination with the point-wise property $\operatorname{div} \boldsymbol{b} \equiv 0$. Integration by parts does not transfer to numerical integration and coercivity of the discrete bilinear form is not clear. To overcome this problem we use integration by parts to derive the equivalent representation of $\mathscr{B}$, namely

$$\mathscr{B}[w, v] = \int_\Omega \nabla v \cdot \boldsymbol{A} \nabla w + \tfrac{1}{2} v\boldsymbol{b} \cdot \nabla w - \tfrac{1}{2} w\boldsymbol{b} \cdot \nabla v + v\, c\, w \, dx.$$

We then apply numerical integration to this representation which features a build-in skew symmetry of the first order term. Now the same rules as in the case $\boldsymbol{b} \equiv 0$ apply.

Aspects for the Implementation

(1) If $\boldsymbol{b} \not\equiv 0$ use integration by parts to derive a representation of the continuous bilinear form featuring a build-in skew symmetry of the first order term.
(2) Use a stable quadrature formula $\hat{Q}$ that is at least exact on $\mathbb{P}_{2p-2}$ for computing element integrals.
(3) Rule of Thump: Integrals that can be computed exactly, should be computed exactly.

Remark 64 (ALBERTA Realization). In ALBERTA the above idea is implemented for assembling the system matrix for 2nd elliptic equations automatically. The problem dependent function

- `LALt` (T, λ_ℓ) that is a realization of $\boldsymbol{A}_T(\lambda_\ell) = |\det DF_T|\, \Lambda \boldsymbol{A}(x(\lambda_\ell)) \Lambda^\top$ on T,
- `Lb0` (T, λ_ℓ) that is a realizations of $\boldsymbol{b}_T(\lambda_\ell) = |\det DF_T|\, \Lambda \boldsymbol{b}(x(\lambda_\ell))$ on T, and

- $\mathtt{c}\,(T\,,\lambda_\ell)$ that is a realization of $c_T(\lambda_\ell) = |\det DF_T|\, c(x(\lambda_\ell))$ on T

have to be supplied. In addition, the assemblage tool needs information about data being piecewise constant. With such information the linear system is assembled automatically in an efficient way; compare with [34, Sect. 3.12].

4.4 Remarks on Iterative Solvers

We finish this section by analyzing the structure of the discrete linear system with main focus on the structure of the residual. The structure of the residual has impact on the choice of the appropriate iterative solver.

Recall the $N \times N$ *system matrix*

$$S := \begin{bmatrix} \mathscr{B}[\Phi_1, \Phi_1] & \dots & \mathscr{B}[\Phi_{\mathring{N}}, \Phi_1] & \mathscr{B}[\Phi_{\mathring{N}+1}, \Phi_1] & & \dots & \mathscr{B}[\Phi_N, \Phi_1] \\ \vdots & \ddots & \vdots & \vdots & & \ddots & \vdots \\ \mathscr{B}[\Phi_1, \Phi_{\mathring{N}}], & \dots & \mathscr{B}[\Phi_{\mathring{N}}, \Phi_{\mathring{N}}] & \mathscr{B}[\Phi_{\mathring{N}+1}, \Phi_{\mathring{N}}] & & \dots & \mathscr{B}[\Phi_N, \Phi_{\mathring{N}}], \\ 0 & \dots & 0 & 1 & 0 & \dots & 0 \\ 0 & \dots & 0 & 0 & 1 & \dots & 0 \\ \vdots & \ddots & 0 & 0 & 0 & \ddots & \vdots \\ 0 & \dots & 0 & 0 & 0 & \dots & 1 \end{bmatrix}$$

and the load vector

$$\boldsymbol{f} := \left[\langle f, \Phi_1\rangle, \dots, \langle f, \Phi_{\mathring{N}}\rangle, g_{\mathring{N}+1}, \dots, g_N\right]^{\mathrm{T}} \in \mathbb{R}^N.$$

Note that even in the case of symmetric $\mathscr{B}$ the system matrix is always *non-symmetric*. We decompose the system matrix $\boldsymbol{S}$ and vectors $\boldsymbol{v} \in \mathbb{R}^N$ according to interior and boundary nodes:

$$\boldsymbol{S} = \begin{bmatrix} \boldsymbol{S}_{\mathrm{II}} & \boldsymbol{S}_{\mathrm{ID}} \\ \boldsymbol{0} & \mathrm{id} \end{bmatrix}, \qquad \boldsymbol{v} = \begin{bmatrix} \boldsymbol{v}_{\mathrm{I}} \\ \boldsymbol{v}_{\mathrm{D}} \end{bmatrix} \in \mathbb{R}^{\mathring{N}} \times \mathbb{R}^{N-\mathring{N}}.$$

The matrix $\boldsymbol{S}_{\mathrm{II}} \in \mathbb{R}^{\mathring{N}\times\mathring{N}}$ couples interior nodes and the matrix $\boldsymbol{S}_{\mathrm{ID}} \in \mathbb{R}^{\mathring{N}\times(N-\mathring{N})}$ couples interior with Dirichlet boundary nodes. We need to solve

$$\boldsymbol{S}\boldsymbol{u} = \boldsymbol{f} \qquad \Longleftrightarrow \qquad \begin{bmatrix} \boldsymbol{S}_{\mathrm{II}} & \boldsymbol{S}_{\mathrm{ID}} \\ \boldsymbol{0} & \mathrm{id} \end{bmatrix} \begin{bmatrix} \boldsymbol{u}_{\mathrm{I}} \\ \boldsymbol{u}_{\mathrm{D}} \end{bmatrix} = \begin{bmatrix} \boldsymbol{f}_{\mathrm{I}} \\ \boldsymbol{f}_{\mathrm{D}} \end{bmatrix}.$$

As we have seen, the system matrix $\boldsymbol{S}$ is a sparse matrix. Direct solvers for sparse matrices need a special sparsity pattern of $\boldsymbol{S}$ in order to perform efficiently. Such a sparsity pattern can be constructed by an appropriate renumbering of basis functions. For $d \leq 2$ there exist nowadays efficient tools for obtaining a suitable sparsity pattern of $\boldsymbol{S}$.

Another class of efficient solvers for a linear system with a sparse matrix are *preconditioned Krylov space solvers*. These are iterative solvers that compute *corrections* based on the residual

$$\boldsymbol{r}(\boldsymbol{v}) := \boldsymbol{f} - \boldsymbol{S}\boldsymbol{v} = \begin{bmatrix} \boldsymbol{f}_{\mathrm{I}} - \boldsymbol{S}_{\mathrm{II}}\boldsymbol{v}_{\mathrm{I}} - \boldsymbol{S}_{\mathrm{ID}}\boldsymbol{v}_{\mathrm{D}} \\ \boldsymbol{f}_{\mathrm{D}} - \boldsymbol{v}_{\mathrm{D}} \end{bmatrix} = \begin{bmatrix} \boldsymbol{f}_{\mathrm{I}} - \boldsymbol{S}_{\mathrm{II}}\boldsymbol{v}_{\mathrm{I}} - \boldsymbol{S}_{\mathrm{ID}}\boldsymbol{v}_{\mathrm{D}} \\ \boldsymbol{g}_{\mathrm{D}} - \boldsymbol{v}_{\mathrm{D}}. \end{bmatrix}$$

Assume an initial guess $\boldsymbol{u}^{(0)}$ with $\boldsymbol{u}_{\mathrm{D}}^{(0)} = \boldsymbol{g}_{\mathrm{D}}$. Then holds for all $\ell \geq 0$

$$\boldsymbol{r}^{(\ell)} = \boldsymbol{r}(\boldsymbol{u}^{(\ell)}) = \begin{bmatrix} \boldsymbol{f}_{\mathrm{I}} - \boldsymbol{S}_{\mathrm{II}}\boldsymbol{u}_{\mathrm{I}}^{(\ell)} \\ \boldsymbol{0} \end{bmatrix} - \begin{bmatrix} \boldsymbol{S}_{\mathrm{ID}}\boldsymbol{g}_{\mathrm{D}} \\ \boldsymbol{0} \end{bmatrix},$$

i. e., there are only corrections for interior nodes that only involve $\boldsymbol{S}_{\mathrm{II}}$. With such an initial guess any iterative solver *utilizes solely properties* of $\boldsymbol{S}_{\mathrm{II}}$. These attributes are completely determined by properties of $\mathscr{B}$ on $\mathring{\mathbb{V}}(\mathscr{T})$.

For given $\boldsymbol{v} = [\boldsymbol{v}_{\mathrm{I}}, \boldsymbol{0}]^\top$ the corresponding finite element function V belongs to $\mathring{\mathbb{V}}(\mathscr{T})$. For *coercive* $\mathscr{B}$ on $\mathring{\mathbb{V}}$ we thus conclude for such $\boldsymbol{v}$

$$\boldsymbol{v} \cdot \boldsymbol{S}\boldsymbol{v} = \boldsymbol{v}_{\mathrm{I}} \cdot \boldsymbol{S}_{\mathrm{II}}\boldsymbol{v}_{\mathrm{I}} = \mathscr{B}[V, V] \geq \alpha \|V\|_{\mathbb{V}}^2 \geq \tilde{\alpha}\, |\boldsymbol{v}|_2^2,$$

using equivalence of norms. The constant $\tilde{\alpha}$ depends on N. Moreover, *symmetry* of $\mathscr{B}$ implies

$$\boldsymbol{S}_{\mathrm{II}} = \big[\mathscr{B}[\Phi_i, \Phi_j]\big]_{i,j=1,\dots,\mathring{N}} = \big[\mathscr{B}[\Phi_j, \Phi_i]\big]_{i,j=1,\dots,\mathring{N}} = \boldsymbol{S}_{\mathrm{II}}^\top,$$

i. e., $\boldsymbol{S}_{\mathrm{II}}$ is symmetric and in combination with coercivity of $\mathscr{B}$ the matrix $\boldsymbol{S}_{\mathrm{II}}$ is spd.

Aspects for the Implementation

(1) For symmetric and coercive $\mathscr{B}$ we can use a preconditioned CG method.
(2) For general $\mathscr{B}$ one has to use GMRes, or BiCGStab, ...

Remark 65 (Decomposition of Discrete and Continuous Solution). The decomposition of the residual implies for the coefficient vector $\boldsymbol{u} = \boldsymbol{u}_{\mathrm{I}} + \boldsymbol{g}_{\mathrm{D}}$ of the discrete solution

$$\boldsymbol{S}_{\mathrm{II}}\boldsymbol{u}_{\mathrm{I}} = \boldsymbol{f}_{\mathrm{I}} - \boldsymbol{S}_{\mathrm{ID}}\boldsymbol{g}_{\mathrm{D}}.$$

This perfectly mimics the construction of the true solution $u = u_I + g$:

$$u_I \in \mathring{\mathbb{V}}: \qquad \mathscr{B}[u_I, v] = \langle v, f\rangle - \mathscr{B}[g, v] \qquad \forall v \in \mathring{\mathbb{V}}.$$

Remark 66 (Initial Guess for an Iterative Solver). Assume the following iteration

$$\text{SOLVE} \longrightarrow \text{ESTIMATE} \longrightarrow \text{MARK} \longrightarrow \text{REFINE/COARSEN}.$$

A *good choice* as initial guess $\boldsymbol{u}_{\mathrm{I}}^{(0)}$ for *any iterative solver* are the coefficients of the *old discrete solution* at *interior nodes* with *boundary values* from the *actual grid*. This needs interpolation resp. restriction of the coefficient vector of the discrete solution during refinement resp. coarsening.

Such interpolation and restriction routines strongly depend on the finite element space and the refinement rule. From the interpolation and restriction of basis functions of $\bar{\mathbb{P}}$ during refinement of $\bar{T}$ interpolation and restriction routines for general finite element functions can be derived. ALBERTA supplies such routines for Lagrange elements; compare with [34, Sect. 1.4.4].

Exercise 67 (2nd Order Elliptic Equation). Write an ALBERTA program for the adaptive solution of the elliptic equation

$$-\nabla\cdot(\boldsymbol{A}\nabla u) + \boldsymbol{b}\cdot\nabla u + c\,u = f \quad \text{in } \Omega = (0,1)^d, \qquad u = g \quad \text{on } \partial\Omega$$

for $d = 2, 3$ with coefficients $\boldsymbol{A} = \varepsilon\,\mathrm{id}_{\mathbb{R}^d}$ $(\varepsilon > 0)$, $\boldsymbol{b} = \left[1,\ldots,1\right]^{\mathsf{T}}$, and $c = d$.

Verify the program with the exact solution

$$u(x) = \prod_{i=1}^{d} x_i\,(1 - e^{(x_i-1)/\varepsilon})$$

using different values $\varepsilon > 0$ and a corresponding right hand side $f(x)$ and boundary values $g(x)$. Use Lagrange elements of order 1–4 and different marking strategies.

As a starting point, an ALBERTA program for the solution of the Poisson problem

$$-\Delta u = f \quad \text{in } \Omega, \qquad u = g \quad \text{on } \partial\Omega$$

is available (the file `ellipt.c` in the `Common` directory). Adjust this code to the above problem by modifying routines for the calculation of element matrices, the routine `r()` for evaluation the lower order term for the estimator `ellipt_est()` (compare with Sect. 5), and exact solution and data.

For the solution of the *non symmetric* linear system use GMRes (solver no. 3 for `oem_solve()`) with restart `10` – `20`. The model implementation of Poisson's equation in the source file `ellipt.c` is described in detail in [34, Sect. 2.1].

5 The Adaptive Algorithm and Concluding Remarks

In this part we discuss the remaining modules of the adaptive algorithm and comment on solver evaluation and choice of an adequate finite element package.

5.1 The Adaptive Algorithm

Let $u \in \mathbb{V}$ be a solution of Problem 54 and let U be its Galerkin approximation in $\mathbb{V}(\mathcal{T})$. Typical a priori error estimates are of the form: If $u \in \mathbb{V}^s$ then

$$\|U - u\|_{\mathbb{V}}^2 \lesssim \sum_{T \in \mathcal{T}} h_T^{2s} \|u\|_{\mathbb{V}^s(T)}^2,$$

where the subspace $\mathbb{V}^s \subset \mathbb{V}$ describes regularity properties of u. This already indicates that one can profit from local refinement of $\mathcal{T}$ by compensating for a large local norm $\|u\|_{\mathbb{V}^s(T)}$ by a small local mesh size h_T. In practice, the true solution u is unknown, and therefore information about $\|u\|_{\mathbb{V}^s(T)}$ is not accessible.

As eluded in the introduction the h-adaptive finite element algorithm is an iteration of the form

$$\mathsf{SOLVE} \quad \longrightarrow \quad \mathsf{ESTIMATE} \quad \longrightarrow \quad \mathsf{MARK} \quad \longrightarrow \quad \mathsf{REFINE}.$$

We have already discussed the modules SOLVE in Sect. 4 and REFINE in Sect. 3. It remains to address the modules ESTIMATE and MARK. The latter one is in general problem independent whereas ESTIMATE strongly depends on the problem under consideration.

Aspects for the Implementation

(1) ESTIMATE has to compute an error bound for the true error that only depends on the discrete solution and given data of the PDE:

$$\|U - u\|_{\mathbb{V}} \lesssim \mathcal{E}_{\mathcal{T}}(U, \mathcal{T}).$$

(2) This bound should be computed from *local quantities* that allow in MARK for decisions about local refinement. This means, the estimator should be given as

$$\mathcal{E}_{\mathcal{T}}^2(U, \mathcal{T}) = \sum_{T \in \mathcal{T}} \mathcal{E}_{\mathcal{T}}^2(U, T)$$

with *error indicators* $\mathcal{E}_{\mathcal{T}}(U, T)$ that are computed from local values.

(3) In practice a stopping test is included in-between the steps ESTIMATE and MARK. This means, given a tolerance $\mathsf{TOL} > 0$ for the estimator $\mathcal{E}_{\mathcal{T}}$, the iteration is terminated if $\mathcal{E}_{\mathcal{T}}(U, \mathcal{T}) \leq \mathsf{TOL}$.

5.1.1 Error Estimation

In this section we comment on the implementation of an error estimator for the model problem

$$-\operatorname{div}(\boldsymbol{A}(x)\nabla u) + \boldsymbol{b}(x)\cdot\nabla u + c(x)\,u = f \quad \text{in } \Omega, \qquad u = 0 \quad \text{on } \partial\Omega.$$

We recall its weak formulation in $\mathbb{V} = H_0^1(\Omega)$

$$u \in \mathbb{V}: \qquad \mathscr{B}[u, v] := \int_\Omega \nabla v^T \boldsymbol{A}\nabla u + v\,\boldsymbol{b}\cdot\nabla u + v\,c\,u\,dx = \int_\Omega f\,v\,dx \qquad \forall\, v \in \mathbb{V}.$$

We suppose that the coefficient satisfy the assumptions of Example 3 that guarantee coercivity of $\mathscr{B}$. In addition, we assume exact integration and exact linear algebra for the computation of the Galerkin approximation U in a finite element space $\mathbb{V}(\mathscr{T}) = \mathsf{FES}(\mathscr{T}, \mathbb{P}_p, \mathbb{V})$ over a conforming triangulation $\mathscr{T}$ of Ω.

Starting point for the a posteriori error analysis is the equivalence

$$\alpha\|U - u\|_{\mathbb{V}} \le \|\mathscr{R}(U)\|_{\mathbb{V}^*} \le \|\mathscr{B}\|\,\|U - u\|_{\mathbb{V}}$$

with the residual $\mathscr{R}(U) \in \mathbb{V}^*$ defined as

$$\langle \mathscr{R}(U), v\rangle := \mathscr{B}[U, v] - \langle f, v\rangle = \mathscr{B}[U - u, v] \qquad \forall v \in \mathbb{V}.$$

The residual is an a posteriori quantity in that it can be computed from the discrete solution and given data of the PDE. Nevertheless, the dual norm $\|\cdot\|_{\mathbb{V}^*}$ is non-computable. Most error estimators provide a computable bound for $\|\mathscr{R}(U)\|_{\mathbb{V}^*}$.

We consider here the residual estimator. We additionally ask that $\boldsymbol{A}$ is piecewise Lipschitz over $\mathscr{T}$, this means $\boldsymbol{A} \in W^1_\infty(\mathscr{T}; \mathbb{R}^{d\times d})$. This allows us to define elementwise the *element residual* as

$$R_{|T} := (-\operatorname{div}(\boldsymbol{A}\nabla U) + \boldsymbol{b}\cdot\nabla U + c\,U - f)_{|T}$$

for all $T \in \mathscr{T}$. For any interior side $S = T_1 \cap T_2$ of $\mathscr{T}$ we define the *jump residual* by

$$J_{|S} := \tfrac{1}{2}[\![\boldsymbol{A}\nabla U]\!]\cdot\boldsymbol{n}_S := \tfrac{1}{2}\big((\boldsymbol{A}\nabla U)_{|T_1} - (\boldsymbol{A}\nabla U)_{|T_2}\big)\cdot\boldsymbol{n}_S.$$

Hereafter, $\boldsymbol{n}_S$ is the unit normal of S pointing from T_1 to T_2. For a boundary side S we set $J := 0$. The *indicators* $\mathscr{E}_{\mathscr{T}}(U, T)$ of the residual estimator are given by

$$\mathscr{E}^2_{\mathscr{T}}(U, T) := h_T^2\|R\|^2_{L^2(T)} + h_T\|J\|^2_{L^2(\partial T)}.$$

Theorem 68 (Global Upper and Local Lower Bound). *There holds*

$$\|U - u\|_{\mathbb{V}}^2 \le C_1 \mathscr{E}_{\mathscr{T}}^2(U, \mathscr{T}) := C_1 \sum_{T \in \mathscr{T}} \mathscr{E}_{\mathscr{T}}^2(U, T).$$

and

$$C_2 \mathscr{E}_{\mathscr{T}}^2(U, T) \le \|U - u\|_{\mathbb{V}(N_{\mathscr{T}}(T))}^2 + \sum_{T' \in N_{\mathscr{T}}(T)} \operatorname{osc}_{\mathscr{T}}^2(U, T').$$

with the oscillation term

$$\operatorname{osc}_{\mathscr{T}}^2(U, T) := h_T^2 \|\bar{R} - R\|_{L^2(T)}^2 + h_T \|\bar{J} - J\|_{L^2(\partial T)}^2 \qquad \forall T \in \mathscr{T}.$$

Hereafter, $\bar{R}$ and $\bar{J}$ are piecewise polynomial approximations to R and J.

Proof. See the books by Verfürth [42] and Ainsworth and Oden [1]. □

Some remarks are appropriate.

(1) Assuming that we can compute L^2 norms, the indicators $\mathscr{E}_{\mathscr{T}}(U, T)$ are computable quantities. The computation of $\mathscr{E}_{\mathscr{T}}(U, T)$ only involves T and its direct neighbors.
(2) Generically, the oscillation term $\operatorname{osc}_{\mathscr{T}}(U, T)$ is of higher order, this means that $\|U - u\|_{\mathbb{V}(N_{\mathscr{T}}(T))}$ is dominant in the lower bound.
(3) The typical choice for the approximations $\bar{R}$ and $\bar{J}$ are the element-wise or side-wise L^2 projection onto polynomials of degree $p - 1$.
(4) Assume that $\mathscr{T}_*$ is a refinement of $\mathscr{T}$ such that $T \in \mathscr{T}$ and its neighbor are "sufficiently" refined. Then there holds a discrete analogue to the continuous lower bound, namely the *discrete lower bound*

$$\tilde{C}_2 \mathscr{E}_{\mathscr{T}}^2(U, T) \le \|U - U_*\|_{\mathbb{V}(N_{\mathscr{T}}(T))}^2 + \sum_{T' \in N_{\mathscr{T}}(T)} \operatorname{osc}_{\mathscr{T}}^2(U, T'),$$

where $U_* \in \mathbb{V}(\mathscr{T}_*)$ is the Galerkin approximation to u in $\mathbb{V}(\mathscr{T}_*)$; compare for instance with [27, 28].

Computation of the Error Estimator. The estimator can be computed by looping over all grid elements and computing the indicators element-wise from local values. For general data $\boldsymbol{A}$, $\boldsymbol{b}$, c, and f the L^2-norms cannot be computed exactly. Therefore, exact integration on T is replaced by numerical quadrature on T and each side $S \subset \partial T$; compare with Sect. 4.3. Note, that in case of piecewise polynomial data an appropriate choice of the quadrature formula results in an exact computation of the L^2-norms.

The element residual for $x \in T$ is

$$R(x) = -\boldsymbol{A} : D^2 U(x) - \operatorname{div} \boldsymbol{A}(x) \cdot \nabla U(x) + \boldsymbol{b}(x) \cdot \nabla U(x) + c(x) U(x) - f(x),$$

where we have used the product rule for the second order term. Therefore, we can compute an approximation to $\|h_T\, R\|^2_{L^2(T)}$ by a quadrature formula having access to functions for the evaluation of data

$$\boldsymbol{A}(x), \quad \operatorname{div}\boldsymbol{A}(x), \quad \boldsymbol{b}(x), \quad c(x), \quad f(x)$$

at given quadrature points $x \in T$. The Galerkin approximation U and its derivatives are evaluated on each element $T \in \mathcal{T}$ by first extracting the local coefficient vector from the global one and then using the local basis representation; compare with Sect. 2.3. Note, that for the evaluation of U, ∇U, and D^2U all derivatives up to 2nd order of the basis functions $\{\bar{\phi}_1, \dots, \bar{\phi}_{\bar{n}}\}$ on $\bar{T}$ have to be accessible. The computation of derivatives of U again involves the Jacobian Λ of the barycentric coordinates on T.

The computation of the jump residual is much more involved since the evaluation of

$$J_{|S} := \tfrac{1}{2}[\![\boldsymbol{A}\nabla U]\!]\cdot \boldsymbol{n}_S := \tfrac{1}{2}\big((\boldsymbol{A}\nabla U)_{|T_1} - (\boldsymbol{A}\nabla U)_{|T_2}\big)\cdot \boldsymbol{n}_S$$

needs values of $\boldsymbol{A}\nabla U$ from both adjacent elements T_1 and T_2 on the common side $S = T_1 \cap T_2$. We thereby have to use a quadrature rule for the $(d-1)$ simplex S. In defining a quadrature formula on a side S we can follow the same ideas explained in Sect. 4.3. This means, we fix a given quadrature rule on a side of the standard element and then use the transformation rule to obtain a quadrature formula for a generic side S.

The barycentric coordinates λ with respect to the common side S have d components. For the evaluation of $(\boldsymbol{A}\nabla U)_{|T_i}$, $i = 1, 2$, they have to be converted into barycentric coordinates λ^{T_1} and λ^{T_2} with respect to T_1 and T_2 having $(d+1)$ components. In doing this one has to fix a unique ordering of the common side's vertices, for instance the ordering given by T_1. Note, that in general the ordering of the vertices of S induced by T_2 differs. Using a different ordering of the vertices of S on T_1 and T_2 results in different world coordinates of the same quadrature points; compare with Fig. 25. After a proper conversion of the quadrature nodes λ on S into barycentric coordinates λ^{T_i} on T_i the evaluation of $(\boldsymbol{A}\nabla U)_{|T_i}$ is standard. This in turn allows then for an approximation of the jump residual $\|h_T^{1/2} J\|^2_S$ by means of numerical quadrature.

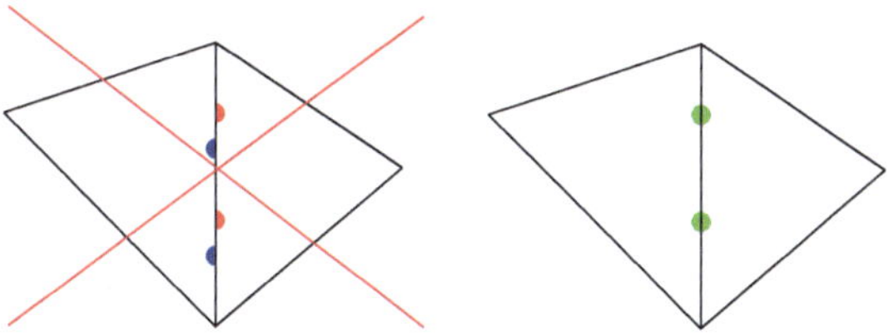

Fig. 25 Quadrature nodes on a common side $S = T_1 \cap T_2$. Different ordering of the vertices of S on T_1 and T_2 (*left*), consistent ordering of vertices of S on T_1 and T_2 (*right*)

Remark 69 (ALBERTA Realization). ALBERTA efficiently computes for constant A the residual estimator when having access to a user defined function evaluating the lower order terms

$$\boldsymbol{b}(x) \cdot \nabla U(x) + c(x)U(x) - f(x).$$

Compare the description of the function `ellipt_est()` in [34, Sect. 3.14.1]. The implementation of this function can serve as a reference when implementing residual type estimators for other problems.

5.1.2 Marking Procedures

Let us motivate marking procedures by recalling the aim of adaptive methods. Given a tolerance TOL for the error $\|U - u\|_{\mathbb{V}}$ and an initial grid $\mathscr{T}_0$ we want to construct a refinement $\mathscr{T}$ of $\mathscr{T}_0$ such that

(1) $\|U - u\|_{\mathbb{V}} = \mathsf{TOL}$,
(2) The number of DOFs in $\mathscr{T}$ is as small as possible.

This is a discrete, constrained minimization problem. We are looking for the optimal (minimal) refinement $\mathscr{T}$ of $\mathscr{T}_0$ such that $\|U - u\|_{\mathbb{V}} = \mathsf{TOL}$. In general, the task to find a solution to this minimization problem is *more costly* than the actual computation of a discrete solution. Therefore, we need some heuristics for constructing a *good* but in general non-optimal grid $\mathscr{T}$. Using techniques from continuous optimization Babuška and Rheinboldt have heuristically characterized optimal meshes [4].

Characterization 70 (Optimal Mesh). *Let $\mathscr{T}$ be a minimal refinement of $\mathscr{T}_0$ such that $\|U - u\|_{\mathbb{V}} = \mathsf{TOL}$. Then holds for all $T \in \mathscr{T}$*

$$\|U - u\|_T \approx \mathsf{TOL}\sqrt{\frac{\#DOFs(T)}{\#DOFs(\mathscr{T})}} = \mathsf{TOL}\, const,$$

i. e., the error is equidistributed over the mesh elements.

Marking strategies therefore aim at the equidistribution of the true error. In practice, the local error is unknown. Hence, marking strategies try to equidistribute the local error indicators $\mathscr{E}_{\mathscr{T}}(U, T)$. Obviously, large error indicators disturb equidistribution. Therefore, elements with large error indicator are selected for refinement. Small indicators also disturb equidistribution. Elements with small indicators can be selected for coarsening, if wanted.

Some remarks about replacing the local error by the error indicator are appropriate. In general oscillation is dominated by the estimator, this means

$$\operatorname{osc}_{\mathscr{T}}(U, T) \le \mathscr{E}_{\mathscr{T}}(U, T).$$

Assume first the generic situation $\operatorname{osc}_{\mathcal{T}}(U,T) \ll \mathcal{E}_{\mathcal{T}}(U,T)$.

(1) The continuous local lower bound then implies: If $\mathcal{E}_{\mathcal{T}}(U,T)$ is large then the local error $\|U-u\|_{\mathbb{V}(N_{\mathcal{T}}(T))}$ is large.
(2) The discrete lower bound for the differences of two finite element solutions implies: If $\mathcal{E}_{\mathcal{T}}(U,T)$ is large we can expect a large local error reduction.

Consider next that oscillation $\operatorname{osc}_{\mathcal{T}}(U,T)$ is proportional to $\mathcal{E}_{\mathcal{T}}(U,T)$. In this case the oscillation term spoils the continuous and discrete lower bound. Oscillation is related to local resolution of data on $\mathcal{T}$. Consequently, if oscillation is large we have to improve local resolution of data. A selection of T with large indicator $\mathcal{E}_{\mathcal{T}}(U,T) \approx \operatorname{osc}_{\mathcal{T}}(U,T)$ for refinement therefore results in an improvement of local data resolution.

General Marking Strategy. Given the estimator $\mathcal{E}_{\mathcal{T}}(U,\mathcal{T})$ and the indicators $\{\mathcal{E}_{\mathcal{T}}(U,T)\}_{T\in\mathcal{T}}$, the most commonly used marking strategies are based on computing first a threshold $\mathcal{E}_{\text{limit}}$ and then marking all elements for refinement, where the indicator is above this threshold, i. e., $\mathcal{E}_{\mathcal{T}}(U,T) \ge \mathcal{E}_{\text{limit}}$. The computation of the threshold $\mathcal{E}_{\text{limit}}$ usually needs additional parameters. This gives the following marking strategy.

Algorithm 71 (General Marking Routine).

function mark$\big(\mathcal{T}, \{\mathcal{E}_{\mathcal{T}}(U,T)\}_{T\in\mathcal{T}}, \ldots\big)$
  compute threshold $\mathcal{E}_{\text{limit}}$;
  $\mathcal{M} := \emptyset$;

  for all $T \in \mathcal{T}$ do
    if $\mathcal{E}_{\mathcal{T}}(U,T) \ge \mathcal{E}_{\text{limit}}$ then
      $\mathcal{M} := \mathcal{M} \cup \{T\}$;
    end if
  end for
  return$(\mathcal{M})$;

It remains to define the threshold in the general marking strategy. Each adaptive iteration requires to solve for the discrete solution, i. e., we have to solve a high dimensional linear or nonlinear system, which in general is costly. Therefore, we have to find a good balance of:

(1) *Selecting only few elements* in order to construct "very good" meshes by picking up frequently improved information about the error.

- *Disadvantage:* Needs many iterations, which implies that we also have to solve frequently for the discrete solution.

(2) *Selecting many elements* in order to reduce the number of iterations.

- *Disadvantage:* The resulting grids may not be optimal, i. e., we are creating too many elements, in particular in early stages of the adaptive iterations.

We next give an overview of the most popular marking strategies.

Equidistribution Strategies. Consider a given grid $\mathscr{T}$ where the estimator meets the tolerance $\mathsf{TOL} > 0$, i. e., $\mathscr{E}_{\mathscr{T}}(U, \mathscr{T}) = \mathsf{TOL}$, and the indicators $\{\mathscr{E}_{\mathscr{T}}(U, T)\}_{T \in \mathscr{T}}$ are equidistributed over $\mathscr{T}$, i. e., $\mathscr{E}_{\mathscr{T}}(U, T) = \text{const.}$ for all $T \in \mathscr{T}$. Then holds

$$\sum_{T \in \mathscr{T}} \mathscr{E}_{\mathscr{T}}^2(U, T) = \#\mathscr{T} \text{const.}^2 = \mathsf{TOL}^2 \qquad \Longrightarrow \qquad \mathscr{E}_{\mathscr{T}}(U, T) = \frac{\mathsf{TOL}}{\sqrt{\#\mathscr{T}}}.$$

This *motivates* to mark all elements with an indicator above this value.

Given a tolerance $\mathsf{TOL} > 0$ and parameter $\theta \in [0, 1]$ define the threshold as

$$\mathscr{E}_{\mathsf{limit}} := \theta \, \frac{\mathsf{TOL}}{\sqrt{\#\mathscr{T}}}.$$

A typical value for the parameter θ is $\theta \approx 0.9$.

The drawback of this marking is the following observation. Using a tolerance for the *absolute* error the threshold is not invariant under scaling of the problem. If TOL is chosen too small, nearly all elements are marked on coarse grids.

Modified Equidistribution Strategy. This disadvantage of the Equidistribution Strategy can be avoided by using the average of the indicators as limit value.

Given parameter $\theta \in [0, 1]$ define the threshold as

$$\mathscr{E}_{\mathsf{limit}} := \theta \, \frac{\mathscr{E}_{\mathscr{T}}(U, \mathscr{T})}{\sqrt{\#\mathscr{T}}}.$$

A typical value for the parameter θ is $\theta \approx 0.9$.

Maximum Strategy. This strategy directly aims at marking only elements with indicators close to the maximal indicator.

Given parameter $\gamma \in [0, 1]$ define the threshold as

$$\mathscr{E}_{\mathsf{limit}} := \gamma \max_{T \in \mathscr{T}} \mathscr{E}_{\mathscr{T}}(U, \mathscr{T}).$$

A typical value for the parameter γ is $\gamma \approx 0.5$.

Dörfler Marking. In the *first convergence proof* for adaptive finite elements [17] Dörfler introduced the idea to control the total estimator by the estimator on the set of marked elements.

Given $\theta \in (0,1]$ select $\mathcal{M} \subset \mathcal{T}$ such that

$$\theta\, \mathcal{E}_{\mathcal{T}}(U,\mathcal{T}) \le \mathcal{E}_{\mathcal{T}}(U,\mathcal{M}).$$

A typical value for θ is $\theta = 0.5$.

Obviously, Dörfler Marking does not fit into the general marking strategy Algorithm 71. In principle, the size of the indicators of single elements in $\mathcal{M}$ does not matter, i. e., $\mathcal{M}$ can contain elements with small error indicator. But then $\#\mathcal{M}$ will be large. For a "good" adaptive method $\mathcal{M}$ should only contain elements with large indicator, such that $\#\mathcal{M}$ is minimal. In fact, for linear, symmetric, and elliptic PDEs it has been proved that minimality of $\mathcal{M}$ implies an optimal error decay in terms of DOFs [15, 23] and [8, 37].

Choosing the minimal subset $\mathcal{M}$ requires a sorting of elements. In practice, this sorting is often avoided by a simple "drawer" algorithm. This results in a threshold $\mathcal{E}_{\mathsf{limit}}$ for the general marking strategy Algorithm 71 and can be interpreted as an adaptive Maximum Strategy for selecting elements into $\mathcal{M}$.

Algorithm 72 (Sorting into Drawers).

function Dörfler_threshold$\big(\mathcal{T},\, \{\mathcal{E}_{\mathcal{T}}(U,T)\}_{T\in\mathcal{T}},\, \theta,\, \nu\big)$

$\mathcal{E}_{\max} := \max_{T\in\mathcal{T}} \mathcal{E}_{\mathcal{T}}(U,T),\ \mathcal{E}_{\mathcal{T}}(U,\mathcal{M}) := 0,\ \gamma := 1$

while $\mathcal{E}_{\mathcal{T}}(U,\mathcal{M}) < \theta \mathcal{E}_{\mathcal{T}}(U,\mathcal{T})$ do
  $\mathcal{E}_{\mathcal{T}}(U,\mathcal{M}) := 0,\ \gamma := \gamma - \nu$
  for all $T \in \mathcal{T}$ do
    if $\mathcal{E}_{\mathcal{T}}(U,T) > \gamma\, \mathcal{E}_{\max}$
      $\mathcal{E}^2_{\mathcal{T}}(U,\mathcal{M}) := \mathcal{E}^2_{\mathcal{T}}(U,\mathcal{M}) + \mathcal{E}^2_{\mathcal{T}}(U,T)$
    end if
  end for
end while

return $\mathcal{E}_{\mathsf{limit}} := \gamma \mathcal{E}_{\max}$.

Typical values for the parameters θ and ν are $\theta \approx 0.5$ and $\nu \approx 0.1$.

Strategies with Coarsening. Adaptive methods allowing for *coarsening of elements* select elements with *small error indicator* for coarsening.

(1) Coarsening is *necessary for instationary problems* with a time-stepping procedure when starting with the *grid from the old time step* as *initial grid* of the current time step.
(2) Adaptive methods simultaneously selecting elements for refinement and coarsening strongly benefit from a *element coarsening indicator* that gives information about a possible error increase after coarsening. Without any coarsening indicator, one has to carefully select parameters in order to avoid cycles of

$$\text{refine} \;\rightarrow\; \text{coarsen} \;\rightarrow\; \text{refine} \;\rightarrow\; \text{coarsen} \;\rightarrow\; \ldots$$

Let $\{\mathcal{E}_{\mathcal{T}}(U,T)\}_{T\in\mathcal{T}}$ be element error indicators and let $\{\mathcal{C}_{\mathcal{T}}(U,T)\}_{S\in\mathcal{T}}$ be element coarsening indicators. Based on suitable thresholds $\mathcal{E}_{\mathsf{limit}}$ for refinement and $\mathcal{C}_{\mathsf{limit}}$ for coarsening, the following algorithm outputs the set $\mathcal{M}^+, \mathcal{M}^-$ of elements marked for refinement respectively coarsening.

Algorithm 73 (Marking Including Coarsening).

function mark$\big(\mathcal{T},\ \{\mathcal{E}_{\mathcal{T}}(U,T)\}_{T\in\mathcal{T}},\ \{\mathcal{C}_{\mathcal{T}}(U,T)\}_{T\in\mathcal{T}},\ \dots\big)$

compute thresholds $\mathcal{E}_{\mathsf{limit}}$ and $\mathcal{C}_{\mathsf{limit}}$;
$\mathcal{M}^+ := \mathcal{M}^- := \emptyset$;

for all $T \in \mathcal{T}$ do
if $\mathcal{E}_{\mathcal{T}}(U,T) \geq \mathcal{E}_{\mathsf{limit}}$ then
$\mathcal{M}^+ := \mathcal{M}^+ \cup \{T\}$;
else if $\sqrt{\mathcal{E}^2(U,T) + \mathcal{C}^2_{\mathcal{T}}(U,T)} \leq \mathcal{C}_{\mathsf{limit}}$ then
$\mathcal{M}^- := \mathcal{M}^- \cup \{T\}$;
end if
end for
return$(\mathcal{M}^+, \mathcal{M}^-)$;

For instance, the Maximum Strategy could select the thresholds $\mathcal{E}_{\mathsf{limit}}$ and $\mathcal{C}_{\mathsf{limit}}$ as follows.

Given parameters $\gamma \in [0,1]$ and $\gamma_c \in [0,\gamma)$ define the thresholds as

$$\mathcal{E}_{\mathsf{limit}} := \gamma \max_{T\in\mathcal{T}} \mathcal{E}_{\mathcal{T}}(U,\mathcal{T}) \qquad \text{and} \qquad \mathcal{C}_{\mathsf{limit}} := \gamma_c \max_{T\in\mathcal{T}} \mathcal{E}_{\mathcal{T}}(U,\mathcal{T}).$$

Typical values are $\gamma \approx 0.5$ and $\gamma_c \approx 0.1$ if coarsening indicators are provided. Otherwise choose $\gamma_c \ll 0.1$.

5.2 *Concluding Remarks*

We conclude by remarks on the ALBERTA philosophy for the implementation of a new solver and some general remarks about selecting the adequate software.

5.2.1 Solver Development in ALBERTA

The basic steps for the implementation of a new adaptive solver in ALBERTA are the following.

(1) Implement an *efficient solver* for the problem under consideration for a given fixed grid $\mathcal{T}$.

- Relies on the assemblage tools discussed in Sect. 4.
- Needs efficient solvers for the resulting linear system, for instance preconditioned Krylov space methods.

(2) *Validate your solver* by rigorous EOC tests on a sequence of uniform refinements for data of your problem computed from a smooth "solution". This means, pick up a suitable smooth function u, apply the differential operator to u to compute the right hand side f of your PDE. Dirichlet boundary data g is directly given by u. We comment on EOC tests below.
(3) Add an *a posteriori error estimator* built from local indicators. Unfortunately there is no black-box solution for estimators up to now. Evaluate the estimator by comparing it with the true error.
(4) Standard *marking routines* are problem independent. Such routines as well as *refinement and coarsening* tools are available in ALBERTA.

Main work has to be done in Step (1), the implementation of the solver. This step enormously profits from a program development in 1d and 2d with rather short run times for tests and very good visualization tools that strongly support debugging. In combination with an estimator this results in an adaptive code also working in 3d.

Experimental Order of Convergence. Let $\{\mathcal{T}_k\}_{k\geq 0}$ be a sequence of uniformly refined meshes, where we have used d bisections for all elements. This yields $h_{\max}(\mathcal{T}_{k+1}) = \frac{1}{2}h_{\max}(\mathcal{T}_k)$ with $h_{\max}(\mathcal{T}) := \max_{T\in\mathcal{T}} |T|^{1/d}$.

We next *assume*, that we have an a priori error estimate of the form

$$\|U_k - u\|_{\mathbb{V}} \approx C\, h_{\max}^s(\mathcal{T}_k)\|u\|_{\mathbb{V}^s}$$

for the sequence of Galerkin approximations $U_k \in \mathbb{V}(\mathcal{T}_k)$. This in turn implies

$$\frac{\|U_k - u\|_{\mathbb{V}}}{\|U_{k+1} - u\|_{\mathbb{V}}} \approx \frac{h_{\max}^s(\mathcal{T}_k)}{h_{\max}^s(\mathcal{T}_{k+1})} = 2^s$$

and allows us to extract *the experimental order of convergence (EOC) s* by

$$s \approx \mathsf{EOC}_k := \log\left(\frac{\|U_k - u\|_{\mathbb{V}}}{\|U_{k+1} - u\|_{\mathbb{V}}}\right) / \log(2).$$

For a "known" solution u, the true error can be computed by means of numerical integration. In this vein, the EOC is a computable quantity that can be calculated on a sequence of uniformly refined grids. If EOC_k does not come close to s after some iterations, then this is a strong indication that the solver is not implemented properly. Possible error sources are bugs in the implementation, improper choice of numerical quadrature formulas and tolerances for iterative solvers, etc. In case of 2nd order elliptic problems we have $s = p$ for $u \in H^{p+1}(\Omega)$.

For adaptively generated grids $\{\mathcal{T}_k\}_{k\geq 0}$ one expects the following error estimate in terms of DOFs

$$\mathcal{E}_k(U_k, \mathcal{T}_k) \approx \|U_k - u\|_{\mathbb{V}} \leq C\, \#\mathrm{DOFs}_k^{-s} \|u\|_{\mathbb{V}^s}$$

including singular solutions u that have some additional regularity $\mathbb{V}^s$ beyond $\mathbb{V}$. For instance, such an estimate is known with $s = p/d$ for the best approximation of u in case $d = 2$ and $p = 1$ if $u \in \mathbb{V}^s := W_q^2(\Omega)$ for some $q > 1$. For linear, symmetric, and elliptic PDEs it has been shown that the standard adaptive algorithm with minimal Dörfler marking using a sufficiently small parameter θ yields an optimal error decay for the Galerkin solution in terms of DOFs [15, 23] and [8, 37].

Following the ideas for computing the experimental order of convergence for uniform refinement we can compute the EOC in case of adaptive refinement by

$$s \approx \mathsf{EOC}_k := -\log\left(\frac{\mathcal{E}_k(U_k, \mathcal{T}_k)}{\mathcal{E}_{k+1}(U_{k+1}, \mathcal{T}_{k+1})}\right) / \log\left(\frac{\mathrm{DOFs}_k}{\mathrm{DOFs}_{k+1}}\right).$$

For a singular solution, i.e., $u \notin H^{p+1}(\Omega)$, that still exhibits some additional regularity we expect $\mathsf{EOC}_k \approx p/d$.

5.2.2 Concluding Remarks

The design of mathematical software significantly profits from specifying the mathematical properties of the problem under consideration. The mathematical language has been developed to precisely describe problems and solutions to problems. The design of the basic data structure should reflect mathematical definitions of important objects. The design of algorithms is based on a mathematical description how a specific operation is executed.

Practical experience strongly suggests that employing mathematical properties simplifies implementation drastically. Nevertheless, the practical implementation is usually much more involved than the mathematical description. But in general even more problems show up without the right mathematical basis.

What is the Right Software? Nowadays there are many free packages on the market to solve PDEs. The design of such packages should follow ideas presented in this course. There are some basic rules for selecting a software package for a specific application.

(1) Do not start to implement any kind of solver from scratch. There are may solutions around!
(2) Check the web for an existing solver for the application at hand:

    a. The solver is documented: Accept
    b. The solver is not documented: Decline

(3) There is no solver available: Check for available general purpose finite element packages like ALBERTA, DEAL, DUNE, PMTLG, ... Choose the best one for your application:

a. The package is documented: Accept
b. The package is not documented: Decline

In this course we have focused on the finite element toolbox ALBERTA. It comprises several advantages: it is fully documented, small but powerful, open source, etc. With advantages come disadvantages. ALBERTA is based on a sequential code, it only allows for simplicial elements, etc. Nevertheless, ALBERTA has been applied successfully to many different kinds of problems at several research institutes distributed all over the world.

6 Supplement: A Nonlinear and a Saddlepoint Problem

In this section we shortly introduce a nonlinear problem and a saddle point problem and describe how to implement solvers for such problems. As we shall see, in both cases we need to solve a sequence of linear elliptic problems as discussed in the previous chapters.

6.1 *The Prescribed Mean Curvature Problem in Graph Formulation*

Hypersurfaces M in $\mathbb{R}^{d+1}$ with mean curvature H/d that are described as the graph of a function $u\colon \Omega \to \mathbb{R}$, i. e., $M = \{(x, u(x)) \mid x \in \Omega\}$, satisfy the *quasi-linear PDE*

$$-\operatorname{div}\left(\frac{\nabla u}{\sqrt{1+|\nabla u|^2}}\right) = H \quad \text{in } \Omega \qquad u = g \quad \text{on } \partial\Omega.$$

The problem is *non-uniformly elliptic*; obviously there are problems with large gradients $|\nabla u|$. Furthermore, the problem may *not be solvable* if the prescribed curvature H is too large; compare with Fig. 26 for constant H.

As for the linear elliptic PDE, we use integration by parts to obtain the variational formulation.

Problem 74 (Prescribed Mean Curvature). Set $\mathbb{V} = H^1(\Omega)$ and $\mathring{\mathbb{V}} = H^1_0(\Omega)$ and solve

$$u \in g + \mathring{\mathbb{V}}: \qquad \int_\Omega \frac{\nabla v \cdot \nabla u}{\left(1+|\nabla u|^2\right)^{1/2}}\,dx = \int_\Omega vH\,dx \qquad \forall v \in \mathring{\mathbb{V}}.$$

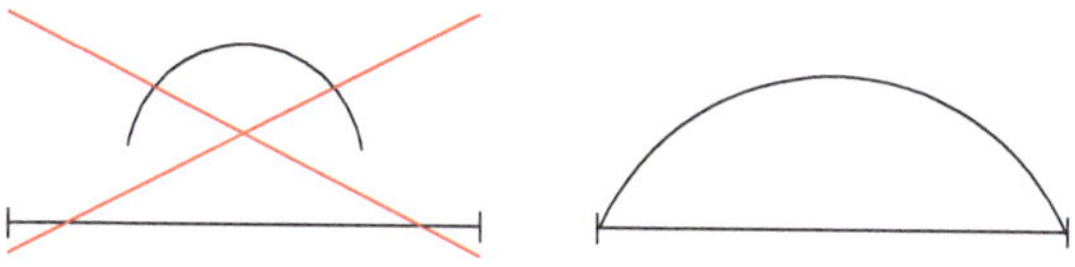

Fig. 26 If H is too large, the prescribed mean curvature problem is not solvable (*left*), whereas it admits a solution for smaller H

Problem 74 are the *Euler-Lagrange Equations* corresponding to the *minimization of the energy*

$$E(w) := \int_\Omega \left(1 + |\nabla w|^2\right)^{1/2} - Hw\,dx \qquad \text{for } w \in g + \mathring{\mathbb{V}}.$$

Problem 74 has at most one solution. A necessary condition for existence of a solution is

$$\exists \varepsilon > 0: \qquad \int_\Omega vH\,dx \le (1-\varepsilon)\|\nabla v\|_{L_1(\Omega)} \qquad \forall v \in \mathring{\mathbb{V}}.$$

See [20, Theorem 16.10] also for sufficient conditions.

Problem 75 (Discrete Prescribed Mean Curvature Problem). For a conforming triangulation $\mathscr{T}$ and given $\bar{\mathbb{P}} \subset C^1(\bar{T})$ set

$$\mathbb{V}(\mathscr{T}) = \mathsf{FES}(\mathscr{T}, \bar{\mathbb{P}}, H^1(\Omega)) \qquad \text{and} \qquad \mathring{\mathbb{V}}(\mathscr{T}) = \mathbb{V}(\mathscr{T}) \cap H_0^1(\Omega).$$

Solve

$$U \in G + \mathring{\mathbb{V}}(\mathscr{T}): \quad \int_\Omega \frac{\nabla V \cdot \nabla U}{\left(1 + |\nabla U|^2\right)^{1/2}}\,dx = \int_\Omega VH\,dx \quad \forall V \in \mathring{\mathbb{V}}(\mathscr{T}),$$

where $G \in \mathbb{V}(\mathscr{T})$ is an approximation to boundary data g.

A necessary and sufficient condition for existence of a unique discrete solution is

$$\exists \varepsilon > 0: \qquad \int_\Omega VH\,dx \le (1-\varepsilon)\|\nabla V\|_{L^1(\Omega)} \qquad \forall V \in \mathring{\mathbb{V}}(\mathscr{T}).$$

See [18, Proposition 2.1].

Defining the nonlinear function $F\colon \mathbb{V}(\mathscr{T}) \to \mathring{\mathbb{V}}(\mathscr{T})^*$ as

$$\langle F(W), V\rangle := \int_\Omega \frac{\nabla V \cdot \nabla W}{\left(1 + |\nabla W|^2\right)^{1/2}} - V\,H\,dx \qquad \forall W, V \in \mathbb{V}(\mathscr{T})$$

Problem 75 is equivalent to finding a root of F, i. e.,

$$U \in G + \mathring{\mathbb{V}}(\mathscr{T}): \qquad F(U) = 0 \qquad \text{in } \mathring{\mathbb{V}}(\mathscr{T})^*.$$

Using a basis $\{\Phi_1, \ldots, \Phi_n\}$ of $\mathbb{V}(\mathscr{T})$ this leads to a nonlinear system of equations in $\mathbb{R}^N$, which can for instance be solved by a *Newton method*.

Algorithm 76 (Newton Method). Start with an *initial guess* $U^{(0)} \in G + \mathring{\mathbb{V}}(\mathscr{T})$. For $\ell \in \mathbb{N}_0$ solve the *linear equation*

$$D^{(\ell)} \in \mathring{\mathbb{V}}(\mathscr{T}): \quad \langle DF\big(U^{(\ell)}\big)\, D^{(\ell)}, V\rangle = -\langle F(U^{(\ell)}), V\rangle \quad \forall\, V \in \mathring{\mathbb{V}}(\mathscr{T})$$

and set $U^{(\ell+1)} := U^{(\ell)} + D^{(\ell)}$. For given $U \in \mathbb{V}(\mathscr{T})$ the Jacobian $DF(U)$ is

$$\langle DF\big(U\big)\, W, V\rangle = \int_\Omega \frac{\nabla V \cdot \nabla W}{\big(1 + |\nabla U|^2\big)^{1/2}} - \frac{\nabla V (\nabla U \otimes \nabla U) \nabla W}{\big(1 + |\nabla U|^2\big)^{3/2}}\, dx$$

for all $V, W \in \mathring{\mathbb{V}}(\mathscr{T})$ with the notation $\nabla U \otimes \nabla U := \big[U_{,x_i} U_{,x_j}\big]_{i,j=1,\ldots,d}$.

Alternatively a Newton Method with inexact Jacobian can be used.

Algorithm 77 (Inexact Newton Method). Start with an *initial guess* $U^{(0)} \in G + \mathring{\mathbb{V}}(\mathscr{T})$. For $\ell \in \mathbb{N}_0$ solve the linear equation

$$D^{(\ell)} \in \mathring{\mathbb{V}}(\mathscr{T}): \quad \langle \tilde{D}F\big(U^{(\ell)}\big)\, D^{(\ell)}, V\rangle = -\langle F(U^{(\ell)}), V\rangle \quad \forall V \in \mathring{\mathbb{V}}(\mathscr{T})$$

and set $U^{(\ell+1)} := U^{(\ell)} + D^{(\ell)}$. For $U \in \mathbb{V}(\mathscr{T})$ the inexact Jacobian $\tilde{D}F(U)$ is given by

$$\langle \tilde{D}F\big(U\big)\, W, V\rangle = \int_\Omega \frac{\nabla V \cdot \nabla W}{\big(1 + |\nabla U|^2\big)^{1/2}}\, dx.$$

for all $V, W \in \mathring{\mathbb{V}}(\mathscr{T})$.

Some remarks about the (inexact) Newton method are in order.

(1) Starting with the *correct discrete boundary values* G, the residual of the nonlinear equation is zero at boundary nodes. Therefore, all corrections $D^{(\ell)}$ of the exact and inexact Newton method belong to $\mathring{\mathbb{V}}(\mathscr{T})$.
(2) The linear sub-problems are a discretization of the linear elliptic PDE

$$-\operatorname{div}\big(\boldsymbol{A}(U)\nabla w\big) = f(U) \quad \text{in } \Omega, \qquad w = 0 \quad \text{on } \partial\Omega.$$

The coefficient matrix $\boldsymbol{A}(U) = \boldsymbol{A}(x)$ is given by

$$\boldsymbol{A}(U) = \frac{\mathrm{id}}{\big(1 + |\nabla U|^2\big)^{1/2}} - \delta \frac{\nabla U \otimes \nabla U}{\big(1 + |\nabla U|^2\big)^{3/2}},$$

where $\delta = 1$ for the exact and $\delta = 0$ for the inexact Newton method. Note, that in general $f(U) \notin L^2(\Omega)$ but $f(U) \in H^{-1}(\Omega) = \big(H_0^1(\Omega)\big)^*$.
(3) Newton's method is converging locally with a quadratic rate, whereas the inexact iteration is converging *at most* with linear rate. But the system matrix of the linearized equation in the simplified iteration has better properties.

Error Estimator. Verfürth has shown that the indicators

$$\mathcal{E}^2(U,T) := C_0^2 h_T^2 \left\| -\operatorname{div}\left(\left(1+|\nabla U|^2\right)^{-1/2}\nabla U\right)\right\|_{L^2(S)}^2 + C_1^2 h_T \left\| \left[\!\left[\left(1+|\nabla U|^2\right)^{-1/2}\nabla U \right]\!\right] \right\|_{L^2(\partial S\cap\Omega)}^2$$

built up an estimator $\mathcal{E}^2(U,\mathcal{T}) := \sum_{T\in\mathcal{T}} \mathcal{E}^2(U,T)$ for the prescribed mean curvature problem with homogeneous boundary data $g = 0$ [41]. This estimator is implemented in the file `Common/pmc-estimator.c`.

We want to remark that this estimator is not *robust* since the problem is not uniformly elliptic. A robust estimator with local conditioning has been derived by Fierro and Veeser [18].

Exercise 68. Implement an ALBERTA program for the solution of the prescribed mean curvature problem. For the solution of the nonlinear discrete system use the Newton solver `nls_newton()`, or the Newton solver `nls_newton_fs()` with step size control:

```
int nls_newton(NLS_DATA *ninfo, int dim, REAL *x);
int nls_newton_fs(NLS_DATA *ninfo, int dim, REAL *x);
```

The detailed description of the `NLS_DATA` data structure and Newton solvers can be found in [34, Sect. 3.15.6]. A model implementation of a nonlinear equation is described in detail in [34, Sect. 2.2].

For the initialization of the data structure `NLS_DATA`, which is the first argument to the Newton solvers, the following functions have to be implemented:

- `update()` assembles for given $U \in \mathbb{V}(\mathcal{T})$ the exact or inexact Jacobian and/or the residual $F(U)$.

  This means, for all basis functions $\Phi_i, \Phi_j \in \mathring{\mathbb{V}}(\mathcal{T})$ compute the system matrix of the linearized problem

  $$\left[\int_\Omega \frac{\nabla\Phi_i \nabla\Phi_j}{(1+|\nabla U|^2)^{1/2}} - \delta \frac{\nabla\Phi_i(\nabla U\otimes\nabla U)\nabla\Phi_j}{(1+|\nabla U|^2)^{3/2}}\,dx\right]_{i,j=1,\dots\mathring{N}},$$

  where $\delta = 1$ is the exact and $\delta = 0$ is the inexact Newton method. The residual is given as

  $$\left[\int_\Omega \frac{\nabla U\nabla\Phi_i}{(1+|\nabla U|^2)^{1/2}}\,dx - \int_\Omega H\Phi_i\,dx\right]_{i=1,\dots,\mathring{N}}.$$

  Furthermore, set the corresponding values of homogeneous boundary values of the corrections.
- `solve()` solves the resulting linear system by a preconditioned CG-method.
- `norm()` computes the H^1-semi-norm $|D|_{H^1(\Omega)}$ for given $D \in \mathbb{V}(\mathcal{T})$.

Stop the iteration if $\|F\left(U^{(\ell+1)}\right)\|$ is sufficiently small in some suitable norm.

As an initial guess for Newton's method use an arbitrary chosen function $U^{(0)} \in G + \mathring{\mathbb{V}}(\mathcal{T})$ on the coarsest grid. Interpolate the coefficient vector of the old solution during refinement and adjust the correct boundary values G on the new grid.

- Test the program with the exact solution

$$u(x) = \sqrt{R^2 - |x|^2} \qquad x \in \Omega = B_r(0) \subset \mathbb{R}^d.$$

  for different $r < R$ and $d = 2, 3$.
- Compute the minimal surface which is given by

$$u(x) = -\ln\left(|x| - \sqrt{|x|^2 - 1}\right) \qquad x \in \Omega = (a, a+5)^2,$$

  where a is chosen to by greater than $\frac{1}{\sqrt{2}}$.
- Compute the minimal surface M for boundary data g given in polar coordinates

$$g(r, \varphi) = \sin k\varphi$$

  on $B_1(0) \subset \mathbb{R}^2$ for different k.

Compare the exact and inexact Newton methods for the above problems.

6.2 The Generalized Stokes Problem

We next slightly generalize the Stokes system describing a stationary, incompressible, viscous flow from Example 4. Assuming that the flow is no longer stationary the corresponding model reads: Find $(\boldsymbol{u}, p)$ satisfying the parabolic PDE

$$\partial_t \boldsymbol{u} - \nu \Delta \boldsymbol{u} + \nabla p = \boldsymbol{f} \quad \text{in } \Omega,\, t > 0, \qquad \operatorname{div} \boldsymbol{u} = 0 \quad \text{in } \Omega,\, t > 0$$

together with boundary and initial values. After applying a time discretization with an implicit Euler discretization and time step size $\tau > 0$ we obtain a sequence of saddle point problems: For $n \in \mathbb{N}$ solve

$$\frac{1}{\tau}\boldsymbol{u}_n - \nu \Delta \boldsymbol{u}_n + \nabla p_n = \boldsymbol{f}_n + \frac{1}{\tau}\boldsymbol{u}_{n-1} \text{ in } \Omega, \quad \operatorname{div} \boldsymbol{u}_n = 0 \text{ in } \Omega, \quad \boldsymbol{u}_n = \boldsymbol{g}_n \text{ on } \partial\Omega.$$

This means, that we have to solve in each time step a *generalized Stokes problem*: For given parameters $\nu > 0$ and $\mu \geq 0$, forcing term $\boldsymbol{f}$, and boundary data $\boldsymbol{g}$ solve

$$\mu \boldsymbol{u} - \nu \Delta \boldsymbol{u} + \nabla p = \boldsymbol{f} \quad \text{in } \Omega, \qquad \operatorname{div} \boldsymbol{u} = 0 \quad \text{in } \Omega, \qquad \boldsymbol{u} = \boldsymbol{g} \quad \text{on } \partial\Omega.$$

Looking at this system we see that a solution $(\boldsymbol{u}, p)$ is never unique, since then $(\boldsymbol{u}, p + c)$ is again a solution for all $c \in \mathbb{R}$. We can rewrite the problem formally as

$$\begin{bmatrix} \mu \mathrm{id} - \nu\Delta & \nabla \\ -\operatorname{div} & 0 \end{bmatrix} \begin{bmatrix} \boldsymbol{u} \\ p \end{bmatrix} = \begin{bmatrix} \boldsymbol{f} \\ 0 \end{bmatrix}.$$

This also indicates that the generalized Stokes problem might be singular. Again formally, $-\operatorname{div}$ is the adjoint operator to ∇ since by Gauß' Divergence Theorem

$$\begin{aligned} \langle \boldsymbol{v}, \nabla q \rangle &= \int_\Omega \boldsymbol{v} \nabla q \, dx = \int_\Omega \operatorname{div}(\boldsymbol{v}\, q) \, dx - \int_\Omega \operatorname{div} \boldsymbol{v}\, q \, dx \\ &= \int_{\partial\Omega} \boldsymbol{v} \cdot \boldsymbol{n}\, q \, do - \int_\Omega \operatorname{div} \boldsymbol{v}\, q \, dx = \int_\Omega - \operatorname{div} \boldsymbol{v}\, q \, dx = \langle -\operatorname{div} \boldsymbol{v}, q \rangle \end{aligned}$$

holds for all $\boldsymbol{v} \in H_0^1(\Omega; \mathbb{R}^d)$ and $q \in H^1(\Omega)$. Consequently, the generalized Stokes problem is *symmetric*. Using Gauß' Divergence Theorem once more we realize that a necessary condition for the existence of a solution is the following *compatibility condition* of boundary data $\boldsymbol{g}$:

$$0 = \int_\Omega \operatorname{div} \boldsymbol{u} \, dx = \int_{\partial\Omega} \boldsymbol{u} \cdot \boldsymbol{n} \, do = \int_{\partial\Omega} \boldsymbol{g} \cdot \boldsymbol{n} \, do.$$

Variational Formulation. We next turn to a weak formulation, which differs on a first glance from the formulation used in Example 4. Set $\mathbb{V} := H^1(\Omega; \mathbb{R}^d)$ and $\mathring{\mathbb{V}} := H_0^1(\Omega; \mathbb{R}^d)$, $\mathbb{Q} := L^2(\Omega)$ and

$$\mathring{\mathbb{Q}} := \Big\{ q \in L^2(\Omega) \mid \int_\Omega q \, dx = 0. \Big\}.$$

The weak formulation is based on *integration by parts* for $-\Delta$ and ∇ defining the *bilinear forms* $a\colon \mathbb{V} \times \mathbb{V} \to \mathbb{R}$

$$a[\boldsymbol{w}, \boldsymbol{v}] := \mu \int_\Omega \boldsymbol{v} \cdot \boldsymbol{w} \, dx + \nu \int_\Omega \nabla \boldsymbol{v} : \nabla \boldsymbol{w} \, dx \qquad \forall \boldsymbol{v}, \boldsymbol{w} \in \mathbb{V}$$

and $b\colon \mathbb{V} \times \mathbb{Q} \to \mathbb{R}$

$$b[\boldsymbol{v}, q] := -\int_\Omega \operatorname{div} \boldsymbol{v}\, q \, dx \qquad \forall \boldsymbol{v}, \in \mathbb{V},\ q \in \mathbb{Q}.$$

Problem 79 (Generalized Stokes Problem). Let $\boldsymbol{f} \in \mathbb{V}^*$ and $\boldsymbol{g} \in \mathbb{V}$ be given and assume that $\int_{\partial\Omega} \boldsymbol{g} \cdot \boldsymbol{n} \, do = 0$ holds. Find $(\boldsymbol{u}, p) \in (\boldsymbol{g} + \mathring{\mathbb{V}}) \times \mathring{\mathbb{Q}}$ such that

$$
\begin{aligned}
a[\boldsymbol{v},\boldsymbol{u}] + b[\boldsymbol{v},p] &= \langle \boldsymbol{v},\ \boldsymbol{f}\rangle \quad \forall \boldsymbol{v} \in \mathring{\mathbb{V}},\\
b[\boldsymbol{u},q] &= 0 \quad \forall q \in \mathring{\mathbb{Q}}.
\end{aligned}
$$

Set $\mathbb{W} = \mathring{\mathbb{V}} \times \mathring{\mathbb{Q}}$, $f = (\boldsymbol{f},0)$, $g = (\boldsymbol{g},0)$, and for $v = (\boldsymbol{v},q), w = (\boldsymbol{w},r) \in \mathbb{W}$ define

$$\mathscr{B}[w,\ v] = a[\boldsymbol{w},\boldsymbol{v}] + b[\boldsymbol{v},r] + b[\boldsymbol{w},q].$$

Then Problem 79 is equivalent to the problem

$$u = (\boldsymbol{u},p) \in g + \mathbb{W}: \quad \mathscr{B}[w,\ v] = \langle f,\ v\rangle \quad \forall v \in \mathbb{W}.$$

Such a variational formulation we have used in Example 4 for the Stokes problem. From Lemma 55 we know that Problem 79 admits a unique solution $(\boldsymbol{u},p) \in \mathbb{W}$ if $\mathscr{B}$ satisfies an inf-sup condition on $\mathbb{W}$. The bilinear form $a\colon \mathbb{V} \times \mathbb{V} \to \mathbb{R}$ is continuous on $\mathbb{V}$ and coercive on $\mathring{\mathbb{V}}$. Therefore, the inf-sup condition of $\mathscr{B}$ is a consequence of the *Ladyshenskaja–Babuška–Brezzi* condition

$$\inf_{q\in\mathring{\mathbb{Q}}} \sup_{\boldsymbol{v}\in\mathring{\mathbb{V}}} \frac{b[\boldsymbol{v},q]}{\|\boldsymbol{v}\|_{\mathbb{V}}\|q\|_{\mathbb{Q}}} = \inf_{q\in\mathring{\mathbb{Q}}} \sup_{\boldsymbol{v}\in\mathring{\mathbb{V}}} \frac{\langle -\operatorname{div}\boldsymbol{v},\ q\rangle}{\|\boldsymbol{v}\|_{\mathbb{V}}\|q\|_{\mathbb{Q}}} \ge \beta > 0. \tag{8}$$

Condition (8) is equivalent to the solvability of the divergence equation, i. e., for all $q \in \mathring{\mathbb{Q}}$ exists $\boldsymbol{v}_q \in \mathring{\mathbb{V}}$ such that

$$-\operatorname{div}\boldsymbol{v}_q = q \text{ in } \Omega \qquad \text{and} \qquad \|\boldsymbol{v}_q\|_{H^1(\Omega;\mathbb{R}^d)} \le \beta^{-1}\|q\|_{L^2(\Omega)};$$

compare with [14] or [19, Theorem III.3.1].

Remark 80 (Saddle Point Structure). Consider $\boldsymbol{g} \equiv \boldsymbol{0}$. For the generalized Stokes problem we can define

$$E[\boldsymbol{w},q] := \tfrac{1}{2}a[\boldsymbol{w},\boldsymbol{w}] + b[\boldsymbol{w},q] - \langle \boldsymbol{w},\ \boldsymbol{f}\rangle \qquad \forall \boldsymbol{w} \in \mathring{\mathbb{V}},\ q \in \mathring{\mathbb{Q}}.$$

This functional is neither bounded from above nor below. The pair $(\boldsymbol{u},p) \in \mathbb{W}$ is a solution of Problem 79 iff

$$E[\boldsymbol{u},q] \le E[\boldsymbol{u},p] \le E[\boldsymbol{w},p] \qquad \forall \boldsymbol{w} \in \mathring{\mathbb{V}},\ q \in \mathring{\mathbb{Q}}.$$

This is equivalent to the

$$E[\boldsymbol{u},p] = \min_{\boldsymbol{w}\in\mathring{\mathbb{V}}} E[\boldsymbol{w},p] = \min_{\boldsymbol{w}\in\mathring{\mathbb{V}}} \max_{q\in\mathring{\mathbb{Q}}} E[\boldsymbol{w},q],$$

i. e., $(\boldsymbol{u},p)$ is a saddle point of the functional E. Furthermore, the velocity is a minimizer, i. e.,

$$u = \arg\min\{E[w,0] \mid w \in \mathring{\mathbb{V}},\ \operatorname{div} w = 0\}.$$

Therefore, p can be interpreted as a *Lagrange multiplier* from constrained optimization.

Problem 81 (Discrete Generalized Stokes Problem). Let $\mathbb{V}(\mathcal{T}) \subset \mathbb{V}$, $\mathbb{Q}(\mathcal{T}) \subset \mathbb{Q}$ be finite element spaces. Define $\mathring{\mathbb{V}}(\mathcal{T}) := \mathbb{V}(\mathcal{T}) \cap \mathring{\mathbb{V}}$ and $\mathring{\mathbb{Q}}(\mathcal{T}) := \mathbb{Q}(\mathcal{T}) \cap \mathring{\mathbb{Q}}$. The *discrete problem* reads: Find $(U, P) \in (G + \mathring{\mathbb{V}}(\mathcal{T})) \times \mathring{\mathbb{Q}}(\mathcal{T})$ such that

$$\begin{aligned} a[U,V] + b[V,P] &= \langle V, f\rangle \quad && \forall V \in \mathring{\mathbb{V}}(\mathcal{T}),\\ b(U,Q) &= 0 && \forall Q \in \mathring{\mathbb{Q}}(\mathcal{T}). \end{aligned}$$

Hereafter, $G \in \mathbb{V}(\mathcal{T})$ is an approximation to g satisfying $\int_{\partial\Omega} G \cdot n \, do = 0$.

Note, that in general for *interpolated boundary values* $G = I_{\mathcal{T}} g$ we have

$$\int_\Omega \operatorname{div} U \, dx = \int_{\partial\Omega} (I_{\mathcal{T}} g) \cdot n \, do \neq 0.$$

Such interpolation operators have to be modified such that $\int_\Omega \operatorname{div} U \, dx = 0$; compare with [13].

The discrete problem has a unique solution, provided the *discrete LBB condition*

$$\inf_{Q \in \mathring{\mathbb{Q}}(\mathcal{T})} \sup_{V \in \mathring{\mathbb{V}}(\mathcal{T})} \frac{b[V,Q]}{\|V\|_{\mathbb{V}} \|Q\|_{\mathbb{Q}}} \geq \beta(\mathcal{T}) > 0$$

holds. A sequence of discrete spaces $\{\mathbb{V}(\mathcal{T}_k), \mathbb{Q}(\mathcal{T}_k)\}_{k\geq 0}$ is called *stable*, iff

$$\inf_{k\geq 0} \beta(\mathcal{T}_k) = \underline{\beta} > 0.$$

The Schur-Complement Operator. Denote by A the system matrix related to the bilinear form a and by B the system matrix related to b. Then Problem 81 is equivalent to the *linear system*

$$\begin{bmatrix} A & B \\ B^T & 0 \end{bmatrix} \begin{bmatrix} u \\ p \end{bmatrix} = \begin{bmatrix} F \\ 0 \end{bmatrix}, \tag{9}$$

where u is the coefficient vector of the discrete velocity U and p the coefficient vector of the discrete pressure P.

Coercivity of a implies invertibility of the matrix A. Hence a block-Gauß elimination yields the following equivalent formulation of (9):

$$S p := (B^T A^{-1} B) p = (B^T A^{-1}) F \qquad \text{and} \qquad A u = F - B p.$$

The matrix $\boldsymbol{S}$ is the *Schur-complement operator*. The discrete LBB condition implies *invertibility* of $\boldsymbol{S}$. Hence, we can solve Problem 81 by inverting $\boldsymbol{S}$, obtaining the discrete pressure $\boldsymbol{P}$ and then inverting $\boldsymbol{A}$ to obtain the discrete velocity $\boldsymbol{U}$.

Remark 82 (Schur-Complement Operator). The condition number of $\boldsymbol{S}$ depends on $\beta^{-1}(\mathscr{T})$. Consequently, stable discretizations are important for having a uniform bound on the condition number of $\boldsymbol{S}$.

The system matrix $\boldsymbol{A}$ is *block diagonal*, this means

$$\boldsymbol{A} = \begin{bmatrix} A & 0 \\ 0 & A \end{bmatrix} \quad \text{in 2d, respectively} \quad \boldsymbol{A} = \begin{bmatrix} A & 0 & 0 \\ 0 & A & 0 \\ 0 & 0 & A \end{bmatrix} \quad \text{in 3d,}$$

where A is the system matrix corresponding to the *scalar differential operator*

$$Lu := \mu u - \nu \Delta u.$$

Although $\boldsymbol{A}$ and $\boldsymbol{B}$ are sparse matrices $\boldsymbol{S}$ is not a sparse matrix and should not be computed. An alternative is to use the *damped Richardson iteration* for solving $\boldsymbol{S}\boldsymbol{P} = (\boldsymbol{B}^T\boldsymbol{A}^{-1})\boldsymbol{F}$. The Richardson iteration is an iterative solver that only requires a matrix-vector multiplication with the system matrix $\boldsymbol{S}$. Any such multiplication in turn then requires the solution of a linear system with system matrix $\boldsymbol{A}$. In summary, this gives the *Uzawa algorithm*.

Algorithm 83 (Uzawa Algorithm). Let $\omega > 0$ be a sufficiently small *damping parameter*, $\mathsf{TOL} > 0$ a *tolerance* for the residual. Starting with an initial guess $P^{(0)} \in \mathring{\mathbb{Q}}(\mathscr{T})$ for the pressure iterate for $m \in \mathbb{N}$:

(1) Solve the *elliptic equation*

$$\boldsymbol{U}^{(m)} \in \boldsymbol{G} + \mathring{\mathbb{V}}(\mathscr{T}): \quad a[\boldsymbol{U}^{(m)}, \boldsymbol{V}] = \langle \boldsymbol{f}, \boldsymbol{V} \rangle - b[\boldsymbol{V}, P^{(m-1)}] \quad \forall\, \boldsymbol{V} \in \mathring{\mathbb{V}}(\mathscr{T}_k).$$

(2) If $\operatorname{div} \mathbb{V}(\mathscr{T}) \not\subset \mathbb{Q}(\mathscr{T})$, *compute the L^2 projection* of $-\operatorname{div} \boldsymbol{U}^{(m)}$, i. e.,

$$R^{(m)} \in \mathbb{Q}(\mathscr{T}): \quad \langle R^{(m)}, Q \rangle_{L^2(\Omega)} = b[\boldsymbol{U}^{(m)}, Q] \quad \forall Q \in \mathbb{Q}(\mathscr{T});$$

otherwise define $R^{(m)} := -\operatorname{div} \boldsymbol{u}^{(m)}$.

(3) *Update the pressure* by

$$P^{(m)} := P^{(m-1)} + \omega \left(R^{(m)} - |\Omega|^{-1} \int_\Omega R^{(m)}\, dx \right).$$

(4) *Stop* if $R^{(m)}$ is sufficiently small, i. e., $\|R^{(m)}\|_{L^2(\Omega)} \le \mathsf{TOL}$.

Some remarks on the Uzawa Algorithm are in order.

(1) The choice of the parameter ω depends on the eigenvalues of $\boldsymbol{S}$. The optimal choice would be

$$\omega_{\text{opt}} = \frac{2}{\lambda_{\max}(\boldsymbol{S}) + \lambda_{\min}(\boldsymbol{S})}.$$

Note, that the eigenvalues $\lambda_{\min,\max}(\boldsymbol{S})$ depend on $\beta(\mathscr{T})$, ν, and μ. The constant $\beta(\mathscr{T})$ is not known explicitly.

(2) The *Taylor-Hood element* of order $\ell \geq 2$, i. e.,

$$\mathbb{V}(\mathscr{T}) = \mathsf{FES}(\mathscr{T}, \mathbb{P}_\ell, \mathbb{V}) \qquad \text{and} \qquad \mathbb{Q}(\mathscr{T}) = \mathsf{FES}(\mathscr{T}, \mathbb{P}_{\ell-1}, C^0(\bar{\Omega}))$$

is stable: $\beta(\mathscr{T}) \geq \underline{\beta} > 0$. Since functions in the pressure space are globally continuous there holds $\operatorname{div} \mathbb{V}(\mathscr{T}) \not\subset \mathbb{Q}(\mathscr{T})$.

(3) The compatibility property $\int_{\partial\Omega} \boldsymbol{G} \cdot \boldsymbol{n} \, do = 0$ of the discrete boundary data is essential for convergence of the Uzawa algorithm. Obviously, $\int_\Omega \operatorname{div} \boldsymbol{U}^{(m)} \, dx \neq 0$ implies $\|R^{(m)}\|_{L^2(\Omega)} = \|\operatorname{div} \boldsymbol{U}^{(m)}\|_{L^2(\Omega)} > 0$ and the algorithm cannot converge.

(4) In the Uzawa algorithm we have avoided to solve for $R^{(m)} \in \mathring{\mathbb{Q}}(\mathscr{T})$. The constraint $\langle q, 1 \rangle = 0$ for functions in $\mathring{\mathbb{Q}}(\mathscr{T})$ is *global*, which does not allow for an easy construction of a *local* basis. All problems to be solved are *well posed* and any discrete pressure fulfills $P^{(m)} \in \mathring{\mathbb{Q}}(\mathscr{T})$.

Error Estimator. Starting with the work by Verfürth [40], several authors have shown that the indicators

$$\begin{aligned}\mathscr{E}_{\mathscr{T}}^2(\boldsymbol{U}, P; T) := {} & C_0^2 h_T^2 \|\mu \boldsymbol{U} - \nu \Delta \boldsymbol{U} + \nabla P - \boldsymbol{f}\|_{L^2(T)}^2 \\ & + C_1^2 h_T \|[\![\nabla \boldsymbol{U}]\!]\|_{L^2(\partial T \cap \Omega)}^2 + C_2^2 \|\operatorname{div} \boldsymbol{U}\|_{L^2(T)}^2\end{aligned}$$

built up an estimator $\mathscr{E}_{\mathscr{T}}^2(\boldsymbol{U}, P; \mathscr{T}) := \sum_{T \in \mathscr{T}} \mathscr{E}_{\mathscr{T}}^2(\boldsymbol{U}, P; T)$ for the generalized Stokes problem with $\boldsymbol{g} = 0$. This estimator is robust and efficient. It is implemented in the file `Common/stokes-estimator.c`.

Exercise 84 (Saddle Point Problem). Implement an ALBERTA program for the solution of the generalized Stokes problem. For the solution of the discrete problem use the Uzawa Algorithm 83. The Uzawa algorithm requires the following routines:

- `build()` for the assembling of the (scalar) matrix

$$\int_\Omega \nabla \Phi_i \nabla \Phi_j$$

for all scalar basis functions of the velocity space (+ Dirichlet boundary values!), the assembling of the mass matrix

$$\int_\Omega \varphi_i \varphi_j$$

for all basis functions in the pressure space and the vector valued load vector

$$\int_\Omega \boldsymbol{f} \cdot \boldsymbol{\Phi}_i \, dx$$

for all basis functions in the velocity space (+ Dirichlet boundary values!). The object for storing vector valued DOF vectors, like the discrete velocity $\boldsymbol{U}$ or the discrete load vector is `DOF_REAL_D_VEC`. For the assembling of this vector the functions `L2scp_fct_bas_d()` and `dirichlet_bound_d()` can be used; compare with [34, Sect. 3.12].

- `add_B_p(P)` adds for given $P \in \mathbb{Q}(\mathscr{T})$

$$-\int_\Omega P \; \operatorname{div} \boldsymbol{\Phi}_i$$

to the vector valued load vector for all *interior* basis functions in the velocity space.

- `add_B_star_v(U)` assembles for given $\boldsymbol{U} \in \mathbb{V}(\mathscr{T})$ the scalar vector

$$-\int_\Omega \varphi_i \; \operatorname{div} \boldsymbol{U}$$

for all basis functions in the pressure space.

- `uzawa()` which solves the discrete saddle point problem for an initial guess $P^{(0)} \in \mathring{\mathbb{Q}}(\mathscr{T})$. The decoupled system in the velocity space can be solved by the function `oem_solve_d()`; compare with [34, Sect. 3.15].

Apply the program to the following problems

(1) $\Omega = (0,1)^2$ with the exact solution

$$\boldsymbol{u}(x) = \begin{bmatrix} (x_1^2 - 2\,x_1^3 + x_1^4)\,(2\,x_2 - 6\,x_2^2 + 4\,x_2^3) \\ -(2\,x_1 - 6\,x_2^1 + 4\,x_1^3)\,(x_2^2 - 2\,x_2^3 + x_2^4) \end{bmatrix},$$

$$p(x) = x_1^2 + x_2^2.$$

(2) $\Omega = (0, 2\,\pi)^2$ with the exact solution

$$\boldsymbol{u}(x) = \sin(x_1)\,\sin(x_2) \begin{bmatrix} -\sin(x_1)\,\cos(x_2) \\ \cos(x_1)\,\sin(x_2) \end{bmatrix},$$

$$p(x) = \cos(x_1) + \cos(x_2) + \cos(x_1)\,\cos(x_2).$$

(3) The *driven cavity* problem on $(0,1)^d$ with boundary data

$$\boldsymbol{g}(x) = \begin{cases} [1,0,\dots]^T, & \text{on } (0,1)^{d-1}\times\{1\},\\ \mathbf{0}, & \text{else.} \end{cases}$$

Note, that with this definition of $\boldsymbol{g}$ the Lagrange interpolant $I_{\mathcal{T}}\boldsymbol{g}$ is well defined. It is easy to see that discrete boundary data $\boldsymbol{G} = I_{\mathcal{T}}\boldsymbol{g}$ satisfies the compatibility condition $\int_{\partial\Omega} \boldsymbol{G}\cdot\boldsymbol{n}\,do = 0$.

References

1. M. Ainsworth, J.T. Oden, *A Posteriori Error Estimation in Finite Element Analysis* (Wiley-Interscience, New York, 2000)
2. D.N. Arnold, A. Mukherjee, L. Pouly, Locally adapted tetrahedral meshes using bisection. SIAM J. Sci. Comput. **22**(2), 431–448 (2000)
3. I. Babuška, Error-bounds for finite element method. Numer. Math. **16**, 322–333 (1971)
4. I. Babuška, W. Rheinboldt, Error estimates for adaptive finite element computations. SIAM J. Numer. Anal. **15**, 736–754 (1978)
5. E. Bänsch, Local mesh refinement in 2 and 3 dimensions. IMPACT Comput. Sci. Eng. **3**, 181–191 (1991)
6. J. Bey, Tetrahedral grid refinement. Computing **55**(4), 355–378 (1995)
7. J. Bey, Simplicial grid refinement: On Freudenthal's algorithm and the optimal number of congruence classes. Numer. Math. **85**(1), 1–29 (2000)
8. P. Binev, W. Dahmen, R. DeVore, Adaptive finite element methods with convergence rates. Numer. Math **97**, 219–268 (2004)
9. S. Boschert, A. Schmidt, K.G. Siebert, in *Numerical Simulation of Crystal Growth by the Vertical Bridgman Method*, ed. by J.S. Szmyd, K. Suzuki. Modelling of Transport Phenomena in Crystal Growth, Development in Heat Transfer Series, vol. 6 (WIT Press, Southampton, 61–96 (36), 2000), pp. 315–330
10. S. Boschert, A. Schmidt, K.G. Siebert, E. Bänsch, G. Dziuk, K.W. Benz, T. Kaiser, in *Simulation of Industrial Crystal Growth by the Vertical Bridgman Method*, ed. by W. Jäger et al. Mathematics – Key Technology for the Future. Joint Projects Between Universities and Industry (Springer, Berlin, 2003), pp. 315–342
11. S. Brenner, R. Scott, *The Mathematical Theory of Finite Element Methods.* Springer Texts in Applied Mathematics 15 (2008)
12. F. Brezzi, On the existence, uniqueness and approximation of saddle-point problems arising from Lagrange multipliers. R.A.I.R.O. Anal. Numer. **R2**, T 129–151 (1974)
13. M.O. Bristeau, R. Glowinski, J. Periaux, Numerical methods for the Navier–Stokes equations. Applications to the simulation of compressible and incompressible viscous flows. Comp. Phys. Rep. **6**, 73–187 (1987)
14. R. Carroll, G. Duff, J. Friberg, J. Gobert, P. Grisvard, J. Nečas, R. Seeley, Equations aux dérivées partielles. No. 19 in Seminaire de mathematiques superieures. Les Presses de l'Université de Montréal (1966)
15. J.M. Cascón, C. Kreuzer, R.H. Nochetto, K.G. Siebert, Quasi-optimal convergence rate for an adaptive finite element method. SIAM J. Numer. Anal. **46**(5), 2524–2550 (2008)
16. P.G. Ciarlet, in *The Finite Element Method for Elliptic Problems.* Classics in Applied Mathematics, vol. 40 (SIAM, PA, 2002)
17. W. Dörfler, A convergent adaptive algorithm for Poisson's equation. SIAM J. Numer. Anal. **33**, 1106–1124 (1996)

18. F. Fierro, A. Veeser, On the a posteriori error analysis for equations of prescribed mean curvature. Math. Comp. **72**(244), 1611–1634 (2003)
19. G.P. Galdi, *An Introduction to the Mathematical Theory of the Navier-Stokes Equations*, vol. 1: Linearized steady problems. Springer Tracts in Natural Philosophy, 38 (1994)
20. D. Gilbarg, N.S. Trudinger, in *Elliptic partial differential equations of second order*. Classics in Mathematics (Springer, Berlin, 2001)
21. I. Kossaczký, A recursive approach to local mesh refinement in two and three dimensions. J. Comput. Appl. Math. **55**, 275–288 (1994)
22. D. Köster, O. Kriessl, K.G. Siebert, Design of finite element tools for coupled surface and volume meshes. Numer. Math. Theor. Meth. Appl. **1**(3), 245–274 (2008)
23. C. Kreuzer, K.G. Siebert, Decay rates of adaptive finite elements with Dörfler marking. Numer. Math. **117**(4), 679–716 (2011)
24. A. Liu, B. Joe, Quality local refinement of tetrahedral meshes based on bisection. SIAM J. Sci. Comput. **16**, 1269–1291 (1995)
25. J.M. Maubach, Local bisection refinement for n-simplicial grids generated by reflection. SIAM J. Sci. Comput. **16**, 210–227 (1995)
26. W.F. Mitchell, Unified multilevel adaptive finite element methods for elliptic problems. Ph.D. thesis, Department of Computer Science, University of Illinois, Urbana (1988)
27. P. Morin, R.H. Nochetto, K.G. Siebert, Data oscillation and convergence of adaptive FEM. SIAM J. Numer. Anal. **38**, 466–488 (2000)
28. P. Morin, R.H. Nochetto, K.G. Siebert, Convergence of adaptive finite element methods. SIAM Rev. **44**, 631–658 (2002)
29. J. Nečas, Sur une méthode pour resoudre les équations aux dérivées partielles du type elliptique, voisine de la variationnelle. Ann. Sc. Norm. Super. Pisa, Sci. Fis. Mat., III. Ser. **16**, 305–326 (1962)
30. R.H. Nochetto, K.G. Siebert, A. Veeser, in *Theory of Adaptive Finite Element Methods: An Introduction*, ed. by R.A. DeVore, A. Kunoth. Multiscale, Nonlinear and Adaptive Approximation (Springer, Berlin, 2009), pp. 409–542
31. R.H. Nochetto, A. Veeser, *Primer of Adaptive Finite Element Methods in Multiscale and Adaptivity: Modeling, Numerics and Applications*, ed. by G. Naldi, G. Russo. CIME-EMS Summer School in Applied Mathematics (Springer, New York, 2011), pp. 125–225
32. M.C. Rivara, Mesh refinement processes based on the generalized bisection of simplices. SIAM J. Numer. Anal. **21**(3), 604–613 (1984)
33. A. Schmidt, K.G. Siebert, ALBERT — Software for scientific computations and applications. Acta Math. Univ. Comen. New Ser. **70**(1), 105–122 (2001)
34. A. Schmidt, K.G. Siebert, *Design of Adaptive Finite Element Software. The Finite Element Toolbox ALBERTA*. Lecture Notes in Computational Science and Engineering 42 (Springer, Berlin, 2005)
35. A. Schmidt, K.G. Siebert, C.J. Heine, D. Köster, O. Kriessl, ALBERTA: An adaptive hierarchical finite element toolbox. `http://www.alberta-fem.de/`. Version 1.2 and 2.0
36. E.G. Sewell, Automatic generation of triangulations for piecewise polynomial approximation. Ph.D. dissertation, Purdue University, West Lafayette, IN (1972)
37. R. Stevenson, Optimality of a standard adaptive finite element method. Found. Comput. Math. **7**(2), 245–269 (2007)
38. R. Stevenson, The completion of locally refined simplicial partitions created by bisection. Math. Comput. **77**(261), 227–241 (2008)
39. C.T. Traxler, An algorithm for adaptive mesh refinement in n dimensions. Computing **59**, 115–137 (1997)
40. R. Verfürth, A posteriori error estimators for the Stokes equations. Numer. Math. **55**, 309–325 (1989)
41. R. Verfürth, A posteriori error estimates for nonlinear problems. Finite element discretizations of elliptic equations. Math. Comp. **62**(206), 445–475 (1994)
42. R. Verfürth, in *A Review of A Posteriori Error Estimation and Adaptive Mesh-Refinement Techniques*. Adv. Numer. Math. (Wiley, Chichester, 1996)

List of Participants

1. Afzal Shehzad
   University of Leoben, Austria
   shehzad.afzal@stud.unileoben.ac.at
2. Antonietti Paola
   Politecnico di Milano, Italy
   paola.antonietti@polimi.it
3. Arcucci Rossella
   University of Napoli Federico II, Italy
   rossella.arcucci@unina.it
4. Artale Valeria
   University of Catania, Italy
   artale@dmi.unict.it
5. Bertoluzza Silvia **(lecturer)**
   IMATI-CNR Pavia, Italy
   silvia.bertoluzza@imati.cnr.it
6. Boffi Daniele
   University of Pavia, Italy
   daniele.boffi@unipv.it
7. Bueno Orovio Alfonso
   Universidad Autonoma de Madrid, Spain
   abucno@caminos.upm.es
8. Ceseri Maurizio
   University of Firenze, Italy
   ceseri@math.unifi.it
9. Chinnici Marta
   ENEA, Italy
   martachinnici@gmail.com
10. Coco Armando
    University of Catania, Italy
    coco@dmi.unict.it

S. Bertoluzza et al., *Multiscale and Adaptivity: Modeling, Numerics and Applications*,
Lecture Notes in Mathematics 2040, DOI 10.1007/978-3-642-24079-9,

11. Coco Salvatore
University of Catania, Italy
coco@diees.unict.it
12. Colli Franzone Piero
University of Pavia, Italy
colli@imati.cnr.it
13. Discacciati Marco
EPFL Lausanne, Switzerland
marco.discacciati@epfl.ch
14. Duci Alessandro
Arivis, Multiple Image Tools GmbH, Germany
alessandro.duci@arivis.com
15. Engquist Bjorn **(lecturer)**
ICES, University of Texas, USA
engquist@math.utexas.edu
16. Ferrandi Paolo Giacomo
MOX - Politecnico di Milano, Italy
paolo.ferrandi@mail.polimi.it
17. Fumagalli Alessio
MOX - Politecnico di Milano, Italy
alessio.fumagalli@mail.polimi.it
18. Gaggero Mauro
University of Genova, Italy
gaggero@diptem.unige.it
19. Gargano Francesco
Univertity of Palermo, Italy
gargano@math.unipa.it
20. Gaudio Loredana
Politecnico di Milano, Italy
loredana.gaudio@polimi.it
21. Geldhauser Carina
University of Tuebingen, Germany
carina.geldhauser@student.uni-tuebingen.de
22. Gonsalves Silveira de Serpa Maria Cristina
Universidade de Lisboa, Portugal
cristinaserpa@hotmail.com
23. Ivanovski Stavro
University of Catania and Astophysical Observatory Catania, Italy
stavro@oact.inaf.it
24. Jourdana Clément
IMATI, CNR, Italy
clement@imati.cnr.it
25. Lapin Vladimir
University of Limerick, Ireland
vladimir.lapin@ul.ie

26. Lesinigo Matteo
    EPFL Lausanne, Switzerland
    matteo.lesinigo@epfl.ch
27. Macciò Danilo,
    CNR, Italy
    ddmach@ge.issia.cnr.it
28. Maldarella Dario
    University of Ferrara, Italy
    maldarella@hotmail.com
29. Mederski Jaroslaw
    Nicolaus Copernicus University, Poland
    mastem@mat.uni.torun.pl
30. Micheletti stefano
    MOX - Politecnico di Milano, Italy
    stefano.micheletti@polimi.it
31. Naldi Giovanni (**editor**)
    University of Milano, Italy
    giovanni.naldi@mat.unimi.it
32. Nochetto Ricardo H. (**lecturer**)
    University of Mariland, USA
    rhn@math.umd.edu
33. Nonnenmacher Achim
    EPFL Lausanne, Switzerland
    achim.nonnenmacher@epfl.ch
34. Obertino Gianluca
    Politecnico di Torino, Italy
    gianluca.obertino@polito.it
35. Passerini Tiziano
    Emory University, USA
    tiziano@mathcs.emory.edu
36. Pedone Massimiliano
    University of Roma La Sapienza, Italy
    massimiliani.pedone@uniroma1.it
37. Perotto Simona
    MOX, Politecnico di Milano, Italy
    simona.perotto@polimi.it
38. Pidatella Rosa Maria
    University of Catania, Italy
    rosa@dmi.unict.it
39. Prignitz Rodolphe
    University of Erlangen, Germany
    prignitz@am.uni-erlangen.de
40. Quarteroni Alfio (**lecturer**)
    MOX, Politecnico di Milano, Italy and EPFL Lausanne, Switzerland
    alfio.quarteroni@epfl.ch

41. Ricciardello Angela
University of Messina, Italy
aricciardello@unime.it
42. Russo Giovanni **(editor)**
University of Catania, Italy
russo@dmi.unict.it
43. Semplice Matteo
University of Insubria, Italy
matteo.semplice@uninsubria.it
44. Siebert Kunibert G. **(lecturer)**
University of Duisburg-Essen, Germany
kg.siebert@uni-due.de
45. Stohrer Christian
University of Basel, Switzerland
christian.stohrer@unibas.ch
46. Turbin Mikhail
Voronezh State University, Russian Federation,
mrmike@math.vsu.ru
47. Verani Marco
Politecnico di Milano, Italy
marco.verani@polimi.it
48. Veeser Andreas **(lecturer)**
University of Milano, Italy
andreas.veeser@unimi.it
49. Weller Stephan
University of Erlangen, Germany
weller@am.uni-erlangen.de
50. Zampini Stefano
University of Milano, Italy
stefano.zampini@gmail.com
51. Zubkov Vladimir
University of Limerick, Ireland
vladimir.zubkov@ul.ie
52. Zvyagin Andrey
Voronezh State University, Russian Federation,
zvyagin.a@mail.ru

LECTURE NOTES IN MATHEMATICS Springer

Edited by J.-M. Morel, B. Teissier; P.K. Maini

Editorial Policy (for Multi-Author Publications: Summer Schools / Intensive Courses)

1. Lecture Notes aim to report new developments in all areas of mathematics and their applications - quickly, informally and at a high level. Mathematical texts analysing new developments in modelling and numerical simulation are welcome. Manuscripts should be reasonably selfcontained and rounded off. Thus they may, and often will, present not only results of the author but also related work by other people. They should provide sufficient motivation, examples and applications. There should also be an introduction making the text comprehensible to a wider audience. This clearly distinguishes Lecture Notes from journal articles or technical reports which normally are very concise. Articles intended for a journal but too long to be accepted by most journals, usually do not have this "lecture notes" character.

2. In general SUMMER SCHOOLS and other similar INTENSIVE COURSES are held to present mathematical topics that are close to the frontiers of recent research to an audience at the beginning or intermediate graduate level, who may want to continue with this area of work, for a thesis or later. This makes demands on the didactic aspects of the presentation. Because the subjects of such schools are advanced, there often exists no textbook, and so ideally, the publication resulting from such a school could be a first approximation to such a textbook. Usually several authors are involved in the writing, so it is not always simple to obtain a unified approach to the presentation.
   For prospective publication in LNM, the resulting manuscript should not be just a collection of course notes, each of which has been developed by an individual author with little or no coordination with the others, and with little or no common concept. The subject matter should dictate the structure of the book, and the authorship of each part or chapter should take secondary importance. Of course the choice of authors is crucial to the quality of the material at the school and in the book, and the intention here is not to belittle their impact, but simply to say that the book should be planned to be written by these authors jointly, and not just assembled as a result of what these authors happen to submit.
   This represents considerable preparatory work (as it is imperative to ensure that the authors know these criteria before they invest work on a manuscript), and also considerable editing work afterwards, to get the book into final shape. Still it is the form that holds the most promise of a successful book that will be used by its intended audience, rather than yet another volume of proceedings for the library shelf.

3. Manuscripts should be submitted either online at www.editorialmanager.com/lnm/ to Springer's mathematics editorial, or to one of the series editors. Volume editors are expected to arrange for the refereeing, to the usual scientific standards, of the individual contributions. If the resulting reports can be forwarded to us (series editors or Springer) this is very helpful. If no reports are forwarded or if other questions remain unclear in respect of homogeneity etc, the series editors may wish to consult external referees for an overall evaluation of the volume. A final decision to publish can be made only on the basis of the complete manuscript; however a preliminary decision can be based on a pre-final or incomplete manuscript. The strict minimum amount of material that will be considered should include a detailed outline describing the planned contents of each chapter.
   Volume editors and authors should be aware that incomplete or insufficiently close to final manuscripts almost always result in longer evaluation times. They should also be aware that parallel submission of their manuscript to another publisher while under consideration for LNM will in general lead to immediate rejection.

4. Manuscripts should in general be submitted in English. Final manuscripts should contain at least 100 pages of mathematical text and should always include
   - a general table of contents;
   - an informative introduction, with adequate motivation and perhaps some historical remarks: it should be accessible to a reader not intimately familiar with the topic treated;
   - a global subject index: as a rule this is genuinely helpful for the reader.

   Lecture Notes volumes are, as a rule, printed digitally from the authors' files. We strongly recommend that all contributions in a volume be written in the same LaTeX version, preferably LaTeX2e. To ensure best results, authors are asked to use the LaTeX2e style files available from Springer's web-server at
   ftp://ftp.springer.de/pub/tex/latex/svmonot1/ (for monographs) and
   ftp://ftp.springer.de/pub/tex/latex/svmultt1/ (for summer schools/tutorials).
   Additional technical instructions, if necessary, are available on request from:
   lnm@springer.com.

5. Careful preparation of the manuscripts will help keep production time short besides ensuring satisfactory appearance of the finished book in print and online. After acceptance of the manuscript authors will be asked to prepare the final LaTeX source files and also the corresponding dvi-, pdf- or zipped ps-file. The LaTeX source files are essential for producing the full-text online version of the book. For the existing online volumes of LNM see:
   http://www.springerlink.com/openurl.asp?genre=journal&issn=0075-8434.
   The actual production of a Lecture Notes volume takes approximately 12 weeks.

6. Volume editors receive a total of 50 free copies of their volume to be shared with the authors, but no royalties. They and the authors are entitled to a discount of 33.3 % on the price of Springer books purchased for their personal use, if ordering directly from Springer.

7. Commitment to publish is made by letter of intent rather than by signing a formal contract. Springer-Verlag secures the copyright for each volume. Authors are free to reuse material contained in their LNM volumes in later publications: a brief written (or e-mail) request for formal permission is sufficient.

Addresses:

Professor J.-M. Morel, CMLA,
École Normale Supérieure de Cachan,
61 Avenue du Président Wilson, 94235 Cachan Cedex, France
E-mail: morel@cmla.ens-cachan.fr

Professor B. Teissier, Institut Mathématique de Jussieu,
UMR 7586 du CNRS, Équipe "Géométrie et Dynamique",
175 rue du Chevaleret, 75013 Paris, France
E-mail: teissier@math.jussieu.fr

For the "Mathematical Biosciences Subseries" of LNM:

Professor P. K. Maini, Center for Mathematical Biology,
Mathematical Institute, 24-29 St Giles,
Oxford OX1 3LP, UK
E-mail : maini@maths.ox.ac.uk

Springer, Mathematics Editorial I,
Tiergartenstr. 17,
69121 Heidelberg, Germany,
Tel.: +49 (6221) 4876-8259
Fax: +49 (6221) 4876-8259
E-mail: lnm@springer.com